Concepts and Applications of Climatology

Concepts and Applications of Climatology

Edited by Loren Gilbert

SYRAWOOD
PUBLISHING HOUSE

New York

Published by Syrawood Publishing House,
750 Third Avenue, 9th Floor,
New York, NY 10017, USA
www.syrawoodpublishinghouse.com

Concepts and Applications of Climatology
Edited by Loren Gilbert

International Standard Book Number: 978-1-68286-650-4 (Hardback)

Cataloging-in-Publication Data

Concepts and applications of climatology / edited by Loren Gilbert.
    p. cm.
Includes bibliographical references and index.
ISBN 978-1-68286-650-4
1. Climatology. 2. Meteorology. 3. Earth sciences. I. Gilbert, Loren.
QC981 .C553 2019
551.6--dc23

TABLE OF CONTENTS

Permissions

List of Contributors

Index

PREFACE

This book has been an outcome of determined endeavour from a group of educationists in the field. The primary objective was to involve a broad spectrum of professionals from diverse cultural background involved in the field for developing new researches. The book not only targets students but also scholars pursuing higher research for further enhancement of the theoretical and practical applications of the subject.

This book elucidates new techniques of climatology and their applications in a multidisciplinary manner. Climatology is the scientific study of climate over a long duration of time. Recent research in climate change and environmental science has contributed to the growth of climatology. This book traces the progress of this field and highlights some of its key concepts and applications of this area of study. Climatology plays a crucial role in weather forecasting which has relevance across a number of other fields such as air transportation, biogeochemistry, etc. This book includes different approaches, evaluations, methodologies and advanced studies with respect to climatology. It aims to serve as a resource guide for students and experts alike and contribute to the growth of this discipline.

It was an honour to edit such a profound book and also a challenging task to compile and examine all the relevant data for accuracy and originality. I wish to acknowledge the efforts of the contributors for submitting such brilliant and diverse chapters in the field and for endlessly working for the completion of the book. Last, but not the least; I thank my family for being a constant source of support in all my research endeavours.

Editor

Impacts of Climate Changes on Management Policy of the Harbors, Land Areas and Wetlands in the São Paulo State Coastline (Brazil)

A. Pezzoli[3]*, P. Alfredini[1], E. Arasaki[1,2], M. Rosso[3] and W. C. de Sousa Jr.[4]

[1]Polytechnic School of São Paulo University, Department of Hydraulic and Environmental Engineering, Harbour and Coastal Area of the Hydraulic Laboratory, Avenida Professor Luciano Gualberto, Travessa 3, n. 380, Cidade Universitaria, CEP 05508-010, São Paulo, Brazil
[2]National Institute of Space Research, Av. Dos Astronautas n. 1758, CEP 12227-010, São José dos Campos, São Paulo, Brazil
[3]Polytechnic of Torino, Engineering Faculty, Department of Environment, Land and Infrastructure Engineering, CorsoDucadegli Abruzzi n. 24, 10129, Torino, Italy
[4]Aeronautic Technology Institute, Civil Engineering Division, Department of Water Resources and Environmental Sanitation; PraçaMarechal Eduardo Gomes n. 50, Vila das Acacias, CEP 12228-900, São Jose dos Campos, São Paulo, Brazil.

Abstract

Santosharbor and São Sebastião Oil Maritime Terminal are the most important oil and gas facility in the São Paulo State Coastline. Santosharbor had, in the last decade, increased rapidly the container handling rate, being the first in Latin America. Santos Metropolitan Region is one of the most important of Brazilian Coastline, also considering the tourism. For that great economic growth scenario it is very important to have wave climate and tidal levels well known considering the sea hazards influence in ship operations.

Since the hind-cast just represents the deep water wave climate, to make time-series of the wave's parameters in coastal waters, for evaluation of sea hazards and ship operations, it is necessary to take into account the variations of those parameters in shallow waters with coastal instrumental data. Analysis of long term wave data-base (1957-2002) generated by a comparison between wave's data modeled by a "deep water model" (ERA40-ECMWF) and measured wave's data in the years 1982-1984 by a coastal buoy in Santos littoral (São Paulo State, Brazil) was made. Validation checking procedures with instrumental measurements of storm surges made in other years than 1982-1984 shows high level of confidence.

These data, obtained from the climatological analysis compared with a data set found from a scale model of the whole area of Santos Bay, Estuary and nearby beaches (Brazil), showed the impact of maritime climate changes, wave climate and tides upon harbor and coastal structures maintenance, beaches stability, tidal inlet saline intrusion and wetlands flooding.

In the same time, the complex environmental system, that characterized the area included between Santos and Caraguatatuba, generates different natural hazard event affecting the maritime activity. A comparative study about the coastal flooding (Santos area) and the fluvial flooding (Juqueriquerê river) was conducted. An analysis about the sea level rise, the wave climate and the flooding risk as well as the sediment transport was developed.

Considering the increasing of the sea hazards, the high values of the facilities and infrastructures in São Paulo State Coastline, it is necessary to mitigate the risks from the point of view of the harbor and coastal structures maintenance and projects purposes increasing defenses procedures. Hence, based on the results obtained by the Authors in previous researches, are highlighted guidelines strategies suggested for Access Channels dimensions, wharves free-board, jetties and breakwaters dimensions, dredging rates, rigid and flexible littoral defenses, saline intrusion and land protection against flooding (including wetlands).

Keywords: Climate change; Geomorphology; Risk assessment; Harbor operations, Waves; Maintenance policy

Introduction

As well know the climate change affect the human activity, the agriculture and the industry [1] as well as the tourism business [2]. However a less bibliography was developed on the effect of the climate change on the maritime navigation. In fact, also if some studies were conducted about the effect of the climate change on the wind conditions and the wave action, the studies about the management and the policies are focused principally about the mitigation of the greenhouse gas emissions (GHG) generate by the navigation [3,4].

Nevertheless it is evident an inadequate bibliography about the effect of the climate change on the maritime navigation. For this reason a less literature is present about the management policy that the Government and the Organizations responsible for the port control can apply to sustain the shipping business due to the climate change effects.

There is the awareness that conditions of bathymetry, tides, winds, currents and waves for next decades shall have climate changes impacts on maritime navigation. The risk is understood, but only in a qualitative way, as composed by Hazard, Exposure and Vulnerability [5].

This paper goal is to overcome the contraposition that it emerges between the defence against the hydraulic risk and the management to preserve the environmental protection for nautical purposes. Moreover, basing on the results obtained by the Authors in the previous published researches, the highlighted guidelines strategies are suggested for access channels dimensions, wharves free-board, jetties and breakwaters dimensions, dredging rates, rigid and flexible littoral defenses, saline intrusion and land protection against flooding (including wetlands).

---

*Corresponding author: A. Pezzoli, Department of Environment, Land and Infrastructure Engineering, Polytechnic of Torino, Torino, Italy
E-mail: d001937@polito.it

The São PauloState (Brazil) Coastline (Figure 1) has around 450 km. The Harbors Areas of Santos and São Sebastião (Figure 2 and 3) concentrates around of 12% Brazilian international trading of 688 million tons per year and around 26% of Brazilian international trading of US$ FOB466million. SantosHarbor is the most important in the Southern Hemisphere and the first in Latin America. São Sebastião Oil Maritime Terminal is the most important Brazilian oil and gas facility. In the last decade important oil and gas reserves were discovered in the SantosOffshoreBasin and São Paulo Coastline received a great demand for supplier boats harbors for the petroleum industry [6]. Santos Metropolitan Urban Region is one of the most important of Brazilian Coastline, also considering the tourism.

For that great economic growth scenario it is very important to have well known the main maritime hydrodynamics forcing processes including climate changes in tidal levels, currents and waves, considering the sea extreme events hazards influence in vessel operations, coastal erosion, land flooding and estuarine mangrove wetlands survival as marine ecosystem [7].

Therefore is essential, due to the differences of the geographic system by a point of view of geomorphology and ecosystem, to analyze the entire area of São Paulo State Coastline. In the same time, as indicate before, this area was divided in two part: Santos and São Sebastião with the Juqueriquerê catchment and waterway.

The first area is affected from the increasing of the storm surges (sea level rise) as well as from the waves generated by the climate change. Instead the second area is influenced by the increasing of frequencies in the storm surges in addition to the flooding risk.

Extending the study at these two different zones, increases the validity of the results proposed in the present research because the phenomena analyzed are different so that the suggested management policies can be applied to a wide number of cases. It is important to note that the geographical system of São Paulo State Coastline have a similarity in the South America in particular in the coastline of the South of the Brazil. For this reason the policies recommended by this research to manage the effect of the climate change on the maritime navigation, can be applied in a large area making the results more general and widely applicable.

It is important to note that the Juqueriquerê Catchment is the major in São Paulo State (Brazil) North Coastline (Figure 4).

The Juqueriquerê Waterway is a 4 km estuarine channel used by small piers and docks. The entrance bar doesn't have any amelioration

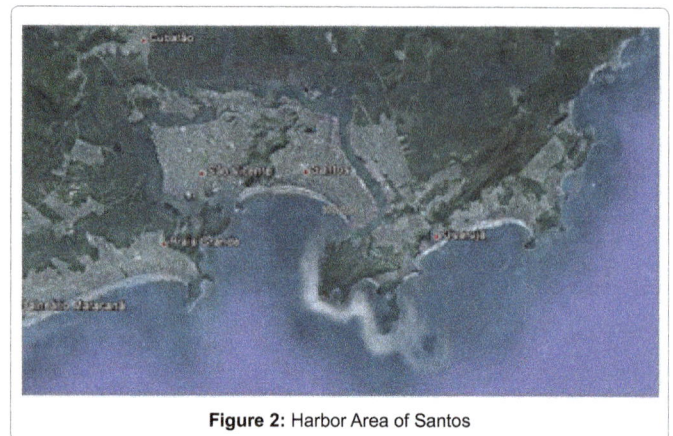

Figure 2: Harbor Area of Santos

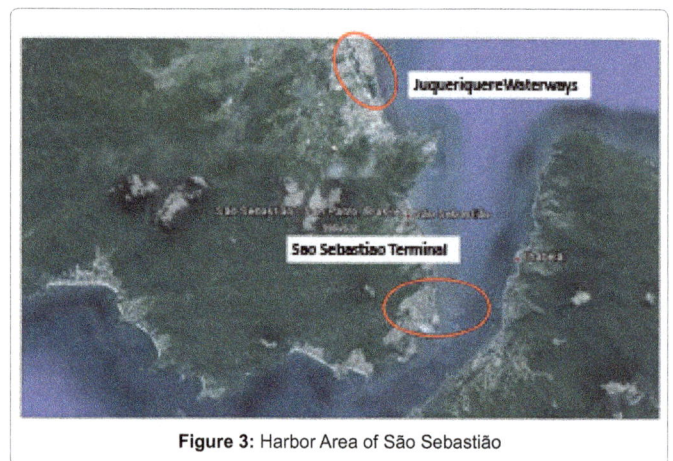

Figure 3: Harbor Area of São Sebastião

works and the boats maneuvers are difficult and dangerous. Example of navigation possibilities in the waterway was the dock operation in the period 1927-1967, the "Fazenda dos Ingleses" (English Farm), has sent the tropical fruits production to England. The railway line of the farm had 120 km, with docks and warehouses in the right bank of the estuary. The Packing House, in the dock area, was considered the second of this type in South America. The cargo boats, more than 20 in the forties decade, had an individual load of 55 dwt. In March 1967, a strong debris-flow, rain more than 600 mm in 2 days (monsoonal rates), combined with storm surge, caused more than 400 casualties and material losses. After that, the docks were closed [8].

In the last four decades, the nautical purposes of leisure and fishing boats increased in both areas mentioned above. It is important to mention the recent interest as supplier area for the offshore LNG and oil. The plant for the gas treatment is located in the left bank of the river and many of the facility heavy cargo equipment used large barges push-pulled by tugs.

Beyond environmental impacts, the cost of the effective improvements consists to bar jetties calibration where this solution means to talk about costs of 5 M €; another possibilities is the permanent maintenance with local dredging works, which costs, in the long term of decades, will be the same.

These examples show as, according to the IPCC forecasting, there is the awareness that conditions of bathymetry, tides, winds, currents and waves for next decades shall have climate changes impacts on the coastal area and in the marine navigation.

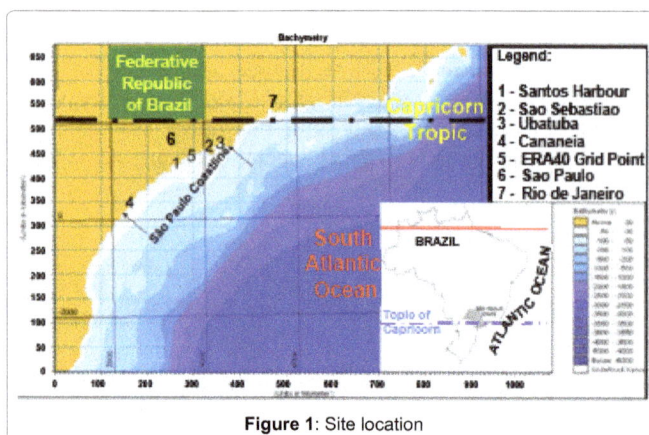

Figure 1: Site location

Material and Methods

The IPCC and PIANC recommendations [4], about the study of the impact on the climate change on the maritime navigation, are to focus on the met ocean variables such as wind, waves, sea level and ice.

Although large-scale climatic processes are driven by the ocean-atmosphere exchange system, very few studies are available on maritime impacts compared to continental impacts due to shorter data series and fewer human consequences [4].

Some analysis about the increasing of the sea level was conducted by Bindoff et al. [9]. The Authors indicates that the global mean sea level increased at an average rate of about 1.7 ± 0.5 mm/year during the twentieth century and that the rate has been slightly higher over the period 1961 to 2003.

In other the climate model prediction elaborated by the IPCC panel [10] shows that the global average rate of rise over the twenty first century will be 25 mm/year, implying that mean sea level will be 0.2 ÷ 0.5 m higher in the 2100 than 2000.

In the same time the waves conditions could be affected by climate changes in a number of aspect. Threnberth et al. [11] reports a statistically significant trend of increasing annual mean and winter mean significant wave height (Hs) for the mid-latitudinal North Atlantic and North Pacific, western subtropical South Atlantic, eastern equatorial Indian Ocean, and the East China and South China Seas. It also reports statistically significant decreases in H_s for western Pacific tropics, the Tasman Sea and the south Indian Ocean. Similar trends are found for the 99% extreme H_s with a maximum increase of winter extreme H_s of 0.4 m per decade in the North Atlantic. The worsening of wave conditions in the north-eastern North Atlantic is most likely connected to a northward displacement of the storm tracks, with decreasing wave heights in the southern North Atlantic [12].

Following these indications, it is possible to understand how the regional analysis of the sea level and the wave climate become important as shows by Debernard and Roed [13] and by Sterl et al. [14].

Considering the lack of bibliography and researches developed in this topic in the South Atlantic and in particular along the coastal line of the South of the Brazil, it was activated in 2010 a joint project called "Rede Litoral" (http://www.redelitoral.ita.br/). The Research Unit, based in the São Paulo University – Polytechnic Institute, has the goal of the research focused on the study of wave and tidal level analysis, maritime climate change, navigation's strategy and impact on the coastal defenses along the São Paulo Coastline Harbor Areas (Brazil).

As well indicated in the Introduction, this paper summarizes the research developed by the Research Unit of the São Paulo University [6-8,15-17] concentrating on the management policies.

This study was developed analyzing three different aspects of the problems (sea level, wave climate and sediments transport), apparently distant from each other, but, in fact, coordinated as well shows by the PIANC report [4].

The long term tidal level variability (high tide, mean sea level and low tide) assessment considering the Santos Dock Company (CDS) tidal variability (Highest High Water, Mean Sea Level and Lowest Low Water) for the last six decades, comprising three moon nodal cycles (58 years) shows a consistent response of relative sea level rise. Those figures were of similar magnitude than the other long term tidal series recorded in São Paulo Coastline, at the tidal gauges of Cananeia (1955-

1992), according to Franco et al [18], 200 km southward, and Ubatuba (1954-2003), 200 km northward (Figure 1).

Then a scale model of the Santos Bay, estuary with the Santos Harbor (Figure 5) was used to evaluate beach erosion and the land and mangrove wetlands flooding (Figure 6). This model was used to modeling tidal cycles and wave climate in the area nearby beaches of Santos [16]. The sea level data obtained by the simulation with the physic model were used to validate the measured data in the hydrometer of Santos Harbor. The high confidence obtained by this comparison, confirmed the validity of the long term data base used for the analysis of the sea level rise.

Considering the problem generated by the short time that the wave data are recorded by the buoy, an innovative methodology is developed in this research.

A long term wave data-base (1957-2002) was made by a comparison between wave's data modeled by the European deep water data base meteorological model ERA-40 Project [19] and measured wave's data in the years 1982-1984 by a coastal buoy in Santos littoral (São Paulo State,

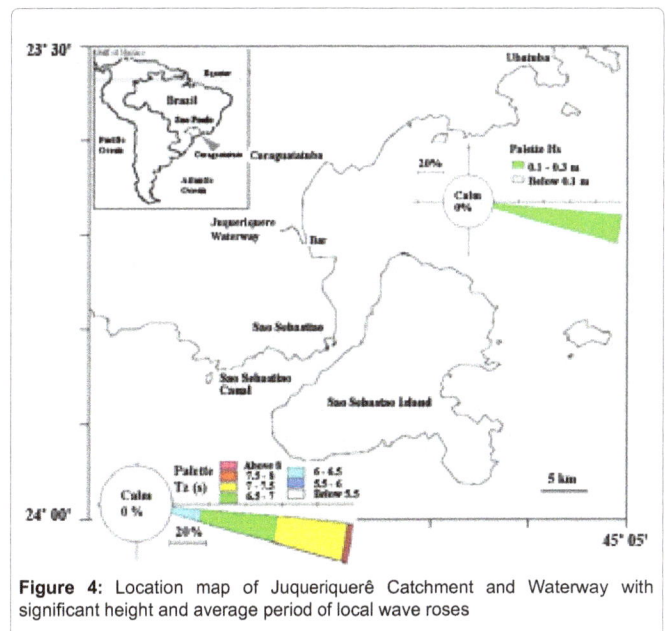

Figure 4: Location map of Juqueriquerê Catchment and Waterway with significant height and average period of local wave roses

Figure 5: Aerial view of Santos Bay, Estuary and Harbor

Brazil). Calibration coefficients according to angular sectors of wave's direction were obtained by the comparison of the instrument data with the modeled ones, and applied to the original scenarios. Validation checking procedures with instrumental measurements of storm surges made in other years than 1982-1984 shows high level of confidence. Finally the significant height (H_s) and the peak period (T_p) obtained by the "virtually" data-base (1957-2002) were analyzed to evaluate the possible effect of the climate change on the sea state [7,15].

Finally, as well indicated in the previous paragraph, was studied the Juqueriquerê Catchment with a particular attention at the sediment transport. It has the following main features: area of 430 km^2, long term average discharge of 11 m^3/s, heavy rainfall rates (around 3000 mm/year) producing high fluvial sediment transport, floods and debris-flows. The last ones are due to the steep slopes and the altitude (~1000 m) of the Serra do Mar mountains near the coast, producing the orographic effect, which rapidly condensates the sea humidity.

The fluvial dynamics is the cause of high solid transport capacity and fluvial and coastal morphology transformations that combined with recurrent and intense flood events, causing riparian and debris-flow on coastal region with important anthropic impact. This event is the cause of extensive risks and damages to population and infrastructures.

Strong debris-flows occur in this region, because events similar to monsoonal rain rates (higher than 300-400 mm per day) occur in multi decadal periods. The region history records shows this type of strong events in 1859, 1919, 1944 and the last and more catastrophic in March 1967.

Results and Discussion

As said in the previous paragraph, it is important summarize the results obtained by the Authors in the researches [6-8,15-17] to define the consequences of the climate change for the navigation purpose and the management policy.

Focusing on the tidal and on the sea level rise [6] and according to the CDS tidal gauge, it was possible to have the annual trends for Highest High Water (HHW), Mean Sea Level (MSL) and Lowest Low Water (LLW) (Figure 7), which shows a generalized sea level rise with the following rates in cm/century: 47, 25 and 45, based on data from 1952 till 2007 (3 moon nodal cycles).

The same evaluation was obtained for the last two moon nodal cycles gives: 57, 39 and 65. It means that there is an increasing rate of sea level rise in the last years and it is possible to reach trend rates from 50 to 100 cm/century in the next decades.

The obtained results about the increasing of the sea level for the São Paulo coastal line are in line with the IPCC scenario [10] as discussed in the Material and Methods paragraph.

According to that scenario, the simulation of scale model showed the flooding of around 50% of the Santos Estuary mangroves [16] and around 100m of the beaches (Figure 8), with the corresponding wave scour. Also in the last century, Santos Harbor wharves free-board (150 cm) lost around 35 cm.

Moreover, the analysis of the wave climate change on the extreme storm surge wave's conditions, using the calibrated ERA-40 1957-2002 data-base, shows an increasing trend, both in the linear and mobile average with a period of 5 years, in the H_s and T_p values and also in the frequency of storm surge events in the last decade (Figures 9 and 10). With the mobile average is possible to see the influence of a warm

episode of El Niño in The Pacific Ocean waters (1991-1993) enhanced with the huge Pinatubo Volcano eruption in The Philippines and a cold one of La Niña (1973-1976).

In the same time, from the Figure 9 and10, we can see that, in the São Paulo coastal line, the H_s have an high frequencies of the estimated values above that 0.7 m. Some and residual cases of the storm's events have the value of the H_s below at 0.7 m associated with the shortest T_p. The agreement between this evaluation of H_s and T_p in São Paulo costal line with the climate conditions calculated from the measured data for the Juqueriquerê Catchment and Waterway (Figure 4 and 6), confirm the validity of the methods used to build the long-term data base that it well representing the sea state conditions of the São Paulo area.

According to that trend it was possible to forecast an H_s increasing since 1957 till 2050 from 1.0 m to 1.4 m, with more than five times the frequency of occurrence of the storm surges since the 1950 and 1960 decades. It is well known that the wave energy per horizontal area and the long shore sand transport in the surf zone of waves is proportional to the square of wave height, meaning an increasing around of 100% per century. Also according to the classical Hudson's Formula, the rubble mound weight of ripraps, breakwaters and jetties are proportional to H_s^3, meaning an increasing of 200% per century for the new design

Figure 6: Scale model of Santos Bay, Estuary and Harbor and nearby beaches

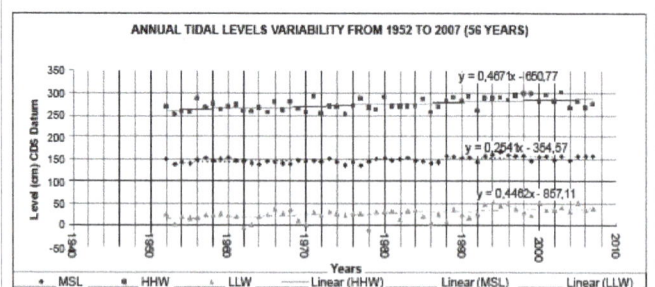

Figure 7: Santos Harbor tidal trends (1952 – 2007)

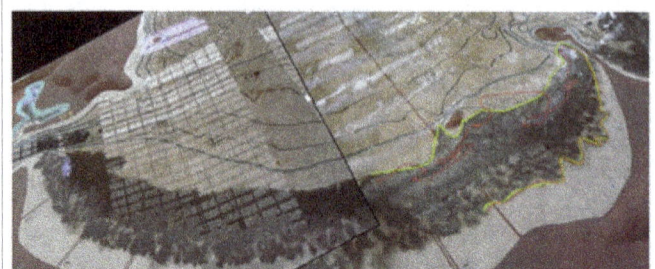

Figure 8: Scale model storm surge test: view of the beach scour

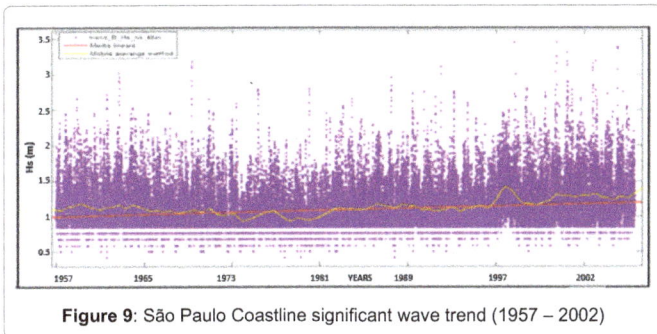

Figure 9: São Paulo Coastline significant wave trend (1957 – 2002)

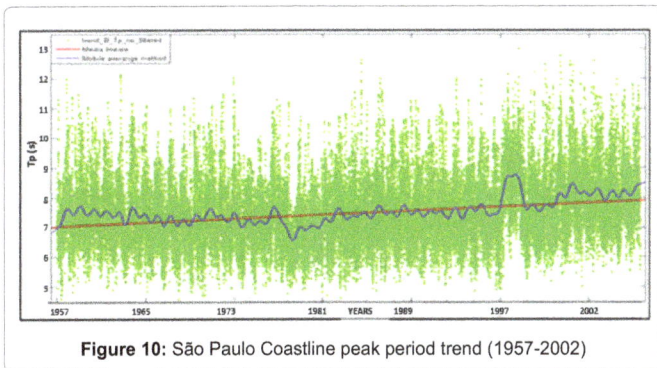

Figure 10: São Paulo Coastline peak period trend (1957-2002)

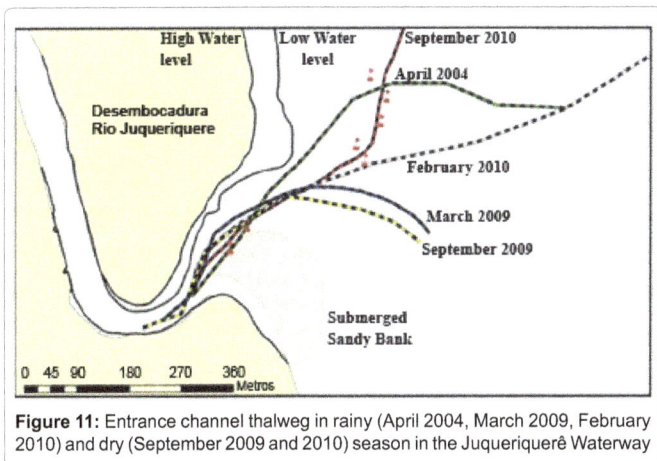

Figure 11: Entrance channel thalweg in rainy (April 2004, March 2009, February 2010) and dry (September 2009 and 2010) season in the Juqueriquerê Waterway

scenario, or an increasing in the damage rate and maintenance costs of those existing structures.

Finally, referring to [6] for what to concern the study and the results about the Juqueriquerê Catchment, it can be summarized that:

- The increasing of the sea level rise is included between 50 to 100 cm/century in the next decades in agreement with the IPCC scenario [10]

- The increasing of the sea level rise generates a flooding of around 50% of the Santos Estuary mangroves and around 100m of the beaches as well demonstrate by the simulation of the physical model

- The climate change impacts on the increasing of the H_s for a 0.4m in the next 50 years as well as in the peak period of the wave

- These calculated increasing of the significant wave height and

of the peak period are in agreement with the wave climate obtained by the measured data for the Juqueriquerê Catchment and Waterway

- In the Juqueriquerê Catchment and Waterway, the composite effect of the increasing of the significant height of the wave with the increasing of the storm and the associated flooding, generates the channel thalweg (Figure 11)

- The increasing of the significant wave height generates a possible increasing of 200% per century for the new design scenario in the harbor's structures

This analysis confirmed how the system is to be considered as a complex system where the effects on the channel, breakwater and harbor structures are generated by sea level rise, the wave climate and the flooding jointly with the sediment transport.

The consequences for navigation purposes, considering depths and channel widths are very complex and are summarized in Figure 12 elaborated considering the obtained results and the related assumptions.

As an example of some possible structural solution, it can be referred at the cost-benefit structural solution to reduce dredging costs in the Santos Access Channel proposed by Arasaki and Alfredini [20] and by Alfredini et al. [21]. The Authors proposed a solution with jetties in the Santos Access Channel which will reduce present maintenance dredging rates and will produce beach enlargement (Figure 13).

Conclusions

The paper summarizes ten years of research about the impact of maritime climate changes in the São Paulo State Coastline, developed by the Authors mainly in the Santos area, where there is the major amount of hydrodynamics data. It was possible to reach the goal of quantifying the magnitude order of tides and wave changes and to correlate them with the impact on maritime structures and the proposed mitigatory measures and structures.

Based on the quantitative assessment made it is possible to present the following and innovative strategic plan and the maintenance policy focusing on the navigation and coastal defenses for São Paulo State Coastline:

1.	Protection	• Wetlands restoration
		• Shoreline enhancement and land preservation
		• Jetties, seawalls, dikes construction/ upgrades against higher design waves and to 1 m of relative sea level rise
2.	Resistance and Resiliency	• Life cycle upgrades: 1 m higher elevation of quays
		• Increasing the external access channel width
		• Increasing maintenance dredging and capital dredging to enlarge external channels and for milder bank slopes
3.	System Management	• Focus investments on lower-risk assets and shift operations away from higher risk assets
4.	Impact Management	• "Green Port" to manage emissions and other impacts
5.	Information and Coordination	• Develop information to support consistent risk assessment, best practice responses, and necessary partnerships between ports and their host regions

Considering the awareness about the importance of climate changes impacts in a coastal area prone to extreme flood and erosion events,

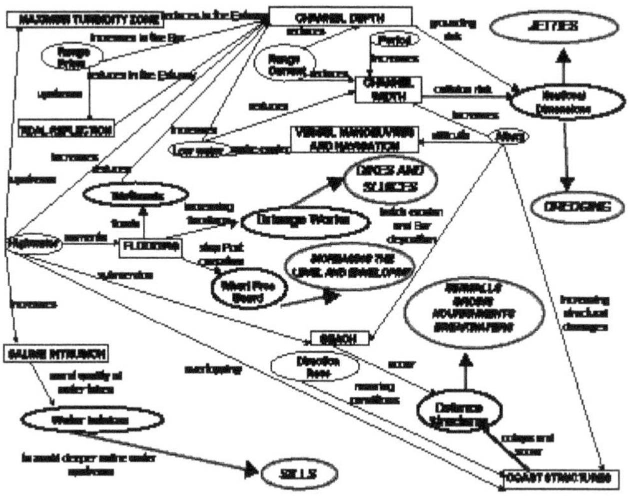

Figure 12: Frame of Santos tidal and wave changes impacts on maritime structures and possible mitigatory works

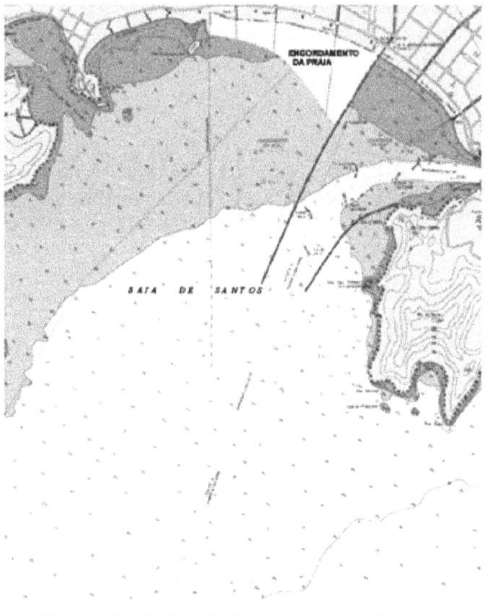

Figure 13: Jetties in Santos Access Channel

important issues to support confidence, or not, to activate structural and rigid solution (i.e. two jetties as shows in Figure 13), maintenance dredging (flexible solution), or non-intervention in the waterway are:

- 1. There is an overall sea level rising trend.

- 2. LLW has the highest rate of linear tidal rising.

- 3. There is an overall tidal range reduction.

- 4. The tidal prism will change, and the tidal currents velocity should increase, if the HHW levels will drown large fluvial areas, compensating the velocity reduction due to the tidal range decreasing.

- 5. Considering the issues above, the bar depth should increase.

- 6. The overall rise of the sea will produce more coastal erosion and littoral drift, in opposition to the outcome of issue 5.

- It is possible to observe a general significant height and average period wave increasing for annual averaged figures over than 1.5 m and 8.0 s (Figure 9 and 10), and the corresponding decadal maximum waves. It means increasing swell. Hence, should be a trend to increase littoral drift, reducing bar depth.

In other, merging the results of the climatological analysis with the result of the physical model, it is possible to made other assumption about the management policies of the São Paulo harbor area and of the Juqueriquerê Catchment and Waterway [6]:

- There are some areas of mud, which may be fluid and sufficient to consider the nautical bottom concept [19], in practice for mud density lower than 1250 kg/m3. In these cases it is possible to reduce the under keel clearance. The analysis of September 2010 and March 2011 survey, with detailed samples of the bar and bathymetry should provide confidence for this answer.

- About the thalweg shifting migration (Figure 11), it is possible to conclude:

- 1. Like for the monsoon weather, the main channel alignment depends upon flood periods, according to rain rate;

- 2. The shifting between two adjacent thalwegs may be produced by extreme river flow conditions, or a storm surge.

Awareness with climate changes impacts importance for the intervention's plan must be considered to obtain a final balanced solution among structures, dredging and non-structural measures for nautical master plan.

It is important to recognize that great natural events are not avoidable, but great disasters are, as the ancient Greek Aristotle (384-322 B.C.) said, "It is probable that the improbable will happen" [5].

References

1. Oliver JE, Hidore JJ, Snow M, Snow R (2010) Climatology: an atmospheric science. Ed Prentice Hall USA 408.

2. Jenkins K, Nicholls S (2010) The impacts of climate variability and potential climate change on tourism business in Torbay, England implications for adaptation. Tourism Analysis 15: 17-30.

3. Den Elzen MGJ, Olivier JGJ, Berk MM (2007) An analysis of options for including international aviation and marine emissions in a post-2012 climate mitigation regime. Netherlands Environmental Assessment Agency (MNP), AH Bilthoven, the Netherlands.

4. AA. VV (2009) Waterborne transport, ports and waterways: a review of climate change drivers, impacts, responses and mitigation. In: Final Report of the EnviCom-Task Group 3, PIANC, and Brussels, Belgium.

5. Kron W (2008) Coasts-The riskiest places on Earth. Proceeding of the 31st International Conference on Coastal Engineering.

6. Arasaki E, Alfredini P, Pezzoli A, Rosso M (2011) Chapter 17: Coastal area prone to extreme flood and erosion events induced by climate changes: study case of Juqueriquerê River Bar navigation, Caraguatatuba (São Paulo State), Brazil. In: Human Resources and Crew Management, CRC Press/Balkema Book, London.

7. Alfredini P, Pezzoli A, Cristofori EI, Dovetta A, Arasaki E (2012) Wave and tidal level analysis, maritime climate change, navigation's strategy and impact on the costal defences – Study case of São Paulo State Coastline Harbour Areas (Brazil). Geophysical Research Abstracts 14: 10735.

8. Arasaki E (2010) Capacidade de adaptação às mudanças climáticas globais: uma abordagem institucional para a gestão de recursos hídricos. São José dos Campos: Instituto Tecnológico de Aeronáutica, Post Doc Thesis.

9. Bindoff NL, Willebrand J, Artale V, Cazenave A, Gregory J et al. (2007) Observations: Oceanic Climate Change and sea level. In "Climate Change 2007: The Physical Science Basis. Contribution of Working Group I to the Fourth Assessment Report of the Intergovernmental Panel on Climate Change" - Eds. Solom S, Qin D, Manning M, Chen Z, Marquis M, Averyt KB, Tignor M, Miller HL, Cambridge University Press, New York, USA.

10. Intergovernmental Panel on Climate Change (2007). Summary for policy-makers. In "Climate Change 2007: The Physical Science Basis. Contribution of Working Group I to the Fourth Assessment Report of the Intergovernmental Panel on Climate Change" - Eds. Solom S, Qin D, Manning M, Chen Z, Marquis M, Averyt KB, Tignor M, Miller HL, Cambridge University Press, New Yourk, USA.

11. Solom S, Qin D, Manning M, Chen Z, Marquis M et al. (2007) Climate Change 2007: The Physical Science Basis. Contribution of Working Group I to the Fourth Assessment Report of the Intergovernmental Panel on Climate Change. Eds Cambridge University Press, New York, USA.

12. Intergovernmental Panel on Climate Change Solom S, Qin D, Manning M, Chen Z, Marquis M et al. (2007) Climate Change 2007: The Physical Sciences Basis Contribution of Working Group I to the Fourth Assessment Report of the Intergovernmental Panel on Climate Change. Eds Cambridge University Press, New York, USA.

13. Debernard JB, Roed LP (2008) Future wind, wave and storm surge climate in the Northern Seas: a revisit. Tellus A 60: 472-438.

14. Sterl A, van de Brink H, de Vries H, Haarsma R, van Meijgard E (2009) An ensemble study of extreme storm surge related to water levels in the North Sea in a changing climate. Ocean Sci 5: 369-378.

15. Dovetta A (2012) Analisi dell'influenza dei cambiamenti climatici sul moto ondoso: applicazione alla zona costiera dello Stato del São Paulo (Brasile). Torino: Politecnico di Torino, MSc Thesis

16. Arasaki E, Alfredini P, Gireli TZ (2008) Greenhouse effect and sea level impacts on Santos Estuary and Bay (Brazil) – Physical model study. Proceeding of the COASTLAB08-IAHR.

17. Alfredini P, Arasaki E, Pezzoli A, de Sousa WC Jr (2013) Impact of maritime climate changes on the harbor, land areas, and wetlands of Sao Paulo coastline (Brazil). In: 6th International Perspective on Water Resource and the Environment – IPWE 2013, January 7-9, 2013, Izmir (Turkey).

18. Franco AS, Kjerfve B, Neves CF (2007) A análise de registros de maré extremamente longos. In: Pesquisa Naval n°19, Estado-Maior da Armada.

19. European Centre for Medium-Range Weather Forecasts (2003) ERA-40 Project. ECMWF, Reading.

20. Arasaki E, Alfredini P (2009) Obras e Gestao de Portos e Costas – A Tecnica Aliada ao Enforque Logistico e Ambiental (2nd Ed). Ed: Edgard Blucher, Sao Paolo, Brazil, ISBN: 9788521204862.

21. AA.VV. (1997) Approach channels - a guide for design. In: Final Report of the Joint Working Group II-30 PIANC-IAPH in cooperation with IMPA and IALA, Brussels (Belgium) and Tokyo (Japan).

2

Observed and Future Climate Variability and Extremes Over East Shoa Zone, Ethiopia

Mequaninta F*, Mitikub R and Shimelesc A

Climate and Geospatial Research, Ethiopian Institute of Agricultural Research, Addis Ababa, Ethiopia

*Corresponding author: Mequaninta F, Climate and Geospatial Research, Ethiopian Institute of Agricultural Research, Addis Ababa, Ethiopia
E-mail: Fasil1190@yahoo.com

Abstract

This study has been conducted with the aim to analyze variability and extremes of daily values of maximum and minimum temperatures, and precipitation. The future data are downscaled using delta method based on outputs from four global climate models (GCMs). The data are simulated for three future 30 year periods, centered at 2030's (2010-2039) and 2050's (2040-2069) and for the two scenarios (A2 and B1). Analysis of the 27 core set of extreme weather indices, which are defined by ETCCDI, is carried out on six selected sites and all the results were reported in detail. In addition comparisons in variability has made between models and values of these indices observed in the base climate period (1981-2010) and values of projected periods.

Among precipitation indicators, significant increasing trends in annual total wet-day precipitation, number of heavy and very heavy precipitation days and decreasing trends in consecutive wet days, simple daily intensity index and precipitation on extremely wet days were found. Yet, the trends show no spatial coverage. In contrast to precipitation indices, the results indicate that most temperature indices showed significant changes in majority of stations. Nevertheless, the trends show less spatial coherence except cold days and summer days indicators.

Regionally, summer days and tropical nights showed significant increases while warm nights, cold days and warmest day showed significant decreases during the base period. On the other hand, for the projected climate, while summer days and warm days would significantly increases but coldest day and cold days would decrease on regional level.

The observed decreasing changes in warm extremes (i.e., warm night and warmest night) which are inconsistent with a warming planet and the heterogeneous behavior identified in most temperature extreme indicators suggested that local factors may play a major role in the study area.

Keywords: Base period; Climate models; Extremes; Indices; Projections; Variability

Introduction

Changes in climate variability and extremes have received increased attention in recent years, since they can have overwhelming impact on environment and society [1]. The large impacts climate extremes can have, and their tendency to change substantially in frequency with even small changes in average climate, mean that changes in extremes can be the first indication that climate is changing [2,3].

The Intergovernmental Panel on Climate Change's Fourth Assessment concluded that some extremes are changing, and will likely continue to change, due to human influences on the weather [4]. It is thus of great interest to analyze climate extremes since it can also be important in understanding future changes [5].

The study of climate extremes is rather complex due to the excessive statistical limitations inherent to extremal analysis (due to the obvious smallness of the samples to be analyzed). To tackle this challenge, the Climate Variability and Predictability (CLIVAR) Expert Team on Climate Change Detection and Indices (ETCCDI), has developed a set of indices on moderate extremes that represents a common guideline for regional analysis of climate.

Many studies have been published on changes in climate extremes around the world using these standard indices [6-18]. However there have been no studies done on precipitation or temperature related extremes in Ethiopia.

The objectives of this paper are to describe temporal trends in extreme weather events and their spatial distribution and to identify indicators which are representative of climate trends across east Shoa zone, Ethiopia. The paper also intends to see how the changes will evolve in the future. This analysis is important for the region which is characterized by high local climate variability.

Hence, the most representative indicators can then be used as predictors in climate-health research. Furthermore, the identification of geographic variation in sensitivity to climate change may also be helpful in resource allocation and policy decisions.

Materials and Methods

Data and quality control

Daily precipitation, maximum and minimum surface air temperature data were obtained from National Meteorological Agency (NMA) from 6 meteorological stations across East Shoa zone, Ethiopia, between 38-40.5°N latitude and 6.75-9.25°E longitude.

The period of record was between 1980-2010. The study area and station locations are shown in Figure 1.

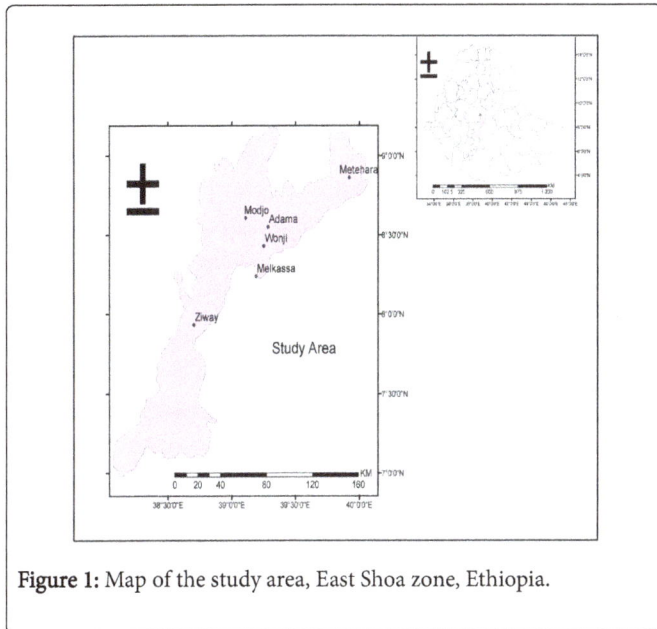

Figure 1: Map of the study area, East Shoa zone, Ethiopia.

In this study an exhaustive data Quality Control (QC) was applied. Initially, data from 10 meteorological stations were available, and after the QC, only stations with less than 10% of missing data for a period of at least 30 years were considered resulting in 6 locations (Table 1).

Station Name	Long(E)	Lat(N)	Elevation(m)
Melkassa	39.19	8.24	1531
Metehara	39.92	8.86	944
Modjo	39.11	8.61	1763
Adana	39.28	8.55	1622
Wonji	39.25	8.43	1540
Ziway	38.7	7.93	1640

Table 1: Meteorological stations used for the analysis.

Methodology

Climate downscaling

Future projections of precipitation, maximum and minimum temperature were generated based on outputs from four Global Climate Models (GCMs) namely: csiroMK3.0, gfdlCM2.1, ncarCCSM3.0 and ukmoHADCM3.

Outputs from the four GCMs were downscaled for all the stations used during the base period using delta method of downscaling [19,20] under two emission scenarios (A2 and B1) and two 30-year future periods centered at 2030 (2010-2039) and 2050 (2040-2069).

In the delta method (also called the constant scaling or perturbation method), the observed station historical daily rainfall series is scaled by a constant factor, determined as the ratio of the mean future GCM grid cell rainfall divided by the mean historical rainfall, to obtain the future daily rainfall series.

For temperature time series, the scaling factor is determined as mean future GCM grid cell temperature minus the mean historical temperature.

Extreme indices analysis

A summary of the temperature and precipitation indicators which were developed by ETCCDI is presented in Table 2. We calculated indices of climate extremes for individual stations for the base period as well as the downscaled future projections. We used linear regression to assess trends in these extreme indicators for each station.

The slopes of the annual trends and their statistical significance to climate indices were calculated based on non-parametric Mann–Kendall test in order to detect trends within the time series. The Mann–Kendall test has proven to be useful in determining the possible existence of statistically significant trends assuming a 95% probability level [5,21]. A trend was termed significant if the t test for the estimate of the slope was significant at α=0.05 level.

The percentile weather indicators were calculated by summing the number of days for which daily values exceed a time of-year-dependent percentile. These percentiles are determined for each day of the year, using data from that day and 2 days on either side of it over the course of the base period. For easy comparison of indices across stations with records of various lengths, the thresholds were computed from a base period, namely 1981–2010 for all stations.

The sample estimates of these indicators in the base years may not be reliable and there may be a discontinuity in the expected rates for the years on the boundaries of the base period [22].

Therefore, the RClimDex program was used to perform a bootstrapping procedure to provide cross-validation of these values [23]. The bootstrapping makes the estimation of the threshold exceedance rate for both the in-base and out-of base periods comparable [24].

The estimated decadal trends for each station were mapped using thematic maps to describe regional trends in temperature and precipitation extremes for the base period. All statistical analyses were done using R open source software (R Foundation for Statistical Computing, Vienna, Austria).

ArcGIS also was used to map the distributions of extreme climate trends across the study area. Finally we calculated zonal wide temporal trends and presented statistically significant trends.

Indices	Name	Definition	Units
CDD	Consecutive dry days	Maximum number of Consecutive days with RR<1 mm	Days
CWD	Consecutive wet days	Maximum number of Consecutive days with RR ≥ 1 mm	Days
PRTOT	Annual total wet day precipitation	Annual total PRCP in wet days (RR ≥ 1 mm)	mm
R10	Number of heavy precipitation days	Annual count of days when PRCP ≥ 10 mm	Days
R20	Number of very heavy precipitation days	Annual count of days when PRCP ≥ 20 mm	Days
SDII	Simple daily intensity index	Annual total precipitation divided by the number of wet days	mm
R95pTOT	precipitation on very wet days	Annual total PRCP when RR ≥ 95th %	mm
R99pTOT	precipitation on extremely wet days	Annual total PRCP when RR ≥ 99th %	mm
RX1day	Max-1 day precipitation amount	Maximum 1-day precipitation	mm
RX5day	Max-5 day precipitation amount	Maximum 5-day precipitation	mm
TN10P	Cold nights	Percentage of days when TN<10th %	Days
TN90P	Warm nights	Percentage of days when TN>90th %	Days
TN10P	Cold days	Percentage of days when TN<10th %	Days
TN90P	Warm days	Percentage of days when TN>90th %	Days
TNn	Coldest nights	Monthly minimum value of daily TN	°C
TNx	Warmest nights	Monthly maximum value of daily TN	°C
TXn	Coldest days	Monthly minimum value of daily TX	°C
TXx	Warmest days	Monthly maximum value of daily TX	°C
DTR	Diumal temperatute range	Monthly mean difference between TX and TN	°C
WSDI	Warm spell duration indicator	Annual count of days with at least 6 consecutive days when TX <90th percentile	Days
CSDI	Cold spell duration indicator	Annual count of days with at least 6 consecutive days when TN <10th percentile	Days
SU25	Summer days	Annual count when TX>25°C	Days
TR20	Tropical nights	Annual count when TN>20°C	Days

Table 2: Definition of extreme weather indicators.

Results

Extremes in the base period (1980-2010)

Temporal trends

Precipitation indicators: Decadal trends across the study area from 1980 to 2010 for the indices of extreme precipitation are presented in Table 3. The bold values represent significant level of 5% (p<0.05). The spatial distribution of trends of precipitation related extreme indicators are also shown in Figure 2. Among specific precipitation indicators, annual total wet-day precipitation (PRTOT), number of heavy (R10) and very heavy precipitation days (R20) showed significant increases (121.3, 4.3 and 2.5 mm per decade respectively) over Modjo.

In terms of duration of dry and wet days, Consecutive Wet Days (CWD) decreased significantly by 1 day/decade over Melkasa station. The Simple Daily Intensity Index (SDII) showed significant decrease (0.8 mm per decade) over Adama. Precipitation on extremely wet days (R99pTOT) indicator also showed significant negative trend of 25.8 mm per decade over Wonji station. Spatially, however, none of the indices showed homogenous significant trends, i.e., either positive or negative.

Temperature indicators: Decadal trends across the study area from 1980 to 2010 for the indices of extreme temperature are presented in Table 4. The bold values represent significant level of 5% (p<0.05). The analysis result evidence that most indices showed significant changes across most stations. Among the indicators, cold nights (TN10P) showed significant decreases of -3.8 and 9.0 days per decade over Adama and Wonji respectively. Similarly, warm nights (TN90P) decreased significantly by 3.5, 5.6 and 15.3 days per decade over Melkassa, Modjo and Wonji respectively, yet it increased over Adama and Metehara by 3.3 days per decade.

Indices	Adama	Melkassa	Metehara	Modjo	Wonji	Ziway
CDD	-247	-2.41	+2.62	-4.8	-5.69	-3.91
CWT	+0.06	-1.04	+0.07	+0.45	-1.00	+0.08
PRTOT	-11.84	+24.937	+11.3	+121.29	+12.58	+12.41
R10	-1.87	+0.69	0.05	+4.27	+0.91	+0.41
R20	-0.6	+0.23	+0.87	+2.51	+0.32	+0.54
R95pTOT	-21.13	-2.14	+10.7	+37.03	-5.58	+7.42
R99pTOT	-13.7	-2.51	-3.94	+16.71	-25.84	+3.69
RX1day	-0.89	-0.17	-0.64	+7.57	-3.67	-1.13
RX5day	-2.32	-7.83	+4.63	+7.34	-2.11	-1.14
SDII	-0.82	+0.31	-0.04	+0.54	+0.11	-0.17

Table 3: Decadal trends in extreme precipitation indicators.

On the other hand, cold days (TX10P) consistent significant decreased by 3.0, 4.6, 7 and 11 days per decade over Modjo, Melkassa, Ziway and Wonji respectively. In terms of warm days (TX90P), the indicator showed significant positive trends of 3.6, 5.1 and 7.7 days per decade over Ziway, Metehara and Modjo respectively, while it decreases significantly over Adama and Wonji (+6.6 and 11.7 days per decade respectively).

Among absolute indicators, coldest night (TNn) is the only indicator which did not show any significant change. Warmest nights (TNx) showed significant negative trend of 0.73, 0.96 and 1.3 degree Celsius per decade over Modjo, Melkasa and Wonji stations respectively. Yet the trend was a significant increase (+0.75°C per decade) over Metehara. Similarly, warmest days (TXx) showed significant decreases of 0.59°C and 1.0°C per decade over Adama and Wonji, respectively; however, it significantly increased over Metehara and Ziway by 0.8°C and 0.7°C decade, respectively. On the other hand, coldest days (TXn) showed significant positive trends of 0.77°C and 0.7°C per decade over Melkasa and Wonji, respectively. Regarding to duration indicators, only warm spell duration indicator (WSDI) showed significant decreases over Wonji by 0.96 days per decade.

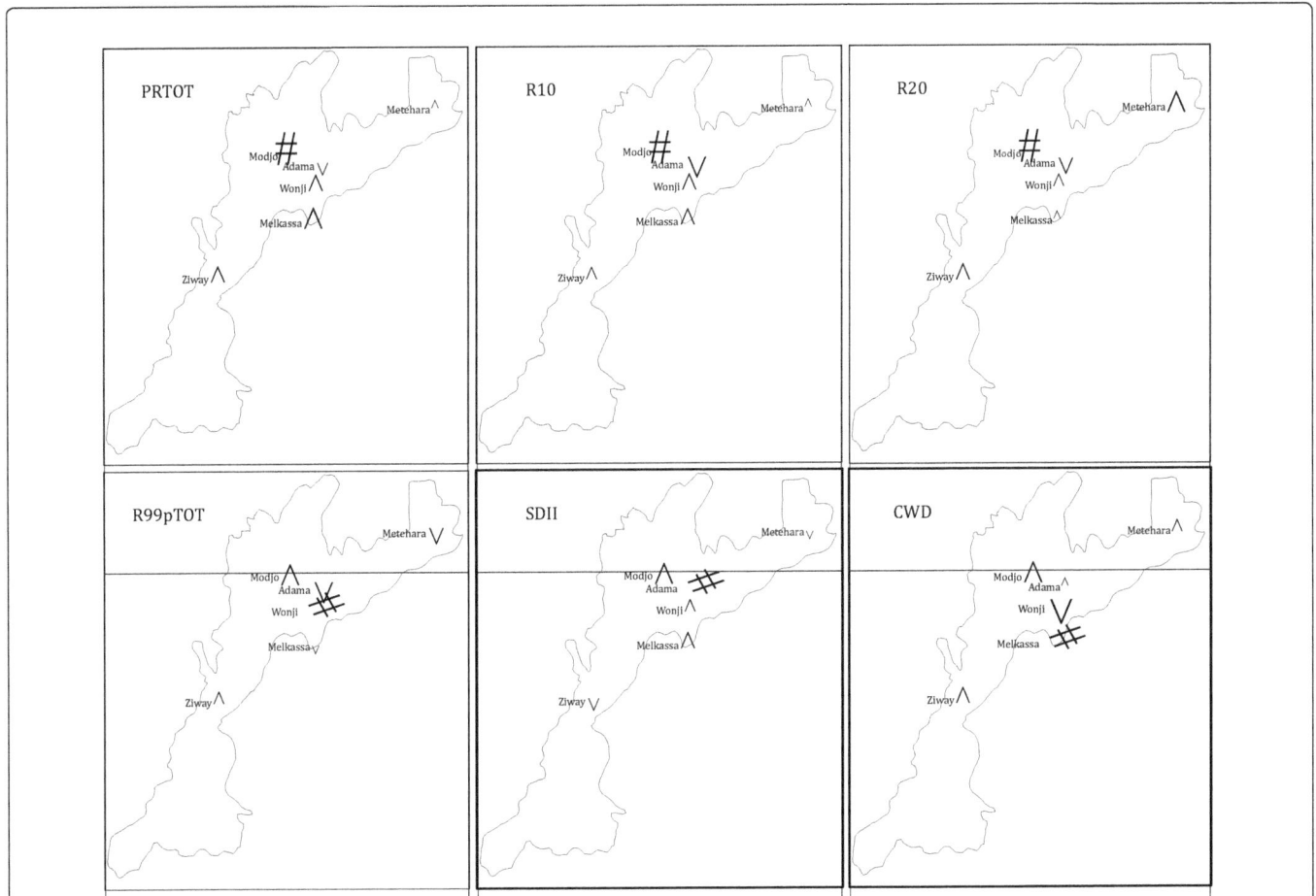

Figure 2: Distribution of trends of precipitation indicators. Upward and downward triangles indicate positive and negative trends respectively. Significant trends represented by the black color and uncolored ones indicate non–significant trends. The size of the triangles represents the magnitude of the trend.

Indices	Adama	Melkassa	Metehara	Modjo	Wonji	Ziway
TN10P	-3.79	-0.27	+0.4	+4.41	-9.01	-2.69
TN90P	+3.3	-3.5	+3.32	-5.58	-15.31	+1.14
TX10P	+1.14	-4.56	+0.83	-3.01	-10.98	-7.01
TX90P	-6.6	-1.81	+5.08	+7.73	-11.73	+3.57
TNn	-0.09	+0.12	-0.22	-0.37	+0.14	+0.11
TNx	+0.09	-0.96	0.75	-0.73	-1.31	+0.37
TXn	-0.67	+0.77	-0.12	+0.09	+0.7	+0.44
TXx	-0.59	-0.54	+0.81	+0.28	-1.02	+0.7
CSDI	-0.27	+0.14	0.02	+0.21	-0.23	+0.07
WSDI	-1.18	-0.02	+0.29	+0.87	-0.96	+0.11

DTR	-0.88	+0.31	+0.08	+1.18	+0.17	+0.4
TR20	-0.26	-0.24	+14.04	-0.04	-0.12	0.04
SU25	-6.94	+8.8	-0.07	+11.49	+14.27	+26.28

Table 4: Decadal trends in extreme temperature related indicators.

With regards to threshold indicators, summer days (SU25) showed consistent significant positive trends of 8.8, 14.3 and 26.3 days per decade over Melkassa, Wonji and Ziway, respectively. On the other hand, tropical nights (TR20) showed significant increase of 14 days per decade over Metehara, yet the trend over Melkasa were negative and small (i.e., -0.24 days per decade) though it were significant. Similarly, diurnal temperature range (DTR) showed significant increase (+1.2°C per decade) over Modjo and significant decrease (-0.88°C per decade) over Adama (Figure 3).

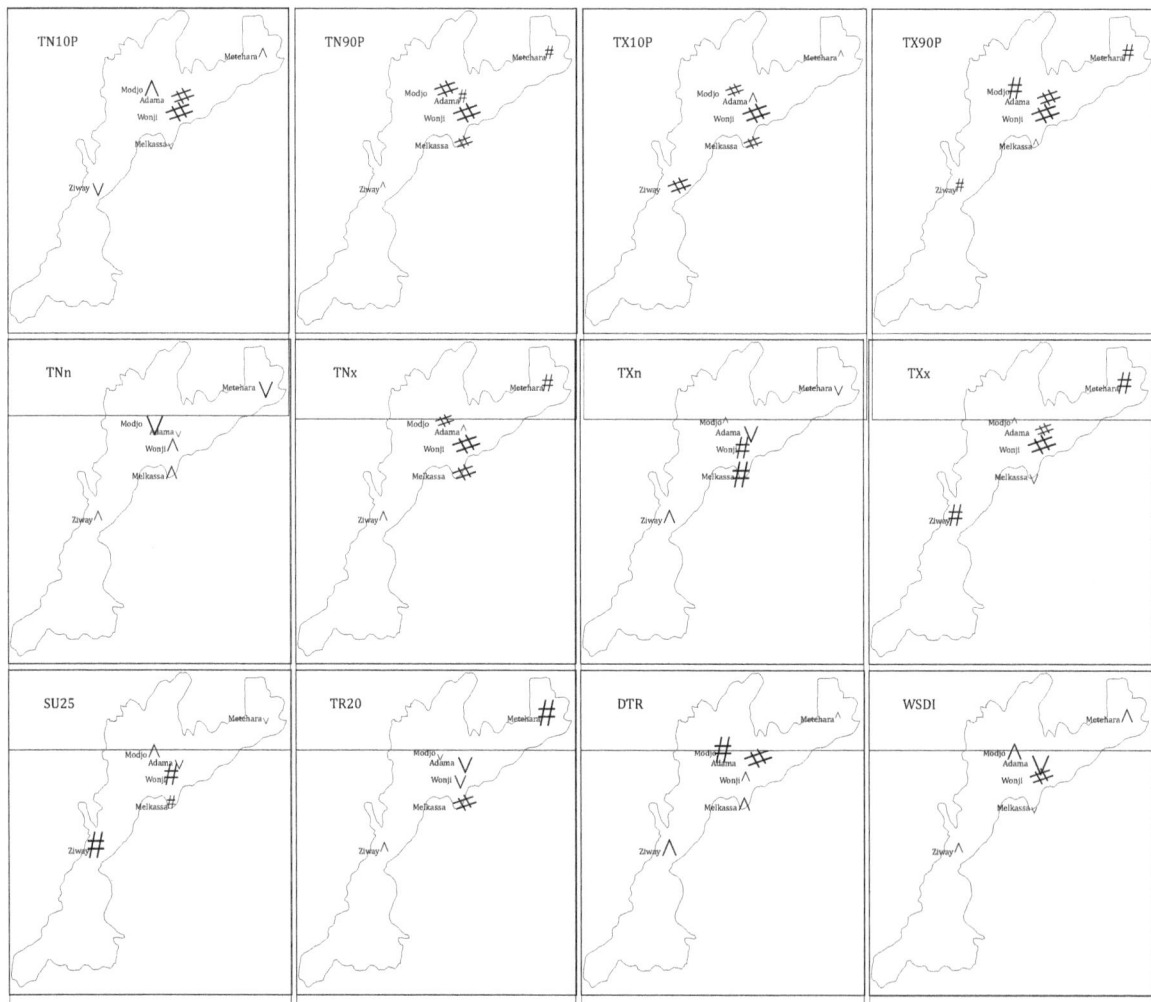

Figure 3: Distribution of trends of precipitation indicators. Upward and downward triangles indicate positive and negative trends respectively. Significant trends represented by the black color and uncolored ones indicate non-significant trends. The size of the triangles represents the magnitude of the trend.

Future extremes

Precipitation indicators: Significant decadal trends of downscaled future daily precipitation extreme indicators across the study area are presented in Table 5. Future projections in all model simulations showed the observed trends in CWD (over Melkasa), PRTOT, R10 and R20 (over Modjo) and SDII (over Adama) would continue in both scenarios. Unlike the base period, projections from gfdlCM2.1 model showed R20 would significantly increase over Metehara (i.e., A2-30's and B1-50's). On the other hand, only projections from gfdlCM2.1 model in A2 scenario showed that the significant negative trend in R99pTOT observed in the base period over Wonji would continue decreasing in the in future (i.e., in 2030's and 2050's).

		csiroMK3_0_A2_30				gfdlCM2_1_A2_30				ncarCCM3_0_A2_30				ukmoHADCM3_A2_30			
		A2		B1		A2		B1		A2		A2		A2		A2	
INDEX	Station	2030	2050	2030	2050	2030	2050	2030	2050	2030	2050	2030	2050	2030	2050	2030	2050
CWD	Melkasa	-0.11	-0.11	-0.11	-0.11	-0.12	-0.11	-0.09	-0.11	-0.11	-0.10	-0.09	-0.11	-0.10	-0.11	-0.10	-0.11
PRTOT	Mojo	13.1	13.5	13.1	13.5	13.4	12.7	13.8	14.6	14.9	14.9	14.7	14.8	13.3	13.5	12.7	13.5
RIO	Mojo	0.47	0.49	0.47	0.49	0.48	0.47	0.49	0.53	0.62	0.54	0.59	0.56	0.52	0.51	0.45	0.53
R20	Metehara	-	-	-	-	0.11	-	-	0.11	-	-	-	-	-	-	-	-
	Mojo	0.32	0.31	0.32	0.31	-	-	0.31	0.30	0.35	0.35	0.33	0.37	0.30	0.29	0.28	0.27
R99pTOT	Wonji	-	-	-	-	-2.75	-3.05	-	-	-	-	-	-	-	-	-	-
SDII	Adama	-0.10	-0.09	-0.1	-0.09	-0.08	-0.08	-0.09	-0.08	-	-	-	-	-0.09	-0.09	-0.09	-0.09

Table 5: Significant decadal trends in precipitation indicators for 2030's (2010-2039) and 2050's (2040-2069) in both A2 and B1 scenarios from the four GCM's. *The table shows only significant trends.

Temperature indicators

Significant decadal trends of downscaled future daily precipitation extreme indicators across the study area are presented in Table 6. While all projections in all models in both scenarios for both periods showed the negative trend in TN10P would continue over Wonji, no model showed significant change over Adama unlike the base period. Similarly, all projections showed the observed trends in the base period would continue in all stations with the exceptions that projections in A2 scenario in 2050's from ukmoHADCM3 model over Modjo and Ziway were not significant. In terms of TN90P, most projections showed the trend over Melkasa, Metehara and Wonji would continue in similar fashion as observed in the base period. Yet, exceptionally, projections from gfdlCM2.1 and ncarCCSM3.0 in A2 scenario in 2050' over Melkasa and projections for the 2050's in both scenarios from all models were not significant. On the other hand, only in 2050's projections from gfdlCM2.1 model (in both scenarios) and from ncarCCSM3.0 model (in A2 scenario) showed that the observed trend would continue over Modjo. Most model projections also showed consistent TX90P trends with the base period (over Metehara, Modjo,

Adama and Ziway stations) except the trends over Adama (in all models) and Metehara (in ukmoHADCM3 under A2 scenario) were not significant in 2050's period. Though, all trends except ncarCCSM3.0 and ukmoHADCM3 (in B1 scenario) were significant, the trend on the reverse would be increasing in the second period, i.e., in 2050's over Wonji.

With respect to Absolute indicators, all model projections showed that the trends in TXx, TXn and TNx observed in the base period would continue consistently except the trend in TXx were not significant in any projections over Adama. Projection also showed that, unlike the base period, TNn would significantly deceases over Modjo with the exception of projections from csiroMK3.0 in 2030's and ukmoHADCM3 in 2050 (i.e., in both A2 and B1 scenarios). Projections showed the observed positive trends in summer days would increase in the future; yet, all trends were not significant in all the projections and in all station. For instance, projections for 2050's period were not all significant except ukmoHADCM3 (over Melkasa) and ncarCCSM3.0 (over Melkasa and Ziway) in projections in B1 scenario.

		csiroMK3.0				gfdlCM2.1				ncarCCSM3.0				ukmoHADCM3			
		A2		B1		A2		B1		A2		A2		A2		A2	
INDEX	Station	2030	2050	2030	2050	2030	2050	2030	2050	2030	2050	2030	2050	2030	2050	2030	2050
TN1OP	Wonji	-0.57	-1135	-57	-1135	-1154	-0.26	-0.57	-0.39	-0.54	-0.28	-62	-0.48	-58	-0.27	-1162	-0.41
TN9OP	Mlelkaia	-0.65	-0.6	-0.65	-0.6	-0.73	-	-0.64	-0.67	-0.69	-	-64	-0.74	-64	-0.68	-0.6	-0.74
	Metehara	0.8	0.75	0.8	0.75	0.86	1166	1180	1196	87	1166	1176	1190	11135	1159	1193	0.99

	Mojo	-	-	-	-	-	-0.91	-	-0.85	-	-0.9	-	-	-	-	-	-
	Adama	0.91	1.11	0.91	1.11	0.97	1198	0.89	1.14	1.01	1196	88	1.06	1188	1.09	1184	1.15
	Wonji	-1.5	-	-1.5	-	-1.54	-	-1.83	-	-1.4	-	-1.66	-1.24	-1.68	-	-1.91	-0.62
TX10P	Melkasa	-0.28	-0.16	-0.28	-0.16	-0.25	-0.11	-0.27	-0.17	-0.26	-0.12	-0.27	-0.23	-0.27	-0.13	-0.29	-0.2
	Mojo	-0.21	-0.17	-0.21	-0.17	-0.2	-0.12	-0.2	-0.16	-	-0.13	-0.22	-0.19	-0.23	-	-0.22	-0.17
	Wonji	-0.62	-0.31	-0.62	-1131	-1160	-0.22	-0.64	-0.37	-	-0.23	-68	-52	-65	-	470	-0.4
	Ziway	-0.46	-0.24	-0.46	-0.24	-43	-0.18	-0.42	-0.27	-0.43		-45	-36	-0.44	-0.17	-0.46	-0.28
TX90P	Melkasa	-	-	-	-	-	-	-	-	-	-	-	-	-	-	-	-
	Metehara	1.18	1.11	1.18	1.11	1.26	0.51	1.22	1.22	1.27	0.75	1.12	1.33	1.25	-	1.16	0.83
	Mojo	1.34	1.58	1.34	1.58	1.38	1.13	1.32	1.49	1.42	1.23	1.28	1.46	1.33	1.32	1.27	1.57
	Adama	-0.87	-	-0.87	-	-0.87	-	-0.86		-0.88		-29	-	-26	-	-0.85	-
	Wonji	-0.87	-	-1187	0.77	-1175	1.05	-	0.58	-0.64	1.07	-1.01	1.1	-90	-	-1.13	1.33
	Ziway	0.93	-	0.93	1.41	1.03	1.43	-	1.33	1.03	1.34	0.94	-	0.92	-	0.85	
TNn	Mojo	-0.09	-1104	-	-1104	-1105	-0.05	-0.05	-0.06	-0.05	-0.05	-0.06	-0.05	-94	-	-4	-
	Melkasa	-	-0.09	-0.09	-0.09	-0.1	-0.1	-0.1	-0.1	-	-0.1	-10	-0.1	-99	-0.09	-0.09	-0.09
	Meehan	-0.08	toe	mos	mos	0.09	ons	ons	1109		1109	1110	1110	1109	1108	1109	1109
	Mojo	-1113	-0.08	-0.08	-0.08	-0.08	-0.08	-0.09	-0.08	-0.08	-0.08	-0.08	-0.08	-0.08	-0.08	-0.09	-0.09
	Wonji	0.08	-1113	-1113	-1113	-0.15	-0.15	-0.14	-0.14	-0.15	-0.15	-0.15	-17	-0.13	413	413	-0.13
lin	Melkaia	0.07	0.08	0.08	0.08	0.08	0.09	0.08	0.09	0.08	0.09	0.1	0.08	0.07	0.07	0.08	0.08
	Wonji	0.09	0.08	0.07	0.08	0.08	1109	1108	1109	1108	1109	1109	1108	1107	1107	1107	1107
	Metehara	-1110	0.09	0.09	0.09	0.09	0.09	0.09	0.09	0.09	0.09	0.09	0.09	0.1	0.1	0.1	0.09
	Wonji	0.08	-1110	-1110	-1110	-1110	-0.1	-0.1	-0.09	-0.1	-0.1	-0.1	-0.1	-0.11	411	410	4.1
	Z nvay	-0.1	0.08	0.08	0.08	0.08	0.08	0.08	0.08	0.08	0.08	0.08	0.08	0.08	0.08	0.08	0.08
	Mellcasa	1.35	-1342	-1110	-0.42	-0.21	-0.84	-0.14	-0.47	-0.19	-0.53	-0.13	-22	-11	-1166	-1110	-0.23
	Metehara	0.26	-	1.35	-	-	-	1.23	-	-	-	-	-	1.22	-	1.23	1.36
	Adama	-0.06	0.62	0.26	0.62	0.36	0.88		0.67	36	0.86		32	0.25	0.85	-	0.46
	WO.*	0.44	-0.37	-0.06	-0.37	-0.1	-0.61	-0.08	-0.34	-0.11	-0.65	-0.05	-0.14	-0.08	-0.53	-	-0.22
SU25	Melia	0.67	-	0.44	-	0.44	-	1146	-	1145	-	1146	0.38	1146	-	0.48	-
	WoMi	1.85	0.3	0.67	0.3	0.73	0.22	0.75	42	0.7	0.23	0.84	0.61	0.79	0.2	0.82	42
	Ziway	-	-	1.85	-	1.77		1.88	-	1.82	-	1.86	1.64	1.84	-	195	122
WSDI	Melkasa	-	-	-	-	-	-	-	-	-	-	-	-	-	-	-	-
DTR	Metehara		-1118		-1118				-0.28	-	-	-	-	-	-	-	-
	Mojo	0.46	-	-	-	-	-	-	-	-	-	-	-	-	-0.2	-	-
	Adama	-0.15	0.48	0.45	0.48	0.49		0.55	54	1147		0.48	0.48	0.47	0.4	1344	-

Wort	-	-0.09	-0.15	-0.09	-	-0.09	-	-0.07	-	-0.09	0.29	-	-	-0.1	0.16	0.11
Ziway	0.13	-	-	-	-	-0.31	-	-	-	-0.29	-	-	-	-0.3	-	-
Mojo	-10	0.13	0.13	0.13	0.13	0.13	0.13	0.13	0.13	0.13	0.13	0.13	0.13	0.13	0.13	0.13
Adama	-	-1110	-1110	-1110	-1110	-0.1	-0.1	-0.1	-0.1	-0.1	-0.1	-0.1	-0.1	-1110	-1110	-1110

Table 6: Significant decadal trends in temperature indicators for 2030's (2010-2039) and 2050's (2040-2069) in both A2 and B1 scenarios from the four GCM's. *The table shows only significant trends.

Similarly, projections were not all consistently significant in TR20 changes specifically over Metehara station. For instance, only 6 out of 16 projections were significant trends over Metehara namely: csiroMK3.0 (for 2030's in both scenarios), gfdlCM2.1 (B1 2030's) and ukmoHADCM3 (except A2 for 2050's). All projections would show the trends in DTR over Modjo and Adama would continue as observed in the base period. Comparisons between the base period in future projections and even among models showed that much difference is apparent with respect to trends in WSDI index. Unlike the only significant decreasing over Wonji, WSDI indicator would show significant trends over all stations except Melkasa. Yet, all models would not show significant trends over all stations except over Adama and Wonji. In fact, among the significant trends over Wonji, WSDI index would increase in ncarCCSM3.0 (in 2030's) and ukmoHADCM3 (in both periods) projected climate under B1 scenario. Furthermore, while the WSDI index would decrease for the rest of the stations, the trend over Adama would be increasing and at the same time relatively higher in magnitude.

Regional trends

The average temporal trends across east Shoa Zone from 1980 to 2010 for statistically significant trends of indices of extreme weather are presented in Table 7. Out of the 26 extreme weather indicators calculated, SU25 and TR20 showed significant increases while TN90, TX10P and TNx showed significant decreases during the base period. However, the trends in TN90P and TR20 indicators are found to be statistically insignificant in the projected future climate. On the other hand, unlike in the historic data, TX90P indicator would increase significantly regionally in the projected future climate. In this regard, most models showed that the TX90P indicator would be significant increase especially in the second period, i.e., in the 2050's. For the 2030's period however, only in gfdlCM2.1 in both scenarios and ncarCCSM3.0 in A2 scenario, TX90P would increase significantly over the region.

Indices	Base	csiroMK3.0				GfdlCM2.1				ncarCCSM3.0				ukmoHADCM3			
		A2		B1		A2		B1		A2		B1		A2		B1	
		30's	50's	30's	50's	30's	50's	30's	50's	30's	50's	30's	50's	30's	50's	30's	50's
SU25	+11.4	+5.2	+2.9	+5.2	+2.9	+5.2	-	+5.3	-	+5.2	+2.5	+5.4	+4.4	+5.5	+2.2	+5.7	+3.3
TY90P	-3.06	-	-	-	-	-	-	-	-	-	-	-	-	-	-	-	-
TNx	-0.39	-0.28	-0.3	-2.8	-0.3	-33	-0.33	-0.34	-0.31	-0.32	-0.31	-0.28	-0.23	-0.27	-0.28	-0.28	-0.27
TR20	+2.11	-	-	-	-	-	-	-	-	-	-	-	-	-	-	-	-
TX10P	-4.7	-2.3	-1.3	-2.3	-1.3	-2.2	-0.9	-2.2	-1.4	-2.2	-1.0	-2.4	-1.9	-2.4	-1.1	-2.5	-1.6
TX90P	-	-	+6.5	-	+6.5	+2.7	+6.0	-2.2	+5.7	+2.9	+6.1	-	+4.0	-	-5.4	-	+4.4

Table 7: East zone average decadal trends. *The table shows only significant trends.

Discussion

Results of the analysis show that no consistent pattern of changes in precipitation extremes could be detected for the majority of precipitation indices. Similarly, among temperature indices, TX10P and SU25 indicators showed relatively homogenous pattern across the region which is in agreement with the regional analysis. The study identified heterogeneous behavior i.e., positive and negative significant trends (p<0.05) observed in TN90P, TX90P, TXx and TNx indicators evidence that changes in routine observing practices may have introduced in homogeneities of non-climatic origin that severely affect the extremes. For instance, Metehara is found in lower altitude and Wonji are found near large sugarcane farms and of course large sugar

factory. The nearby lake to Ziway station and the higher topography which encircle Adama station are also other local factors which can impact the local climates of the stations.

The analysis for climate extreme indices for the projected climate also evolves quite in similar fashion as observed in the base period despite little differences among the GCM's in simulating individual climate extreme indices in some stations. The only significant difference observed among the models was the variations in simulating the trends in WSDI indices. This due to the downscaling method we used was simple statistical downscaling method.

The zonal average analyses, SU25 and TR20 showed significant increases while TN90, TX10 and TNx showed significant decreases during the base period. For the projected climate, SU25, TNx and TX10P would show similar trends as observed in the base period. Nevertheless, the trends TR20 and TN90P were not statistically significant in all projected climate. Additionally, we found significant increases in TX90P index in most projected climates in the zonal average analysis.

Conclusion

In this paper, we have examined the spatial and temporal distribution of trends in climate extremes using standard indices to describe for the last quarter of the 20th century when the largest changes occurred. In addition temporal trends of climate extremes in projected future climate for two periods centered at 2030 and 2050 under two scenarios A2 and B1are also investigated. Our study found neither consistent trend in precipitation extreme indicators nor homogenous trends across the study area. Similarly, the study identified TX10P and SU25 as representative indicators across the study area. The heterogeneous behavior identified in most temperature extreme indicators suggested that local factors play major role in the study area. Furthermore, the observed decreasing changes in warm extremes (i.e., TN90P and TNx) which are inconsistent with a warming planet evidenced the impact of these nonclimatic factors. It should be noted that the results presented in this paper are from small area and the downscaling method was also simple statistical downscaling method i.e., delta method. Therefore, we recommend further research using these representative indicators for longer period of data over larger area and using other downscaling method to better understand the impact of changing climate extremes.

References

1. Meehl GA, Karl T, Easterling DR (2000) An introduction to trends in extreme weather and climate events: Observations, socioeconomic impacts, terrestrial ecological impacts, and model projections. Bull Am Meteorol Soc 81: 413-416.

2. Nicholls N, Alexander L (2007) Has the climate become more variable or extreme? Progress 1992-2006. Prog Phys Geogr 31: 77-87.

3. Simolo C, Brunetti M, Maugeri M (2010) Understanding climate change-induced variations in daily temperature distributions over Italy. J Geophys Res 115: D22110.

4. IPCC (2007) Climate Change 2007: Synthesis report. Working Group I, II and III contributions to the Fourth Assessment Report, report, Geneva, Switzerland, p: 104.

5. Dufek AS, Ambrizzi T (2008) Precipitation variability in Sao Paulo State, Brazil. J Theor Appl Climatol 93: 167-178.

6. Aguilar E, Peterson TC, Ramirez Obando P (2005) Changes in precipitation and temperature extremes in Central America and northern South America 1961-2003. J Geophys Res 110: D23107.

7. Moberg A, Jones PD (2005) Trends in indices for extremes in daily temperature and precipitation in central and western Europe 1901–1999. Int J Climatol 25: 1149-1171.

8. Alexander LV, Zhang X, Peterson TC (2006) Global observed changes in daily climate extremes of temperature and precipitation. J Geophys Res 111: 5.

9. Moberg A, Jones PD, Lister D (2006) Indices for daily temperature and precipitation extremes in Europe analyzed for the period 1901-2000. J Geophys Res 111: D22106.

10. New M, Hewitson B, Stephenson DB (2006) Evidence of trends in daily climate extremes over southern and west Africa. J Geophys Res 111: D14102.

11. Tank AMGK, Peterson TC, Quadir DA (2006) Changes in daily temperature and precipitation extremes in central and south Asia. J Geophys Res 111: D16105.

12. Alexander LV, Hope P, Collins D (2007) Trends in Australia's climate means and extremes: a global context. Aust Met Mage 56: 1-18.

13. Brunet M, Sigro J, Jones PD (2007) Long-term changes in extreme temperatures and precipitation in Spain. Contr Sci 3: 331-342.

14. Peterson TC, Zhang X, Brunet-India M (2008) Changes in North American extremes derived from daily weather data. J Geophys Res 113: D07113.

15. Toreti A, Desiato F (2008) Changes in temperature extremes over Italy in the last 44 years. Int J Climatol 28: 733-745.

16. Carlos ACDS (2011) Trends in indices for extremes in daily air temperature over UTAH, USA. Rev Bras Met 26: 19-28.

17. Vincent LA, Aguilar E, Saindou M (2011) Observed trends in indices of daily and extreme temperature and precipitation for the countries of the western Indian Ocean 1961-2008. J Geophys Res 116: D10108.

18. Insaf TZ, Lin S, Sheridan SC (2012) Climate trends in indices for temperature and precipitation across New York State 1948-2008. Int J Air Qual Atmos Health 6: 247.

19. Fowler HJ, Blenkinsop S, Tebaldi C (2007) Review: Linking climate change modelling to impacts studies: recent advances in downscaling techniques for hydrological modelling. Int J Climatol 27: 1547-1578.

20. Mpelasoka FS, Chiew FHS (2009) Influence of Rainfall Scenario Construction Methods on Runoff Projections. J Hydromet 10: 1168-1183.

21. Haylock MR, Peterson TC, Alves LM (2006) Trends in Total and Extreme South American Rainfall 1960-2000 and Links with Sea Surface Temperature. J Clim 19: 1490-1512.

22. Von Storch H, Zwiers FW (2003) Statistical Analysis in Climate Research. Cambridge University Press, NY, USA, p: 484.

23. Zhang XB, Yang F (2004) RClimDex (1.0) User Manual, pp: 1-23.

24. Zhang XB, Hegerl G, Zwiers FW (2005) Avoiding inhomogeneity in percentile-based indices of temperature extremes. J Clim 18: 1641-1651.

3

Review on Influence of Climate Alterations on Corals and Associated Fishes for Indian Scenario

Uthaya Siva M*, Selvakumar P and Sakthivel A

CAS in Marine Biology, Faculty of Marine Science, Annamalai University, Parangipettai-608502, India

Abstract

Global warming is progression in which the earth temperature and the temperature on the atmosphere layers that are close to earth rise artificially as a result of the intense increase in some gases that occur in consequence of various human activities and that are qualified as greenhouse gases in the atmosphere. As to global climate change, it is the phenomenon where other climatic factors change as well depending upon global warming. Marine ecosystems are not in a steady state, but are affected by the environment, which varies on many spatial and temporal scales. Generally due to climate change and global warming the small and beneficial microorganism such as algae, bacteria, phyto and zooplanktons are getting disturbed or due to increasing of temperature and sea level raise it powerless to survive in the water surroundings. Detailed information is reported on the main Indian reef areas, including the Lakshadweep Islands, Andaman and Nicobar Islands, and Gulf of Mannar, with limited and largely anecdotal information for reefs elsewhere in the country. A fish population cannot be tolerant of high temperature changes in the area where it is distributed in a certain time interval. If these changes are within a certain temperature boundary and slow, it generally causes migration of fish specifically ornamental clown fishes. Temperature takes important physiological phenomena such as feeding, respiration, osmoregulation, growth and reproduction under control. If the individuals of population cannot adjust themselves according to the sudden and strong changes in temperature, one or some of their metabolism activities may deteriorate and mass deaths may occur and the changes in the sea water level will endanger the coastal habitat and species such as the sea turtle which uses the beaches as reproduction areas and lays eggs there will be unfavorably affected since their reproduction areas will become narrower.

Keywords: Climate Alterations, Osmoregulation, Global warming, Marine ecosystems, Temperature

Introduction

Now it is widely accepted statement that the climate change is no longer simply a potential threat, it is unavoidable; a consequence of 200 years of Excessive Greenhouse Gas (GHG) emissions from fossil fuel combustion in energy generation, transport and industry, deforestation and intensive agriculture [1]. IFAD and other development agencies have recognized the climate change as one the greatest threats facing mankind today [2,3] and tomorrow, it's have highlighted fact, that the poorest and most vulnerable will be disproportionately affected by the impacts of global warming [4]. Its defines the process in which the earth's temperature and temperature on the atmosphere layers that are close to earth rise artificially as a result of the intense increase in some gases that occur in consequence of various human activities and that are qualified as greenhouse gases in the atmosphere of Earth. As to global climate change, it's the major phenomenon where other climatic factors were changed as well depending upon global warming. Tekinayve Guroy clarified that Oceans and sea waters are mostly affected by the process of climate change instigated by global warming. Since they constitute a large portion of our planet and have rich biodiversity of copious of organisms. A temperature have increased only a few degrees does not only because an increase in the temperature of large water masses such as oceans, seas, lakes, and ponds but it also causes hydrological events that cause a change in the physical and chemical characteristics of water. Water temperature is the most important environmental parameter that affects the life cycle, physiology and behaviors of aquatic living beings [5]. With global warming, the waters of oceans are also warming up though there are considerable variations in different geographical regions and at different times. Warming has been more intense in surface waters, and there are evidences for deep water warming too. The world's oceans are also affected by changes in precipitation, wind and currents, which

are the result of geographical differences in temperature and humidity of the atmosphere. Thus, important oceanic weather systems such as the El Nino Southern Oscillation (ENSO) and Indian Ocean monsoon will be affected by global warming. Other direct effects of warming on aquatic systems include changes in precipitation, evaporation, river flows, groundwater, lakes and sea levels. These changes have altered the energy balance in the atmosphere, resulting in a warming effect. Later, Kennedy et al. described that Marine ecosystems are not in a steady state, but are affected by the environment, which varies on many spatial and temporal scales. Generally due to climate change and global warming the small and beneficial microorganism such as algae, bacteria, phyto and zooplanktons are getting disturbed or due to increasing of temperature it unable to survive in the water environment. These beneficial microbes are playing a vital role in the form of primary production in the ecosystem by which the fishes and other organisms getting feed. Changes in temperature are related to alterations in oceanic circulation patterns that are affected by changes in the direction and speed of the winds that drive ocean currents and mix surface waters with deeper nutrient rich waters. These processes in turn affect the distribution and abundance of plankton, which are food for small fish [6]. Hoegh-Guldberg et al. reported that Coral reefs are highly diverse and economically important to approximately

*Corresponding author: Uthaya Siva M, Ph.D., Research Scholar, CAS in Marine Biology, Faculty of Marine Science, Annamalai University, Parangipettai-608502, India
E-mail: uthayasiva.m1@gmail.com

100 million people in the tropics, yet their future is being threatened by rapid climate change [7]. Sutherland et al. stated that Identifying and prioritizing management of coral reefs in areas of lowclimate stress or reefs that are resilient to climate change is a leading rationale for informing the placement and implementation of challenging management interventions, such as no-take or closed areas [8]. The current status of coral reefs, coral reef research and Government policy towards conservation and management of reefs in India. Vineeta Hoorn reviewed more recent information is cited as available and analyzed to show the most recent trends in resource condition, use and conservation [9]. Due to global warming followed by sea level rise is more dangerous to affect coral reefs. When the corals are affected continually followed by the associated organisms such as fishes, mollusk, and warms are getting eradicated from ecosystem.

Macroalgae (Seaweeds, Sea grass and Coral reef)

Schaffelke et al. and Ritson-Williams et al. explained Macroalgae is a collective term used for seaweeds and other benthic marine plants that are generally visible to through naked eye. Seaweeds and sea grass are broadly comprise species from three different phyla, Rhodophyta (red algae), Heterokontophyta (predominantly Phaeophyceae, the brown algae), and Chlorophyta. They are clearly distinguished from microalgae, which require an associate microscope to be observed as phytoplankton, benthic and pelagic diatoms, free-living din flagellates, cyanobacteria (blue-green algae) and the symbiotic zooxanthellae that live within coral tissue. In some cases, benthic microalgae, such as some cyanobacteria and Chrysophyta, form large colonies that resemble thalli of Macroalgae. Such colony-forming cyanobacteria are often common components of turf algal assemblages [10,11]. Coral reefs are very important organisms living in the sea sores and zooxanthellae are occupied the tissue allthe corals and coral polyps as well with a symbiotic relationship. More than 90% of the corals gettingenergy come from the associate zooplankton of zooxanthellae, so corals are extremely dependent on symbiotic relationship of zooxanthellae. Coral bleaching will occurs when the coral host ejects zooxanthellae. Photosynthetic pigments are come from zooxanthellae and its gives colour. Therefore, the tissue of the coral appears transparent and the coral's bright white skeleton is revealed.

Marine Ornamental Fish

Dey and Ajit Kumar et al. defined as Ornamental fishes can be consider as an attractive in world over with colourful fishes of peaceful nature that are kept as pets in confined spaces of an aquarium or a garden pool with the purpose of enjoying their beauty for fun and fancy and those are visually exciting objects [12,13]. Because of they may have unique shapes, colouration, body forms and movements. Ornamental fishes are also called 'living jewels' for their beautiful colours and playful behavior. Ornamental fishes are typically small sized, colorful and most often bizarre shaped in appearance [13]. However, these fishes need not necessarily be always colorful. In fact, certain Fish species loved by aquarist are quite ugly; in such cases the peculiar appearance is a source of attraction for the aquarium lovers and naturalist [14], with the inspiring popularity of aquarium keeping in households in many parts of the world, ornamental has become an important part in international trade.

The world trade of ornamental fish is one of the most valued at about US $ 9.0 billion (FAO 2004). In Indian domestic trade in ornamental fish is conservatively estimated at Rs.10 crores. It is growing at the rate of 20 percent annually and the present domestic demand is higher than the supply [15]. Ornamental fish trade was

increasing frequently because of demand. This is the main reason of collecting marine ornamentals from the wild, so few years afterward the biodiversity will be totally eradicate due to overfishing so now very necessary to produce hatchery bred for conserve the natural stock. In natural conditions, plenty of foods of different types are available to suit the requirements of different fishes living in a particular environment. Aquarists have a great role in learning and understanding nature, especially aquatic life. Coral reef fishes (also called clown fishes and anemone fishes) are members of the family, Pomacentridae; subfamily Amphiprioninae. Among the commercially traded families of reef fishes, Pomacentridae dominate, accounting for nearly 43%. The damsels contain approximately 235 species worldwide [16]. They are unique because all of the species have an obligate symbiotic relationship with sea anemones [17], formation of a group consisting monogamous pairs and protandrous hermaphrodites [18]. Anemone fishes are one of the most popular and best marine ornamental fishes and have been bred successfully in captivity [19]. Among freshwater ornamental fishes, around 90% are farmed and reproduced while 10% are collected from the wild, but in the case of marine ornamental fish, more than 95% is collected from the wild [20]. Ajith Kumar et al. reported the commercially important marine ornamental fishes are clowns, damsels, angels, wrasses, butterflies and sea horses [13]. Sale stated most coral reef fish species disperse widely as pelagic larvae but after settling to a reef, they adopt a much more sedentary habit within a small territory [21]. Booth and Wellington demonstrated that habitat preferences of these fishes match well with their effects on subsequent survival, growth and reproduction [22]. Coral reef fishes are very sensitive to accumulate in the marine ecosystem and corals as well. If the corals are getting bleached automatically the associate fishes will go away. So it's very much necessity to study the corals and marine ornamental fishes as well.

Coral Reefs Ecosystem in South Asia

Previous reports are demonstrating that Climate changes are the main regional threat to coral reefs in the sea sores of South Asia. Effects from higher temperature are more variable precipitation; more extreme weather events. Sea level rise are already being felt in South Asia and it is increasing gradually. Due to this fact the coastal peoples are affecting all the years. A reduction in calcification rates caused by rising of ocean acidification may be equally severe or even more so. However, reefs are also facing severe straight human stresses from over-fishing as well as destructive fishing, coastal development, runoff from land and increased sedimentation. These are the main drivers of reef degradation as well as coral bleaching: areas where human effect and use of reef resources has been minimal are presently comparatively healthier (e.g. Chaos and around the Jaffna Peninsula in northern Sri Lanka) than other areas. Poor management of coastal areas, including many MPAs and NGOs, as well as intensive reef resource use, remains a concern in all countries [23]. Coastal and marine ecosystems are the main resources providing large benefits to the countries of the region through fisheries and tourism, which are highly important economic sectors in the Maldives, India and Sri Lanka. Many people throughout the region are directly dependent on reef and fisheries resources, however, poverty is widespread, especially among coastal populations (which are also dependent on other natural resources). Moreover the millions of peoples rely heavily on coastal and marine ecosystems and resources for employment, income, and food protein [23]. For example, in the Andaman Nicobar and Lakshadweep Islands as much as 90% of the protein intake for poor households comes from reef fishing and gleaning from the Arabian Sea and Bay of Bengal. Not only because of fisheries, at the same time is the tourism also the important factors to improve the economic status of the country as well as coastal peoples.

Coral Reefs Ecosystem in India

The detailed information was reported on the main Indian reef beds, including the Lakshadweep Archipelago, Andaman and Nicobar Islands, and Gulf of Mannar, with limited and largely anecdotal information for corals elsewhere in the country (Figure 1). The patterns of coral reefs recovery are described previously as continuing in the Lakshadweep Archipelago with coral cover was increased at the most of the coral reef sites. Recovery is faster on west-facing than eastern sites, largely due to differences in settlement patterns and substrate stability in the sea sores. There has been a reduction in algal turf and macro-algal cover compared to earlier studies, possibly explained by healthy populations of algal eating fish, particularly scarids and acanthurids, which are reported to facilitate coral recovery in the sea sores. The former dominant Acropora species, such as A. abrotanoides, are returning to dominance. An increase in coral bleaching at levels higher than normal summer bleaching was observed in April 2007;the extent and mortality is not known but appears to be limited. The impact of the 2004 tsunami was examined in detail on the Andaman and Nicobar Islands in 2005 and 2006. More than 100 km of shallow reef area was damaged in the Andaman Islands with most of this due to tectonic uplift and aerial exposure, as well as by the tsunami waves carving channels between islands. Many reef areas, especially in the Andaman Islands, were moderately to slightly affected and coral cover remains between 30 and 70%. A significant reduction in coral cover occurred at North Reef, Northern Andaman, and Interview Island, Middle Andaman. More than 200 km of reef was damaged in the Nicobar Islands due to tectonic activity, the tsunami and consequent sedimentation. Subsidence of the islands changed beach profiles and high erosion and sedimentation continued for more than 8 months after the tsunami. Severe damage in these areas extended to more than 20 m depth. Diversity of hermatypic corals in the Indian Ocean is present in the Figure 2. Coral cover at Car Nicobar is now about 5% after suffering more than 90% mortality [23]. Good coral larval recruitment has been reported from the Andaman Islands, while in the Nicobars it remains negligible. In some parts of the Nicobar Islands hard corals are facing competition from soft corals.

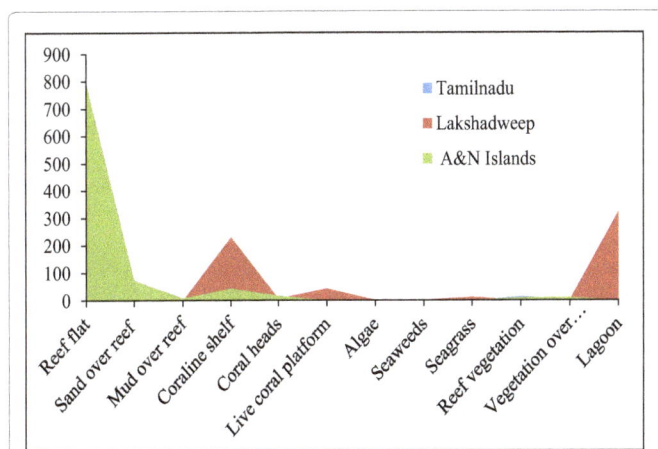
Figure 1: Area Estimates of Coral Reefs in the Country (Km2) in three regions of Indian Ocean [28].

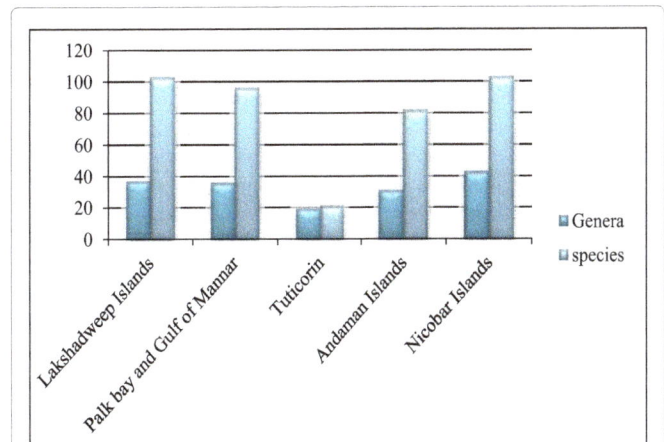
Figure 2: Diversity of hermatypic corals in the Indian Ocean [28].

Coral Reefs Ecosystem in Tamilnadu:

Vineeta Hoon stated that coral reefs in the Tamil Nadu coast are available to see with naked eye at Palk Bay near Rameshwaram and in the Gulf of Mannar. Palk Bay is the place separated from the gulf of Manner by Mandabam peninsula and Rameshwaram Island. The reef is centered on 9°17'N and 79°15'. There is only one fringing reef in the Palk bay, which lies in an east-west direction along the mainland from the Pamban channel at the Pamban end of the bridge to Rameshwaram Island. This reef is 25-30 km long, and generally less than 200 m wide; maximum depth is around 3 m. Visibility is poor around 1 meter and it is badly affected by the north east monsoon. The reef flat is relatively broad from Pamban channel to the southern end near Ramnad and narrow from Pamban to south of Rameshwaram. Diversity in the Palk bay reef consists of common sea grasses, sixty five species of coral have been recorded with a large number in the family Acroporidae. Turtles and Dugongs are found in the area. Squid breeding grounds occur near Rameshwaram [9]. In presently the nowadays reef growth is poor and it is not in a pristine condition since it was quarried in the sixties [24]. Satellite data shows that the reef flat is barren and is followed by sandy beach on the landward side. A small patch of reef fringes at the Dhanushkodi tip [25]. The Gulf of Mannar reefs on the other hand are developed around a chain of 21 islands that lie along the 140 km stretch between Tuticorin and Rameshwaram [26]. These islands are located between latitude 8°47'N and 9°15'N and longitude 78°12'E and 79°14'E. A detailed account of each Island is provided by Krishnamurthy and Deshmukh and Venkatramani (1995). The islands lie at an average of about 8 km from the main land. They are a part of the Mannar Barrier reef which is about 140 km long and 25 km wide between Pamban and Tuticorin [27]. Different types of reef forms such as shore platform, patch, coral pinnacles and atoll type are also observed in the Gulf of Mannar. The islands have fringing coral reefs and patch reefs around them. Narrow fringing reefs are located mostly at a distance of 50 to 100 m from the islands. On the other hand patch reefs arise from depths of 2 to 9 mt and extend to 1 to 2 km in length with width as much as 50 meters. Reef flat is extensive in almost all the reefs in the Gulf of Mannar. Reef vegetation is richly distributed on these reefs. The total area occupied by reef and its associated features is 94.3 sq km. Reef flat and reef vegetation including algae occupies 64.9 and 13.7 sq km, respectively [28]. Visibility is affected by monsoons, coral mining and high sedimentation load. The reefs are more luxuriant and richer than the reefs of Palk bay. Presently Pillai presents an overview of the status of coral reefs in Mannar and the species diversity [24,29-31]. His publication [29] cites the example of Manali Island in the Gulf of Mannar, and lucidly presents the situation in the gulf before the 1960's coral mining activities and the situation after the mining had taken place. He feels that coral growth in the Gulf of Mannar will be irretrievably stunted since the bottom is sandy and the planulae will not be able to settle [24]. Recent underwater surveys conducted by Kumaraguru are more encouraging [32]. They reveal that there is luxuriant coral growth around the Manali Island and that the overall condition of the reef patches in the Gulf of Mannar is not too alarming.

The Impacts of Global Warming and Climate Change on Coral Ecosystem

The effects of the global warming has created and will probably create on aquatic ecosystem can be listed as increase in water temperature and drying of the lakes, regression of glaciers, increase in the sea level, degradation in coastal ecosystem, change in precipitation amounts and models, change in the frequency and density of extreme weather phenomena, change in streams, increase in the extinction of species, increase in the distribution areas of ailment vectors, coral bleaching, and coral corruption. The main impact of global warming and climate change will be seen on coasts and also the sea sores, the most productive areas of seas. Experts conducting studies on coasts have calculated that each rise of one cm in the sea level may cause a horizontal regression with a width of 1m on sandy coastal lines due to erosion [32] and sue to this fact the corals are unable to get sunlight. The changes in the sea water level will endanger the coastal habitat and species such as the sea turtle which uses the beaches as reproduction areas and lays eggs there will be unfavorably affected since their reproduction areas will become narrower. Mainly the associated organisms are product the corals from the predators and both are maintaining a symbiotic relationship between corals and associates, so when the symbolism is damage automatically the total ecosystem will damage.

The Impacts of Global Warming and Climate Change on Fisheries

Water temperature comes of the first in the list of most determinant factors since it is essential for the reproduction of ornamentals fishes which was being nearby corals and the formation of an ideal living environment. Fish population can be tolerant of temperature changes in the area where it is distributed in a certain time interval. If these changes are within a certain temperature boundary and slow, it generally causes migration of fish population to other reef areas. So automatically the environment will come an end. Temperature takes important physiological phenomena such as feeding, respiration, osmoregulation, growth and reproduction under the control. If the individuals of population cannot adjust themselves according to the sudden and strong changes in the temperature, one or more of their metabolism activities may deteriorate and mass deaths may occur. If long term temperature increases are observed in a region, in any stock that is prevalent in that areas, shifting of the southern border of the ovulation area towards the north, changing of ovulation areas horizontally with any increase in the bottom water temperature, preference of northern latitudes as new areas of feeding and growth, increase in food salts and amounts with the increase in temperature on higher latitudes and changes in the present currents, prolongation of the growth period in a year, and shifting of the limit within which larvae may live towards northern latitudes may be observed [33]. Graeme Mac fadyen and Edward Allison reviewed that Countries such

as Russia, Peru and Columbia are sensitive to climate changes due to their high catches and reliance on exports or high employment from fisheries, but their larger economies and higher human development indices mean they are likely to have a higher adaptive capacity to deal with potential impacts [34].

References

1. IPCC (2007) Climate Change 2007: The Physical Science Basis. Summary for Policymakers. Intergovernmental Panel on Climate Change 18.

2. IFAD (2007) IFAD Strategic Framework 2007–2010. IFAD, Rome, Italy, 35 pp.

3. World Bank (2010) World Development Report 2010: Development and Climate Change. The World Bank, Washington, DC, USA, 424 pp.

4. IFAD (2008) Climate Change: A Development Challenge. IFAD, Rome, Italy, 4 pp.

5. Tekinay A, veGüroy D(2007) ĐklimdeğiikliğiTürkiyebalıküretimininasıletkileye cek? I.TürkiyeDklimDeğiikliğiKongresi, BildirilerKitabı: 329-334. 11-13 Nisan 2007, Đstanbul.

6. Kennedy VS, RR Twilley, JA Kleypas, JH Cowan Jr and SR Hare (2002) Coastal and marine ecosystems and global climate change. Pew Center on Global Climate Change, Arlington, USA, 52 pp.

7. Hoegh-Guldberg O, Mumby PJ, Hooten AJ (2007) Coral reefs under rapid climate change and ocean acidification. Science, 318, 1737–1742.

8. Sutherland WJ, Adams WM, Aronson RB (2009) Conservation Biology, ENFA and regression-kriging. Ecological Modeling, 220, 3499–3511. 23, 557–567.

9. Hoon V, Sriskanathan G, Townsley P, Cattermoul B (2008) Socioeconomic Monitoring Guidelines for Coastal Managers of South Asia, SocMon South Asia IUCN/CORDIO.

10. Schaffelke B, Heimann K, Marshall PA, Ayling AM (2004) Blooms of Chrysocystis fragilis on the Great Barrier Reef. Coral Reefs 23:514

11. Ritson-Williams R, Paul VJ, Bonito V (2005) Marine benthic cyanobacteria Overgrow coral reef organisms. Coral Reefs 24:629

12. Dey, VK (1996) Ornamental fishes and Handbook of Aqua farming. The Marine Products Export Development Authority, Cochin.

13. Ajith Kumar T T, Marudhupandi T, Balamurugan J & Balasubramanian T (2012) Clown fish (Amphiprion sandaracinos, (Allen, 1972) and A. melanopus (Bleeker, 1852)). Success in Hatchery Seed Production by the Centre of Advanced Study in Marine Biology, Annamalai University. Fishing Chimes**32** (3).

14. Abhisek basu, Dibyendu dutta and Samir banerjee (2012) Indigenous ornamental fishes of west bengal. Recent research in science and technology 2012, 4(11): 12-21

15. Threatened freshwater fishes of India (2010) and IUCN, 2011.

16. Allen GR (1991) Damsel fishes of the world, Mergus publishers, Mello, Germany. pp: 271.

17. Fautin D G and Allen G R (1997) Anemone fishes and their host sea anemones. Western Australian Museum 160.

18. Ross R M (1978) Territorial behaviour and ecology of the anemone fish Amphiprion melanopus on Guam. Z. Tierpsychol., 46: pp. 71-83.

19. Hoff FH (1996) Conditioning, spawning and rearing of fish with emphasis on marine clown fish. Dade city: Aquaculture Consultants, 212.

20. Tarlochan sing and VK Dey (2006) Trends in world ornamental fish trade. Souvenir, Ornamentals Kerala 2006, Dept. of Fisheries, Govt. of Kerala: 3-8.

21. Sale PF (1980) The ecology of fishes on coral reefs. Oceanogr Mar Biol Ann Rev 18: 367-421.

22. Booth DJ, Wellington G (1998) Settlement preferences in coral-reef fishes: Effects on patterns of adult and juvenile distributions, individual fitness and population structure. Aust J Ecol 23:274–279

23. Jerkertamelander and Arjanrajasuriya(2008) Status of Coral reefs in south Asia: Bangladesh, Chagos, India, Maldives and Srilanka. Status of Coral Reefs of the World: 2008.

24. Pillai CSG (1996) "Coral reefs of India: Their Conservation and Management, in (Pillai CSG and Menon N.G. eds) "Marine Biodiversity, Conservation and Management," CMFRI, Cochin, India.

25. Bahuguna. A, S Nayak (1994) "Coral reef Mapping of Tamil Nadu using satellite data" SAC (ISRO), Ahmedabad, India.

26. Krishnamurthy K (1987) "The Gulf of Mannar Biosphere Reserve: Project document-5, Ministry of Environment & Forests, Government of India.

27. Venketesan K.R (n.d). Gulf of Mannar Marine National Park (Proposed), unpublished report of the Chief wildlife warden, K.R Venketesan, Tamilnadu.

28. DOD & SAC (1997) "Coral reef maps of India," Department of Ocean Development and Space Application Centre, Ahmedabad, India.

29. Pillai CSG (1975) An assessment of the effects of Environment and Human interference on the Coral Reefs of Palk Bay and the Gulf on Mannar, along the Indian Coast in Seafood export Journal Vol VIII No 12.

30. Pillai CSG (1986)"Recent Corals from South-East Coast of India" in PSBR James (ed "Recent Advances in Marine Biology ', Today and Tomorrow Printers and Publishers, India.

31. Kumaraguru AK (1997) "Project title: Ecology of ornamental fishes of export value in the Gulf of Mannar", Project report for the period ending 31-1-97", School of energy, environment and natural resources, Madurai Kamaraj University, Madurai, sponsored by Department of ocean development, GOI.

32. Avar D (2005) Balikcılık Biyolojisive Populasyon Dinamiği. Nobel KitabeviYayınları: 332, Adana.

33. Çalta M ve Ural M (2001) AynaliSazan (Cyprinuscarpio L., 1758) yumurtalarininacilmasivelarvalarinyaamaoraniuzerinesuyun pH degerininetkisi. E.U. Su Urunleri Dergisi 18 (3-4): 319 – 324.

34. Graeme Macfadyen and Edward Allison (2009) Climate Change, Fisheries, Trade and Competitiveness: Understanding Impacts and Formulating Responses for Commonwealth Small States. Graeme Macfadyen, Poseidon Aquatic Resource Management Ltd, Geneva, Switzerland, in association with Dr. Edward Allison, World Fish Centre, Malaysia.

4

How Well is the Tropical Africa Prepared for Future Physiologic Stress? The Nigerian Example

Eludoyin OM*

Department of Geography and Planning Sciences, AdekunleAjasin University, Akungba-Akoko, Ondo State, Nigeria

Abstract

The huge literature gap in the knowledge of physiologic climatology on tropical Africa indicates poor awareness to the issue of physiologic stress in the region. This study examined the variability in the physiologic comfort over Nigeria using both annual and hourly patterns of unitary (temperature and relative humidity) and integrative indices (effective temperature, temperature-humidity and relative strain indices), and assessed the perceptions of a randomly selected Nigerians in 18 tertiary institutions across the country. Results indicated thermal stress in Nigeria, and showed that both heat and cold stress varied temporally (annually and hourly) and spatially (1200-1500 Local Standard Time, LST as the most thermally uncomfortable period of the day, and ≤ 0900 and around 2100 LST were more comfortable). Perception of the comfortable climate exhibits variation based on the latitudinal location of the respondents but the coping strategies vary with the wealth of individuals. The study indicated that whilst many parts of Nigeria could be vulnerable to physiologic stress, indigenous and modern know-how to cope with future physiologic stress is largely unknown. The study therefore recommends significant improvement in climate-oriented policies, especially in the areas of healthcare and infrastructure.

Keywords: Day-time thermal comfort; Coping strategies; Effect of extreme weather; Climate education

Introduction

Physiologic climatology is a field of scientific study that is concerned with the effects of climatic elements and patterns on the physiologic behaviour of man and other warm blooded animals [1-3]; and with classification or regionalisation of climatic environments based on measurable human psychological and physiologic reactions [4]. Researches on the effects of climate on human's feelings and behaviour have increased in recent times, probably because of the increased concerns for extreme climate conditions and their consequences on human health and livelihood [5,6]. Available social, health, economic and technological facilities in many vulnerable communities are also inadequate to mitigate the negative effects of extreme climate conditions [7,8].

Two classes of approaches- unitary and integrative- are generally available in literature as indicators of physiologic comfort. The unitary approach considers certain elements of climate (temperature, humidity, radiant energy and air movement or wind) as suitable indicators of physiologic climate per unit time [1]. The setback to the unitary concept is that the climate factors that are usually linked with human physiology are rarely excusive but are rather integrative [9,10]. The integrative approach demonstrates that humans respond physiologically to more than one element of climate at a time [11,12], and as such indicators which combine, extrinsically, at least two elements of climate are more realistic. A better approach has been to combine both unitary and integrative approaches, complementarily. For example, Terjung [4,13] characterised the physiologic comfort in the African and North American (contiguous) continents from two main integrative indices - the comfort index (an integration of dry bulb temperature and relative humidity) and wind effect index (a combination of solar radiation and wind chill)- complementarily; the wind effect index complemented the comfort index where wind was considered to be of little or no effect. Similarly, Gregorczuk and Cena [14] mapped the physiologic comfort of the world using the effective temperature index (ETI). The ETI is one of the oldest indices for illustrating the physiologic comfort, though previously designed for indoor condition, that have gained interests

among scholars in the tropics that, probably because they have found the index relevant to the region [14-17]. More recent studies [18,19] have indicated the use of more integrative indices in Nigeria - a typical tropical country- and these also include the temperature-humidity index (THI) and relative strain index (RSI), as applied in other countries such as Hungary [11] and Argentina [20]. Other indices such as apparent temperature (AT), predicted mean vote (PMV) are among the over one hundred integrative indices that have been used in literature [21-24].

Whilst information generated from the analysis of physiologic comfort have been found useful for planning of holidays, migration, tourism and building in many countries [25,26], African countries have appeared to show less interest in the aspect of climatological research; most African researches in climate science and applications have focused on ensuring food security. On the other hand, the increasing population and urbanisation rate in Africa, as well as indications of temperature increase due to anthropogenic factors (such as gas flaring and oil exploration activities) are justified reasons to study the change in the physiologic comfort in the area [27].

Few studies [17,28] have however provided information on the daytime variability of physiologic comfort in any part of Africa- the studies are on Nigeria- and these studies require updating [28] was a microscale study on Ilorin in the guinea savanna part of Nigeria). The present study compares the 1971 and 2001 weather of selected meteorological stations in Nigeria as a complement to the study of the

*Corresponding author: Eludoyin OM, Department of Geography and Planning Sciences, AdekunleAjasin University, Akungba-Akoko, Ondo State, Nigeria E-mail: baynick2003@yahoo.com

59 year (1951-2009) average data of physiologic comfort. The study also sought insight into the indigenous perception of both indoor and outdoor workers in selected tertiary institutions across Nigeria. Perception on physiologic comfort is known to be subjective factors such as types of clothing wears used to cover the skin covers, previous weather experience and certain adaptation factors, including culture and body type [29-31]. The overall goal is to determine the average change in physiologic stressed based on daytime variation and assess the knowledge of efficient coping strategies in case of an endangering physiologic stress.

Study Area

Nigeria is located within 4-14°N and 3-15°E in the southeastern edge of the West African region, and is characterized by dry season (usually accompanied by the tropical continental airmass influenced dust–laden or Harmattan wind from the Sahara desert) and rainy season (which is strongly influenced by the tropical maritime from the Atlantic Ocean. The Nigerian climate can be grouped into: the tropical rainforest climate, tropical savanna climate and highland climate or montane climate [32]. The tropical rainforest climate ('Af' by Köppen climate classification) characterises the southern region, and can be sub–grouped into the tropical wet and tropical wet and dry climates while the Tropical savanna climate comprises the guinea, sudan and sahel savanna, and characterizes most of the central and northern regions. The guinea belt occupies the limits of tropical rainforest climate, and extends to the central part while the northern fringe is occupied by the sudan tropical savanna climate. The north–eastern fringes exhibit the sahelian climate while the montane climate occurs in settlements on high altitude (especially above 1520 m as in the Plateau Mountains) (Figure 1). A little less than 50% of the above 100 million Nigerian populations live in the urban areas as 2010 (Table 1). The rural population, which contains about 50% of the remaining population in Nigeria is however characterised by poor infrastructure, including poor access to healthcare system, education and communication facilities. The poor electricity in Nigeria also made many people to be vulnerable to heat stress and other environmental impact, typical of developing countries in Africa, Asia and Latin America [8].

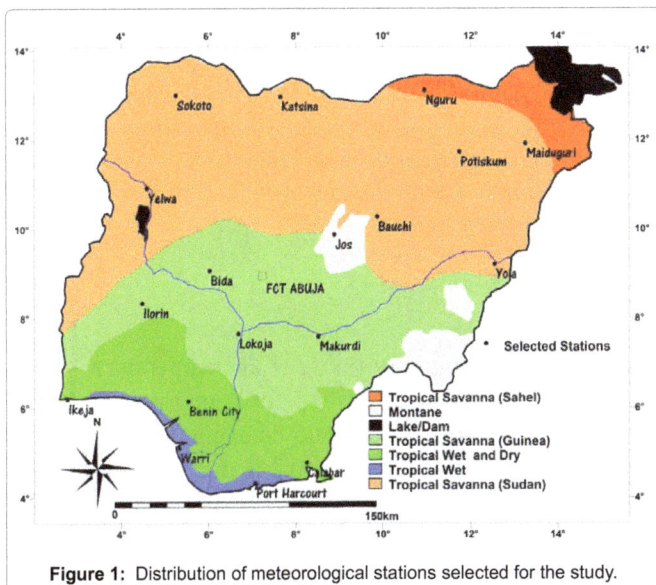

Figure 1: Distribution of meteorological stations selected for the study.

Variables	Specifics	Rate
Landuse	Deforestation	4000 sq. km per year
	Reforestation	10 sq. km per year
	Forested area (2008)	10.8%
Urban Population	Annual growth	3.8%
	Urban Population in 2004, 2010	45%, 48.9%
Rural Population	Annual growth	1.8%
Total population	Population density in 2004, 2009	137.6, 167.5 persons per km
	Annual growth	2.5%
Total fossil fuels emission	1951	460'000 metric tons
	1980	18,586'000 metric tons
	2008	26113'000 metric tons

Table 1: Some information about land area and population of Nigeria (2011 World Statistics Country Profile).

Materials and Methods

Two types of data sources were used for this study. The first sets of data were the temperature and relative humidity data, which were obtained from the Nigerian Meteorological (NIMET) Stations' office, Lagos. The NIMET office in Lagos is a custodian of the quality assured climate data obtained from all the meteorological stations across Nigeria. The meteorological stations, located at least in each State and in other areas of specialized government (and colonial) interests make up of about 56 stations out of which less than 20 possessed temperature and relative humidity data from 1951, although most of the stations have long-term (pre-1951) rainfall data. Selected integrative indices required data on temperature and relative humidity, hence only meteorological stations with these data were selected. For representativeness, a network of 2° by 2° grid was overlayed on the map of Nigeria, and a representative station was selected for each grid, where possible. Since only daytime (0600, 0900, 1200, 1500, 1800 and 2100 hour) temperature and relative humidity data were available (because data were collected at 3 hour interval), this study used only the daytime (0600-2100 Local Standard Time, LST) relative humidity and temperature data for 20 meteorological stations to compute the selected thermal comfort indices (effective temperature index, temperature-humidity index and relative strain index) (Figure 1). Selected thermal comfort indices are described in Box 1. The 20 stations were selected to represent the different sub-climate types (tropical wet (Port Harcout, Warri); tropical wet and dry (Ikeja, Benin and Calabar); guinea savanna (Bida, Ilorin, Lokoja, Makurdi); sudan savanna (Katsina, Sokoto, Maiduguri, Potiskum, Bauchi, Yola and Yelwa); sahel savanna (Nguru); and montane (Jos)) (Figure 2). Day-time thermal comfort for 1971 and 2001 was also compared to provide experience for typical cool and warm year thermal comfort for Nigeria (the choice of 1971 and 2001 was based on the global temperature model [33] and data availability).

Values for the investigated meteorological stations and periods were interpolated using the moving averages technique to plot the descriptive maps for Nigeria, and the monthly day-time thermal comfort for selected stations with standard geographic information (ILWIS, version 3.4, and a third party, SURFER) software. Classification of the physiologically comfortable (ETI values of 18.9°-25.6°C or THI values of 15°-24°C or RSI values of 0.1-0.2 (ratio, no unit)) and regions that exhibited cold or heat stress condition (cold stress occurs at values below the minimum while the maximum threshold marks the beginning of a heat stress condition) were determined based on the information in Box 1.

Secondly, preference and responses of Nigerians on thermal

Effective Temperature Index
The effective temperature index (ETI) was originally intended for indoor conditions in industry and mines and not for open air conditions (Gregorczuk and Cena, 1967) but has later gained wider applications in comfort and climatological analyses in north America and Europe. The results of ETI represent the thermal sensations of a man insulated from air movement and from solar radiation. Values of ETI is usually estimated as (1-2).

$$T_{eti} = t - 0.4(t-10)(1-\frac{H_{rh}}{100}) \quad (1), \qquad T_{eti} = 0.4(t_{dry} + t_{wet}) + 4.8 \quad (2)$$

Where

T_{eti} = effective temperature, t = air temperature, H_{rh} = relative humidity, t_{dry} = dry bulb temperature, t_{wet} = wet bulb temperature.

Uncomfortable situations due to cold stress occur with ETI ≤ 18.9°C and heat stress > 25.6°C (Ayoade 1978). Challenges with ETI are; that the basic observations on the original model development were made on a specific group of subjects; healthy young white men and women living under American conditions of climate, housing and clothing, the observation relate only to sedentary conditions; and that the index in its original form gives too much weight to changes in humidity at the lower end of the Comfort index scale.

Temperature-Humidity Index (THI)
The THI was developed by Thom (1959) for providing a broad approximation of stress changes in a city over time and for developing useful design guidelines for cities (Jáuregui and Soto, 1967). The original index, which combined the wet and dry bulb temperatures to produce the THI, has been modified by McGregor and Nieuwolt (1998) to use air temperature and relative humidity (3)

$$T_{thi} = 0.8 * t + \frac{H_{rh} * t}{500} \qquad (3)$$

Where

T_{thi} = Temperature-humidity index, t *is* the air temperature (°C) and H_{rh} *is* the relative humidity (%).
Nieuwolt (1977) suggested the following THI classification, based on the climate of the United States of America and suggestion of the United States Weather Bureau (1959).

      a. 21 - 24 °C = 100% of the subjects felt comfortable
      b. 24 - 26 °C = 50% of the subjects felt comfortable
      c. THI > 26 °C = 100% of the subjects felt uncomfortably hot

Relative Strain Index (RSI)
In order to take account of the effects of clothing and net radiation, the relative strain index (RSI) was developed for a sedentary standard man dressed in a business suit (healthy, 25 years old and not acclimatized to heat) (Giles et al., 1990; Lee, 1953). The RSI is the ratio of the rate at which heat accumulates in the body of one organism to the maximum possible rate at which it can be removed. The RSI is expressed in the equations (3-5)

$$T_{rsi} = \frac{(t-21)}{(S8-e)} \qquad (3), \quad e = \frac{(H_{rh} \times H_{vp})}{100} \quad (4), \qquad RSI = \frac{[10.7 + 0.74(T-35)]}{44 - 0.0075\,HV} \qquad (5)$$

$$V = 6.11 \times 10^{\frac{7.5T}{237.7+T}} \quad (6)$$

Where

T_{rsi} = Relative Strain Index, H_{vp}=Vapour pressure (hPa), H_{rh} = Relative humidity (%), t = air temperature (°C), and e = Dew point temperature. The RSI threshold for physiological failure is 0.5 for young people and 0.3 for people older than 65 years (Giles et al., 1990). A quarter of the people would be uncomfortable at an RSI of 0.2 and no one is comfortable at an RSI of 0.3. For elderly and ill people, 0.2 is the threshold above which they are subject to heat stress.

Box 1: Information about selected integrative indices

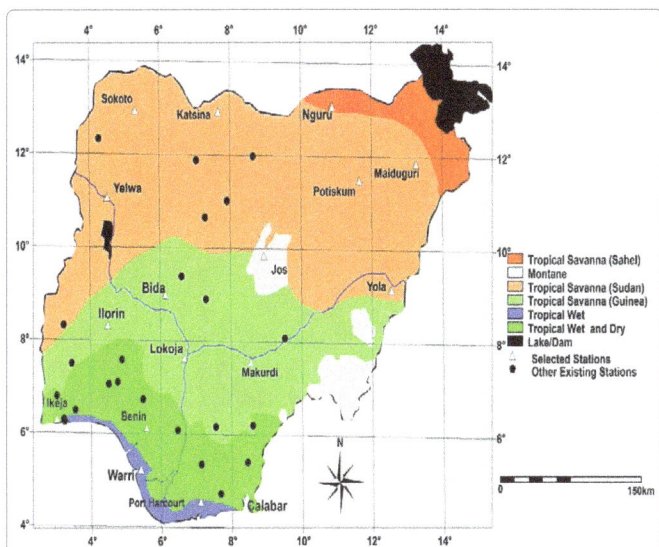

Figure 2: Climate sub-regions in Nigeria showing the location of selected meteorological stations.

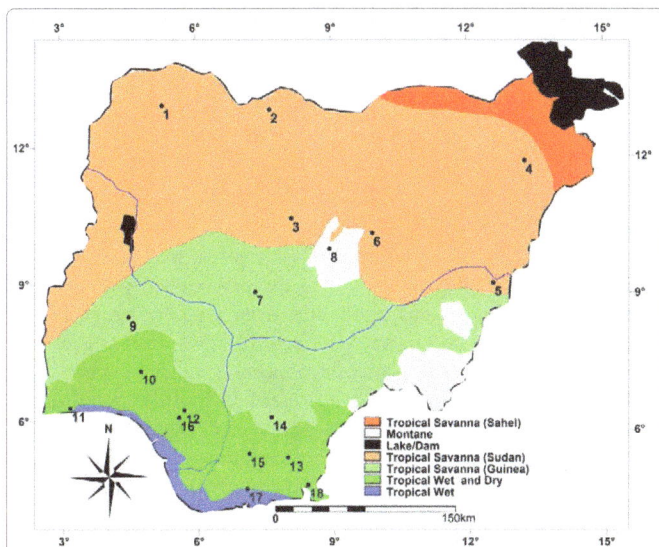

1. Usman Dan Fodio University Sokoto
2. Hassan Usman Katsina Polytechnic
3. Kaduna Polytechnic, Kaduna
4. University of Maiduguri
5. Federal University of Technology, Yola
6. Federal Polytechnic, Bauchi
7. University of Abuja
8. University of Jos
9. University of Ilorin
10. Obafemi Awolowo University, Ile-Ife
11. University of Lagos
12. Adekunle Ajasin University, Akungba-Akoko
13. Abia State University
14. Enugu State University
15. Imo State University
16. University of Benin
17. University of Port Harcourt
18. University of Calabar

Figure 3: Locations where questionnaire were administered for this study.

comfort was assessed by administering a set of questionnaire to about 200 randomly selected workers from each of purposively selected 18 tertiary (University, Polytechnic, or College of Education) locations in Nigeria (Figure 3).

Crocombe and Malama (1989) had argued that the responses from a tertiary institution could sometimes be stronger than that of a community, probably because those in schools are likely to be more conscious and inquisitive than others within the entire community. To determine the sampled population the Slovin's formula (equation 7)

was applied for a targeted population of 10,000 in each institution.

$$\text{Sample population} = \frac{N}{1 + Ne^2} \qquad (7)$$

N= total targeted population, e=confidence level (0.05 for 95%)

A sample size of 200 was finally accepted because less than the targeted sample sizes for most institutions were returned.Almost 60% of the respondents worked indoor while 40% worked outdoor. Most of the respondents were adults between the ages of 18 and 60 years, who worked either on full or part time basis at the sample locations. The 18-60 years age group makes up the most productive set of the population. More than 70% of the respondents had lived in the sample location for at least 5 years and this gives the group an advantage of fair understanding of the climate of the sample location. More than 99% of the respondents had attained at least primary education status (and 67.5% have had tertiary or post-secondary education), suggesting that they would easily understand the content of the questionnaire and its purpose. At least, 40% of the workers claimed to make a minimum of five hundred thousand (500,000 Naira) per annum; an equivalent of about US $9 per day as at the time of the research.

Results

Characterisation of the Nigerian thermal climate conditions

i. Mean values and variability

The 59-year (1951-2009) average minimum, maximum and mean temperatures in Nigeria (based on the selected 20 stations) are 21.4, 32.8 and 27°C, respectively (Table 2). Table 2 also indicated that minimum temperatures were generally below the overall mean in the tropical savanna (except at Sokoto and Yola) and montane regions while mean maximum temperature in all stations within the savanna was higher than its overall mean, except at Bauchi and Ilorin. The montane, tropical wet and dry and tropical wet climate regions, however, had lower (than the overall average) mean maximum temperature. The mean relative humidity varied between 36.5 and 85.1%, with 62% as the mean. Variations in both temperature and relative humidity increased from south towards the north, except for few stations in the guinea and sudan savanna which exhibited higher variability than the sahel. Lowest mean annual ETI occurred at the montane region (Jos, 19.4°C) while the highest values occurred at Warri in the tropical wet climate (26.1°C). Annual ETI values at most of the stations within the sudano–sahelian and montane (Jos) climate regions are lower than the overall average (24.3°C). Stations within the guinea savanna, and tropical rainforest on the other hand, exhibited higher ETI than the overall average. Highest mean THI (26.3°C) occurred in Lokoja (Guinea savanna) while the montane station exhibited the lowest (19.6°C) mean THI. Highest RSI ratio (0.2) occurred at Warri in the Tropical wet climate while the montane region exhibited the smallest ratio (0.01) (Table 2).

ii. Trends in the unitary (temperature, relative humidity) and integrative indices

The results of linear regression analysis and their level of significance for each variable and at the investigated stations indicated that temperature variables and the integrative indices exhibited significantly increasing trend at most stations, except the montane and few stations in the savanna. Both minimum and mean temperatures exhibited decreasing trends at the montane region, while maximum temperature at Yelwa has increased.

Climate region	Meteorological Station	Temperature (°C)						Relative Humidity (%)		ET (°C)		THI (°C)		RSI (ratio)	
		Minimum		maximum		Mean									
		M	SD	M	SD	M	SD	M	SD	M	SD	M	SD	M	SD
Tropical Savanna (Sahel)	Nguru	21.2	4.4	35.3	3.6	28.2	3.5	36.5	21.1	23.1	1.5	24.2	1.5	0.17	0.05
Tropical Savanna (Sudan)	Katsina	19.4	4.3	33.7	3.4	26.5	3.5	38.9	23.5	22.5	0.6	23.3	0.6	0.13	0.02
	Sokoto	22.0	3.6	35.2	3.2	28.6	3.0	42.7	23.6	24.4	0.6	25.4	0.6	0.19	0.03
	Maiduguri	20.1	4.6	35.3	3.6	27.7	3.6	39.9	20.7	23.5	0.6	24.4	0.6	0.16	0.02
	Potiskum	19.7	4.5	34.4	3.3	27.0	3.4	39.5	23.9	22.4	1.3	23.3	1.2	0.13	0.10
	Yelwa	21.3	3.6	34.1	3.0	27.7	2.3	60.7	19.1	24.9	0.5	25.5	0.7	0.18	0.06
	Bauchi	19.0	3.5	32.7	2.8	25.8	2.6	46.7	23.1	22.4	0.7	23.1	0.7	0.09	0.10
	Yola	21.7	3.2	34.7	3.1	28.2	2.4	55.2	23.5	24.9	0.9	25.7	0.9	0.21	0.05
Montane	Jos	16.0	2.3	27.6	2.3	21.8	1.7	50.0	25.9	19.4	0.3	19.6	0.4	0.01	0.01
Tropical Savanna (Guinea)	Bida	22.8	1.8	33.7	2.8	28.3	1.9	64.4	18.1	25.5	0.8	26.1	0.6	0.21	0.06
	Ilorin	21.3	1.7	32.2	2.6	26.8	1.6	74.4	11.6	24.4	0.5	24.9	0.5	0.17	0.02
	Lokoja	22.8	2.0	33.0	2.4	27.9	1.7	73.4	8.2	25.9	0.7	26.3	0.6	0.22	0.04
	Makurdi	22.3	2.5	33.3	2.6	27.8	1.8	69.8	14.2	25.4	0.9	25.9	0.7	0.21	0.04
Tropical Wet and Dry	Ikeja	23.1	1.3	30.9	1.9	27.0	1.4	82.6	5.8	25.3	0.6	26.1	0.6	0.21	0.04
	Benin	23.0	0.8	31.3	2.0	27.1	1.4	83.9	5.7	25.9	0.6	26.1	0.6	0.21	0.04
	Calabar	22.8	1.0	30.6	1.8	26.8	1.2	85.1	5.0	25.7	0.9	25.9	0.8	0.19	0.09
Tropical Wet	Warri	23.1	0.9	31.4	1.8	27.3	1.2	83.9	4.8	26.1	0.4	26.3	0.9	0.22	0.02
	Port Harcourt	22.4	1.2	31.1	1.8	26.7	1.1	83.4	5.5	25.6	0.4	24.8	2.1	0.20	0.03
Overall mean		21.4	3.4	32.8	3.4	27.0	2.8	62.0	24.8	24.1	0.9	24.8	1.8	0.20	0.18

Table 2: Descriptive statistics of temperature, relative humidity and thermal climate at selected meteorological stations in Nigeria (M represents mean, and SD is the standard deviation).

On the other hand, trends of temperatures, relative humidity has decreased in most regions, especially in the tropical rainforest region (b ≥ -0.07; p<0.05). Bauchi in the sudan savanna region however exhibited significant increase (b=0.17; p<0.05) within the study period. Similarly, ETI, THI and RSI exhibited increased trend at most of the stations, especially within the sudan and sahelian savanna as well as tropical rainforest regions (Table 3a and 3b).

iii. **Daytime thermal comfort**

Evaluated thermal indices (THI, ETI and RSI) showed temporal and spatial variations. Peak of the heat stress condition occurred at 1500 Local Standard Time (LST) while the early morning time (before 0900 LST) were more conducive (Figure 4). Figure 4 also shows that the monthly variations of thermal comfort at selected settlements in Nigeria exhibited spatial variations. The results of ETI and RSI were similar, and therefore only ETI and THI are presented. In both ETI and THI, Lokoja (in the guinea savanna) showed most day-time hours of heat stress (1100-1700 LST) while Jos (montane region) exhibited the least number of hours with day-time thermal stress condition. Calabar and Benin (tropical wet) exhibited heat stress in the afternoon (1200-1700 LST) in June-October. Heat stress at the tropical wet regions occurred for more hours (1000-1700 LST) between November and April.

When compared, thermal comfort in 1971 exhibited slight difference from that of 2001, especially at 0900 LST and 2100 LST (Figure 5). Some of regions (in the northwest) that were mapped to experience cold stress in 1971 were comfortable (some have also exhibited heat stress) in 2001. The tropical wet region also have shown increased level of heat stress around 0900 LST. The ETI map suggests that cold stress condition extended further around the montane region. The results of the comparison also indicated a prevalence of cold stress in a typical cool year, and heat stress in a warm year.

Responses of Nigerians to thermal climate

Whilst the perceptions of sampled Nigerians varied on the seasonal distributions of heat stress (p ≤ 0.05 for dry season, p ≥ 0.05 for Harmattan and rain season), about 50% from southern (tropical wet) and northern (savanna) regions described the dry season as generally warm and characterized by heat-related morbidity, including headache and heat rash. The Harmattan was described by most respondents (>70%) as dusty, windy and linked with dry eyes, dry skin and dry throat while the rainy season is generally cool (and in some cases, cold) and linked with severe cold, headache and cough. About 90% of the sample populations responded to heat stress condition mainly by alternating their modes of dresses (from thick to thin layered dress in dry season or vice versa in the rainy and Harmattan season. Majority (>64%) also cover their head and arms as response to the dusty Harmattan or severe cold in both Harmattan and rainy season) (Figure 6). While alternating different dressing mode to cope with a prevailing weather is a practice worldwide, people whose income is above the poverty level (the poor is defined here as those whose concern is primarily on feeding themselves and families) prefer to install air conditioners in their car, house and office (or at least in one of them) in the present study.

Figure 7 shows the preference for meeting the thermal discomfort challenges in the offices and houses. At least 70% of the entire respondents were however not disturbed about the seasonal variations in the thermal comfort as more than half of the respondents (51%) largely attributed climate change to 'what only god can change' or 'what god uses to punish the people where he is angry with them'.

Discussion

Main objective of this study was to assess physiologic comfort across Nigeria and determine if Nigerians are well prepared to cope with future physiologic stress. The study indicated significant variability and

Table 3a: Linear trends in the unitary (temperature and relative humidity) (3a) and integrative indices (3b) in selected stations in Nigeria.

Climate Region	Stations	Temperature (°C)			Relative Humidity (%)
		Minimum	Maximum	Mean	
Tropical Savanna (Sahel)	Nguru	19.98+0.04(x)*	34.67+0.02(x)	27.32+0.03(x)*	36.75-0.01(x)
	Katsina	18.82+0.02 (x)	33.11+0.02(x)*	25.97+0.02(x)*	41.65+-0.1(x)
	Sokoto	20.38+0.01(x)*	34.6+0.02(x)*	27.47+0.03(x)*	40.76+0.06(x)
	Maiduguri	19.34+0.02(x)	35.04+0.01(x)	26.85+0.03(x)	42.81-0.09(x)
Tropical Savanna (Sudan)	Potiskum	18.45+0.04(x)*	33.42+0.03(x)*	25.95+0.03(x)*	40.76-0.04(x)
	Yelwa	20.96+0.01(x)	34.81-0.03(x)*	26.89-0.01(x)	60.02+0.03(x)
	Bauchi	17.82+0.04(x)*	32.17+0.02(x)*	24.99+0.03(x)*	41.97+0.17(x)*
	Yola	20.77+0.04(x)*	34.36+0.01(x)	27.56+0.02(x)*	53.99+0.04(x)
Montane	Jos	17.19-0.03(x)*	27.01+0.02(x)*	21.85-0.002(x)	44.72+0.16(x)
	Bida	22.27+0.02(x)	33.05+0.02(x)*	22.66+0.02(x)*	66.76-0.09(x)
Tropical Savanna (Guinea)	Ilorin	20.42+0.03(x)*	32.74-0.02+(x)	26.58+0.01(x)	75.99-0.05(x)
	Lokoja	22.45+0.01(x)*	32.59+0.02(x)*	27.52+0.01(x)*	72.44-0.07(x)*
	Makurdi	21.18+0.03(x)*	32.52+0.02(x)*	27.19+0.02(x)	71.55-0.05(x)
	Ikeja	21.65+0.05(x)*	30.25+0.02(x)*	25.95+0.04(x)*	86.28-0.13(x)*
Tropical Wet and Dry	Benin	22.63+0.01(x)*	30.68+0.02(x)*	26.26+0.03(x)*	84.87-0.03(x)*
	Calabar	21.79+0.04(x)*	29.96+0.02(x)*	26.29+0.02(x)*	86.55-0.04(x)*
Tropical Wet	WWarri	22.39+0.03(x)*	31.16+0.01(x)*	26.77+0.02(x)*	84.85-0.03(x)*
	Port Harcourt	21.79+0.02(x)*	30.17+0.03(x)*	25.98+0.02(x)*	84.56-0.03 (x)

Linear trend of the asterisked (*) row is significant at the corresponding station within 95% confidence level (p≤0.05)

Table 3b: Linear trends in thermal comfort indices in selected stations in Nigeria.

Climate Region	Stations	Effective temperature (°C)	Temperature-humidity index, THI (°C)	Relative strain index, RSI (no unit)
Tropical Savanna (Sahel)	Nguru	21.46+0.06(x)*	22.64+0.05(x)*	0.17+0.0001(x)
	Katsina	22.45+0.001(x)	25.11+0.005(x)	0.13-0.0003(x)
	Sokoto	23.44+0.03(x)*	24.27+0.033(x)*	0.15+0.001(x)*
	Maiduguri	23.49+0.0001(x)	24.27+0.003(x)	0.14+0.001(x)*
Tropical Savanna (Sudan)	Potiskum	20.70+0.06(x)*	21.71+0.05(x)*	0.04+0.003(x)
	Yelwa	24.90-0.01(x)	25.67-0.007(x)	0.21-0.0001(x)
	Bauchi	21.61+0.03(x)*	22.18+0.03(x)*	0.07+0.001(x)
	Yola	24.07+0.03(x)*	24.72+0.03(x)*	0.15+0.0002(x)*
Montane	Jos	19.31+0.003(x)	19.41+0.006(x)	0.01+0.0001(x)
	Bida	25.64-0.01(x)	25.98+0.003(x)	0.22+0.0001(x)
Tropical Savanna (Guinea)	Ilorin	24.77-0.01(x)	25.12-0.006(x)*	0.17+0.0001(x)
	Lokoja	25.90+0.001(x)	26.27+0.003(x)	0.22-0.0001(x)
	Makurdi	25.59-0.006(x)	25.87+0.001(x)	0.21+0.0001(x)
	Ikeja	24.59+0.03(x)*	25.26+0.03(x)*	0.16+0.002(x)*
Tropical Wet and Dry	Benin	25.14+0.03(x)*	25.25+0.03(x)*	0.17+0.002(x)*
	Calabar	25.89-0.01(x)	25.98-0.002(x)	0.21+0.0001(x)
Tropical Wet	Warri	25.62+0.02(x)*	25.82+0.02(x)*	0.20+0.001(x)*
	Port Harcourt	24.89+0.02(x)*	22.06+0.10(x)*	0.15+0.0001(x)*

Linear trend of the asterisked (*) row is significant at the corresponding station within 95% confidence level (p≤0.05)

showed potential for physiologic stress at many meteorological stations in Nigeria. Features in the Nigerian urban centres which encouraged heat related physiologic stress include the extensive transport and commercial activities in the regions [34]. Studies have equally indicated that the Nigerian climate is often affected by the movements of the Intercontinental tropical discontinuity, ITD, prevailing air masses, relief, continentality, proximity of river bodies, and anthropogenic factors, including urbanisation, gas flaring activities among others [28,32,35,36]. Commercial activities, transportation and industry are growing in many parts of Nigeria, and with them administration and the human population, all of which can also impact the local climate in different regions. The day-time and night-time discrepancy in the savanna and tropical wet is supported by the difference in the relative humidity in these regions. The tropical wet regions are characterised by thick cloud, which can prevent penetration of solar energy and maximise thermal comfort in the tropical wet, especially in day-time. The savanna region exhibits lower relative humidity, and this suggest less cloud cover with the consequent high radiation (heat) in the region. Conversely, re-radiation of heat can be delayed by the thick cloud cover in the tropical wet region than will be delayed in savanna region.

In addition, the results of this study indicated that 1200-1500 LST was the most thermally stressful in Nigeria. This is typical of the tropical region where the sun is known to be directly overhead at noon before the heat accumulates and peaks shortly after. Samendra and Ayesha [37] also showed that temperatures and heat conditions usually peak at 1500

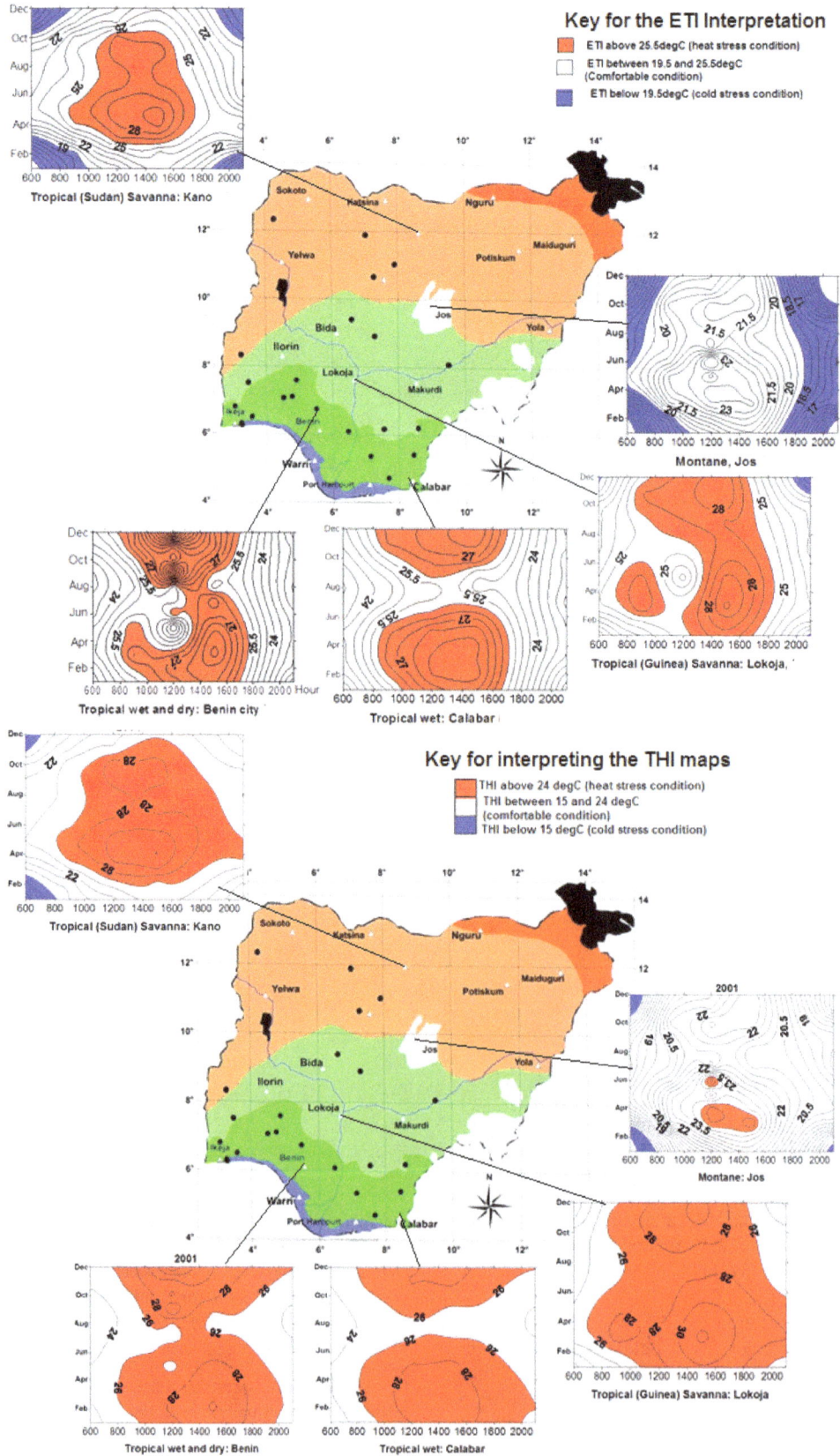

Figure 4: Representative pattern of physiologic comfort in Nigeria based on ET (a) and THI (b).

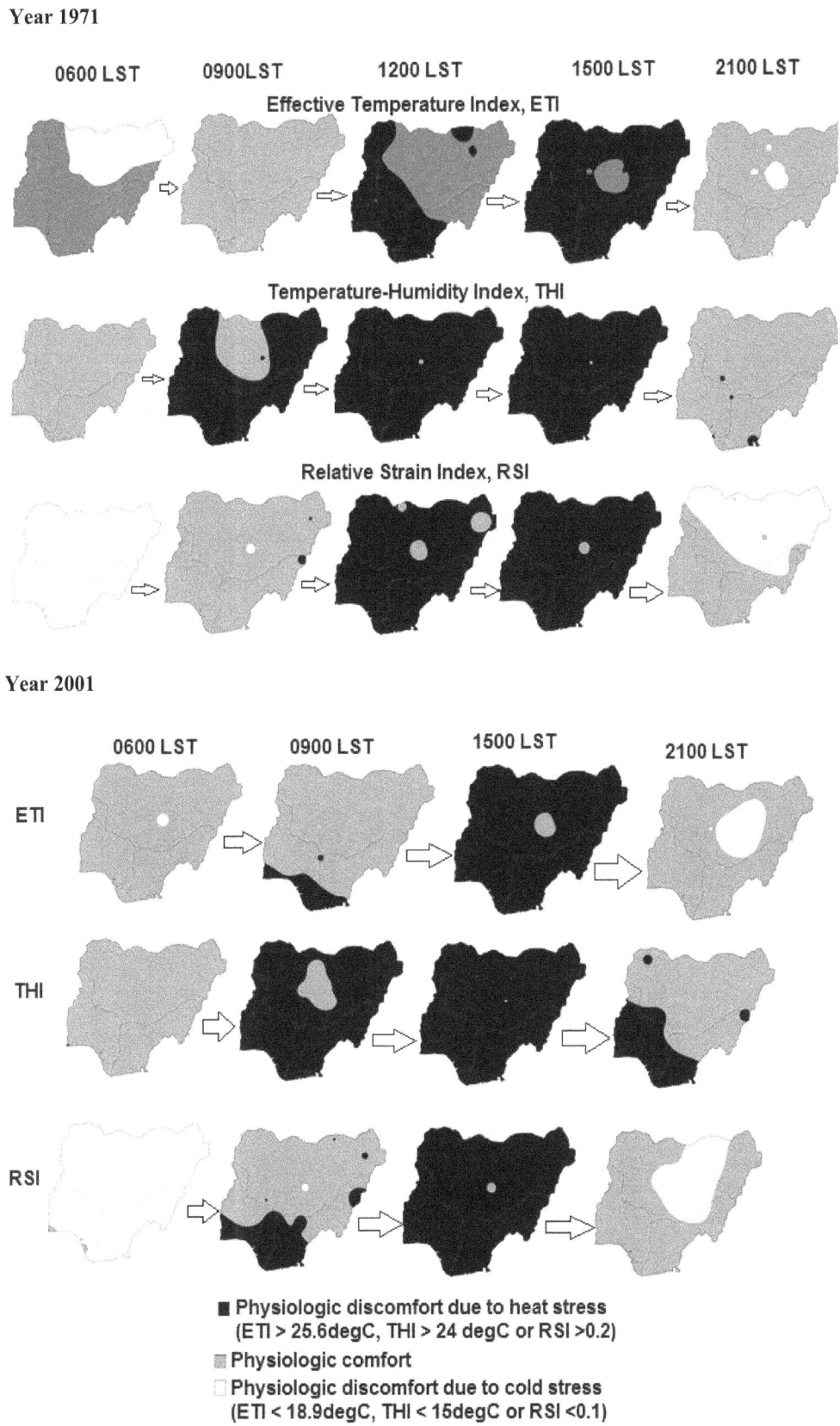

Figure 5: Mean hourly patterns of physiologic comfort in Nigeria for 1971 and 2001 (Note: Data was not available for most stations at 1200 LST in 2001, and was therefore not mapped for the period).

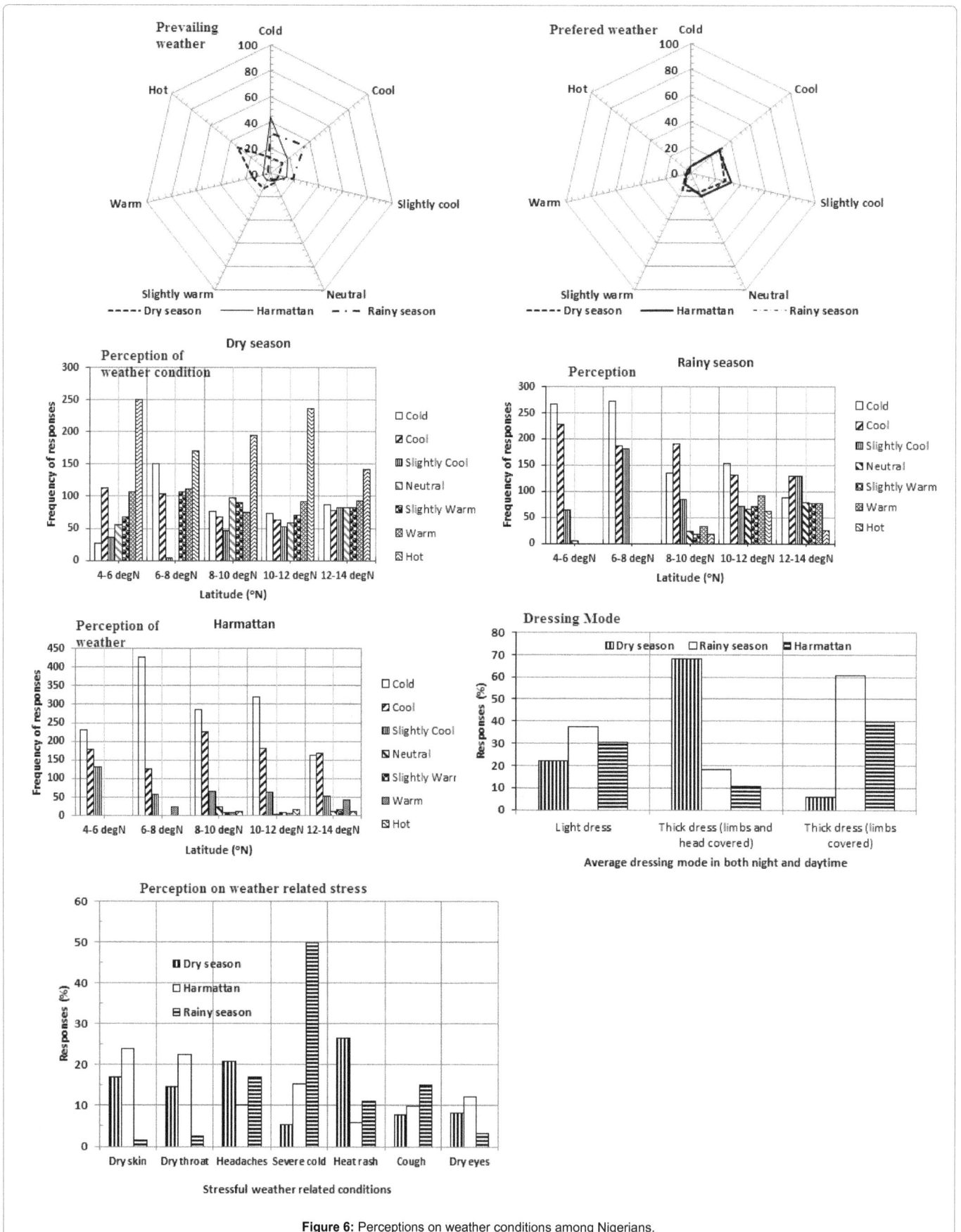

Figure 6: Perceptions on weather conditions among Nigerians.

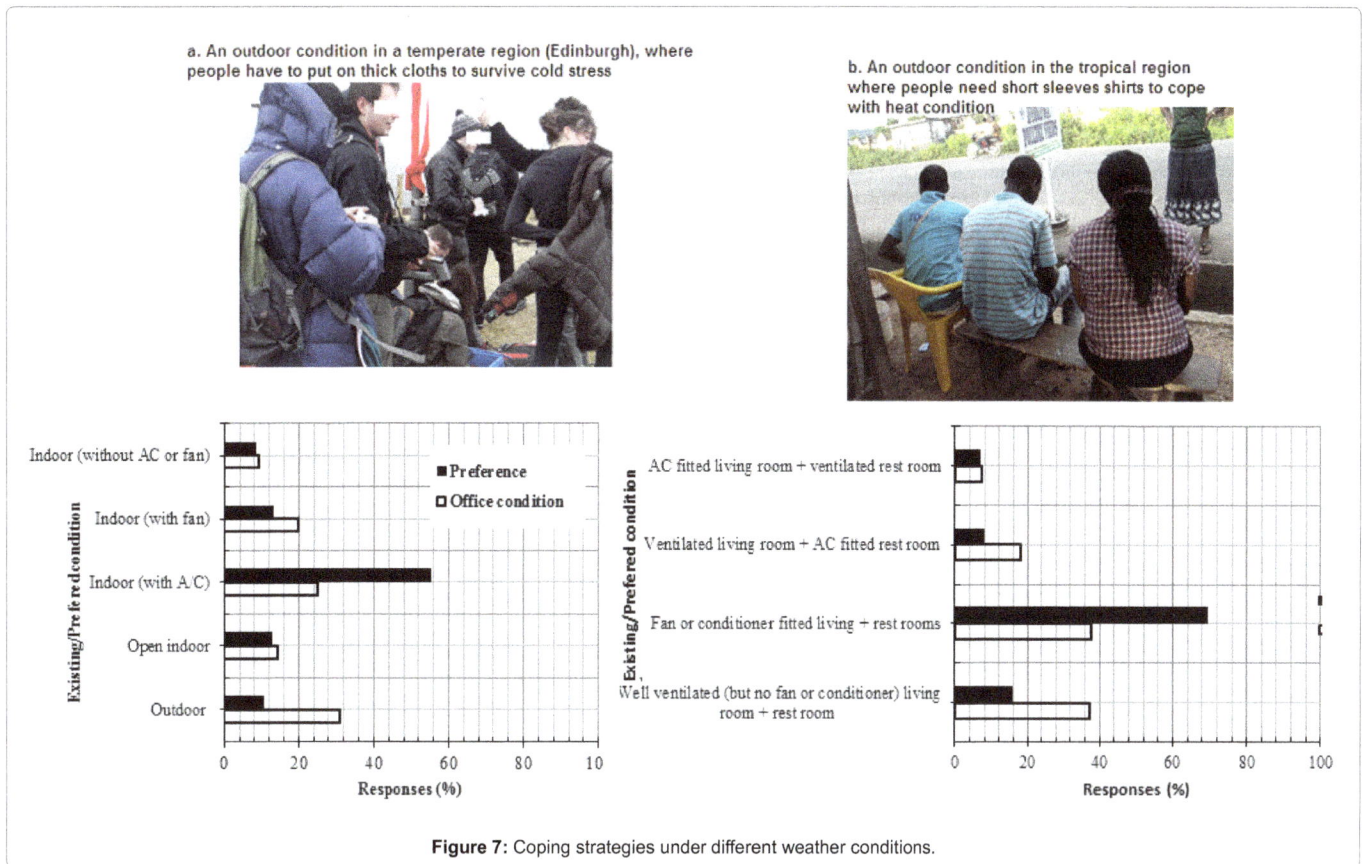

Figure 7: Coping strategies under different weather conditions.

hour in Dhaka, Bangladesh, and that more people feel uncomfortable around this hour than other hours of the day. Runnals and Oke [38] also showed that maximum heat condition occurred around this time (afternoon), and argued that the morning and night are usually more comfortable. Furthermore, the responses indicated low awareness to the morbidity and mortality consequences of extreme climate effects. Sawa and Buhari [39] has attributed the outbreak of measles and meningitis in Zaria, Nigeria in the last decade to extreme temperature while many cases of mortality were recorded as a result of heat waves in North America [22,39-41].

However, unlike the other regions which have mapped weather health-response plans [22,40], Nigeria does not (as at the period of this paper) have a documented plan or any infrastructure to respond to extreme weather events, other than the National Emergency Management Agency, that has often been criticized for its inefficiency [42]. Many reports have indicated that Nigeria, like most developing countries is not prepared for the challenge of extreme climate and climate change, especially because of poor technology and poor resource allocation [8,43,44]. Allen [45], in a study of the assessment of the millennium development goal scores of Ondo State (in the southwest Nigeria) indicated that most states in Nigeria will poorly perform in the areas of sanitation, water supply and health. Given the results of the peoples' perceptions on thermal climate in Nigeria, it can be generally deduced that poor climate education and sensibility, poor technology (for cost effective, cheap and energy saving devices), unequal distribution of financial wealth, poor social welfare schemes (and where they exist, some of the programmes are poorly implemented and severely infested by corrupt or discriminatory practices) are the bane of social infrastructure in Nigeria.

Conclusions and Recommendations

This study has examined the thermal comfort in Nigeria and the responses to thermal stress. The study showed that thermal stress will increase in many parts of Nigeria as indicated by results of the linear regression on 59 years data, probably due to increasing rate of urbanization, population and the global temperature increase. The study however does not indicate adequate indigenous approach of physiologic stress among the people within the community of tertiary education, and there is no evidence to indicate any preparation to cope with future physiologic stress. The study recommends institutional approach towards a high level preparation for future heat stress, and this including developing a climate-oriented healthcare systems in Nigeria. This study shows the typical issues on future physiologic stress in many developing countries in the tropical region.

References

1. Lee DHK (1953) Physiological climatology as a field of study. Annals of the Association of American Geographers 43: 127-137.

2. Alcamo J, Moreno JM, Novaky B, Bindi M, Corobov R, et al. (2007) Climate Change 2007: Impacts, Adaptation and Vulnerability. In: Parry ML, Canziani OF, Palutikof JP, van der Linden PJ, Hanson CE (eds.), Contribution of Working Group II to the Fourth Assessment Report of the Intergovernmental Panel on Climate Change. Cambridge University Press, pp. 541-580.

3. Smith JB, Schneider SH, Oppenheimer M, Yohe GW, Mastrandrea MD, et al. (2009) Assessing dangerous climate change through an update of the Intergovernmental Panel on Climate Change (IPCC) reasons for concern. Proceedings of the National Academy of Sciences 106 : 4133-4137.

4. Terjung W (1967) The geographical application of some selected physio-climatic indices to Africa. International Journal of Biometeorology 11: 5-19.

5. Jauregui E (1993) Urban bioclimatology in developing countries. Experientia 49: 964-968.

6. Matzarakis A, Amelung B (2008) Physiological Equivalent Temperature as Indicator for Impacts of Climate Change on Thermal Comfort of Humans. Seasonal Forecasts, Climatic Change and Human Health 30: 161-172.

7. Boko M, Niang I, Nyong A, Vogel C, Githeko A et al. (2007) ClimateChange 2007: Impacts, Adaptation and Vulnerability, Contribution of Working Group II. In: Canziani OF, Parry ML, Palutikof JP, van der Linden PJ, Hanson CE (eds.), Fourth Assessment Report of the Intergovernmental Panel on Climate Change. Cambridge University Press, Cambridge, pp. 433-467.

8. United Nations Population Fund (2007) UNFPA. The state of world population 2007: Unleashing the potential of urban growth, UNFPA, USA, pp. 32.

9. De Dear R, Fountain M (1994) Field experiments on occupant comfort and office thermal environments in a hot-humid climate. ASHRAE Transactions 100: 457-475.

10. Wong NH, Khoo SS (2003) Thermal comfort in classrooms in the tropics. Energy and Buildings 35: 337-351.

11. Unger J (1999) Comparisons of urban and rural bioclimatological conditions in the case of a Central-European city. International Journal of Biometeorology 43: 139-144.

12. Smoyer KE, Rainham GC, Hewko JN (2000) Heat-stress-related mortality in five cities in Southern Ontario: 1980–1996. International Journal of Biometeorology 44: 190-197.

13. Terjung WH (1966) Physiologic Climates of the Conterminous United States: A Bioclimatic Classification Based on Man. Annals of the Association of American Geographers 56: 141-179.

14. Gregorczuk M, Cena K (1967) Distribution of effective temperature over the surface of the earth. International Journal of Biometeorology 11: 145-149.

15. Peel C (1958) Thermal conditions in traditional mud dwellings in northern Nigeria. The Journal of Tropical Medicine and Hygiene 61: 189-203.

16. Peel C (1961) Thermal comfort zones in Northern Nigeria: an investigation into the physiological reactions of nursing students to the thermal environment. The Journal of Tropical Medicine and Hygiene 64: 113-121.

17. Ayoade J (1978) Spatial and seasonal patterns of physiologic comfort in Nigeria. Theoretical and Applied Climatology 26: 319-337.

18. Eludoyin OM, Adelekan IO (2013) The physiologic climate of Nigeria. Int J Biometeorol 57: 241-264.

19. Eludoyin OM, Adelekan IO, Webster R, Eludoyin AO (2014) Air temperature, relative humidity, climate regionalization and thermal comfort of Nigeria. International Journal of Climatology 34: 2000-2018.

20. Alessandro A, de Garín A (2003) A study on predictability of human physiological strain in Buenos Aires City. Meteorological Applications 10: 263-271.

21. Iordache MC, Cebuc I (2009) Analysis of the impact of climate change on tourism in some European countries. Analele Stiintifice ale Universitatii" Alexandru Ioan Cuza" din Iasi-Stiinte Economice 56: 270-286.

22. Smoyer KE (1998) A comparative analysis of heat waves and associated mortality in St. Louis, Missouri--1980 and 1995. Int J Biometeorol 42: 44-50.

23. Rainham DGC, Smoyer-Tomic KE (2003) The role of air pollution in the relationship between a heat stress index and human mortality in Toronto. Environmental Research 93: 9-19.

24. Holmer I (2004) Cold but comfortable? Application of comfort criteria to cold environments. Indoor Air 14: 27-31.

25. de Freitas CR (2003) Tourism climatology: evaluating environmental information for decision making and business planning in the recreation and tourism sector. Int J Biometeorol 48: 45-54.

26. Emmanuel R (2005) Thermal comfort implications of urbanization in a warm-humid city: the Colombo Metropolitan Region (CMR), Sri Lanka. Building and Environment 40: 1591-1601.

27. United Nations Statistics Division UNSD (2013) World Statistics Pocketbook, Country Profile, Nigeria.

28. Olaniran O (1982) The physiological climate of Ilorin, Nigeria. Theoretical and Applied Climatology 31: 287-299.

29. Busch JF (1992) A tale of two populations: thermal comfort in air-conditioned and naturally ventilated offices in Thailand. Energy and Buildings 18: 235-249.

30. Davis RE, Knappenberger PC, Michaels PJ, Novicoff WM (2004) Seasonality of climate-human mortality relationships in US cities and impacts of climate change. Climate Research 26: 61-76.

31. Githeko A, Woodward A (2003) International consensus on the science of climate and health: the IPCC Third Assessment Report. Climate change and human health: risks and responses: 43-60.

32. Iloeje NP (2001) A new geography of Nigeria. New Revised Edition, Longman Nig. Ltd, pp. 200.

33. Le Treut H, Somerville R, Cubasch U, Ding Y, Mauritzen C (2007) Historical overview of climate change. Earth Chapter 1, Cambridge University Press, pp. 93-127.

34. Akinbode OM, Eludoyin AO, Fashae OA (2008) Temperature and relative humidity distributions in a medium-size administrative town in southwest Nigeria. J Environ Manage 87: 95-105.

35. Adebayo Y (1991) Day-time effects of urbanization on relative humidity and vapour pressure in a tropical city. Theoretical and applied climatology 43: 17-30.

36. Ayoade J (1983) Introduction to Climatology for the Tropics. Wiley 4: 258.

37. Samendra K, Ayesha K (1994) The variability and probability extremes of some climatic elements over Dhaka. Reports of the Technical Conference on Tropical Urban Climates, Dhaka, Bangladesh.

38. Runnals KE, Oke TR (2000) Dynamics and control of the near-surface heat Island of Vancouver, British Columbia. Physical Geography 21: 283-304.

39. Sawa B, Buhari B (2011) Temperature Variability and Outbreak of Meningitis and Measles in Zaria, Northern Nigeria. Research Journal of Applied Sciences, Engineering and Technology 3: 399-402.

40. Kalkstein LS, Jamason PF, Greene JS, Libby J, Robinson L (1996) The Philadelphia hot weather-health watch/warning system: development and application, summer 1995. Bulletin of the American Meteorological Society 77: 1519-1528.

41. Smoyer-Tomic KE, Rainham DG (2001) Beating the heat: development and evaluation of a Canadian hot weather health-response plan. Environ Health Perspect 109: 1241-1248.

42. Omodanisi ES, Eludoyin AO, Salami AT (2014) A multi-perspective view of the effects of a pipeline explosion in Nigeria. International Journal of Disaster Risk Reduction 7: 68-77.

43. United Nations Environment Programme (2002) UNEP. Global Environment.

44. World Health Organisation WHO (2011) Regional consultation on health of the urban. Proceedings of the 2010 Regional cponsultation of Mubai, India, Regional Office for South East Asia, UNFPA, pp. 82.

45. Allen AA (2013) Population dynamics and infrastructure: meeting the millennium development goals in Ondo State, Nigeria. African Population Studies 27: 229-237.

5

Hydrological Response to Climate Change of the Upper Blue Nile River Basin: Based on IPCC Fifth Assessment Report (AR5)

Sintayehu Legesse Gebre[1]*, Fulco Ludwig[2]

[1]Department of Natural Resources Management, Jimma University, P.o.box 307, Ethiopia
[2]Department of Earth System Science, Wageningen University, P.o.box 47 6700AA, The Netherlands

Abstract

Climate change is likely to affect the hydrology and water resources availability of upper Blue Nile River basin. Different water resource development projects are currently existed and under construction in the region. In order to understand the future impacts of climate change, we assessed the hydrological response of climate change of four catchments (Gilgel Abay, Gumer, Ribb, and Megech) of the upper Blue Nile River basin using new emission scenarios based on IPCC fifth assessment report (AR5). Five biased corrected 50 kms by 50 kms resolution GCMs (Global Circulation Model) output of RCP 4.5 and RCP 8.5 emission scenarios were used. The future projection period were divided in to two future horizons of 2030`s (2035-2064) and 2070`s (2071-2100). The Hydrologic Engineering Center-Hydrological Modelling System (HEC-HMS) was calibrated and validated for stream flow simulation. All the five GCMs projection showed that, maximum and minimum temperature increases in all months and seasons in the upper Blue Nile basin. The change in magnitude in RCP 8.5 emission is more than RCP 4.5 scenario as expected. There is considerable average monthly and seasonal precipitation change variability in magnitude and direction. Runoff is expected to increase in the future, at 2030`s average annual runoff projection change may increase up to +55.7% for RCP 4.5 and up to +74.8% for RCP 8.5 scenarios. At 2070`s average annual runoff percentage change increase by +73.5% and by +127.4% for RCP 4.5 and RCP 8.5 emission scenarios, respectively. Hence, the increase in flow volume in the basin may have a significant contribution for the sustainability of existed and undergoing water development projects. Moreover; it will help for small scale farmer holders to harness water for their crop productivity. However, a precaution of mitigation and adaptation measures ought to be developed for possible flooding in the flood plains area of the River basin.

Keyword: BlueNile; GCM; HEC-HMS; RCP; Scenario

Introduction

In these days the awareness of the effect of climate change due to human activities has been accelerating. Climate change and variability has many significant effects on the hydrological cycle and thus also on hydrology and water resources system. The Intergovernmental panel on climate change [1] has addressed this realization. Green house gasses have played a great role in changing the climate change at global as well as regional level. The release of these gases to the atmosphere has been disturbing the normal composition of the atmosphere [2].

Nowadays there is strong scientific evidence that indicates the average temperature of Earth surface is increasing due to greenhouse gas emissions. For example, the average global temperature has increased by about 0.6°C since the late 19th century and the latest IPCC [3], scenarios project temperature rises of 1.4 - 5.8°C and sea level rise of up to 100 cm by 2100.

Global Warming and precipitation are expected to vary considerably from region to region. Average change in climate, changes in frequency and intensity of extreme weather events are likely to have major impacts on natural and human systems [4]. With respect to hydrology, climate change can cause significant impacts on water resources by resulting changes in the hydrological cycle. For example, the changes on temperature and precipitation can have a direct consequence on the quantity of evapotranspiration and on runoff component. Consequently, the spatial and temporal availability of water resource can be significantly changes which in turn can affect agriculture, industry, and urban development [5].

The Blue Nile River is a main water resource for different transboundary countries which is already under immense pressure due to various competitive uses as well as social, political, and legislative conditions. In addition to these, previous studies show that many parts of the Nile Basin are sensitive to climatic variations [6-9] implying that climate change could have considerable impacts on the water resource availability. Therefore, it is necessary to analyse the possible changes in the available water resource aspect under the changing climatic conditions. However, due to variable climatic regions this impact might not be similar throughout the basin. Hence, dividing the basin into different regions will be a convincing and proficient approach when studying climate change impact.

Despite the fact that the impact of different climate change scenarios projected at a global scale, the exact type, and magnitude of the impact at a catchment scale is not investigated in most parts of the world [10]. Hence, identifying local impacts of climate change at a catchment level is quite important. The Upper Blue Nile River catchments are the main sources for the Blue Nile River basin and their water resources are an important input for the different water development projects and the livelihood support of the people/communities in the basin. Currently, different multipurpose water resources development structures are proposed and under constructions in the river basin. It is very critical

*Corresponding author: Sintayehu Legesse Gebre, Department of Natural Resources Management, Jimma University, P.o.box 307, Ethiopia
E-mail: sintayehulegesse@gmail.com

to determine the hydrological response to climate change for the sustainability of the projects and looking for the possible mitigation measures otherwise all the cost indebted will be lost in failing to meet the objectives. In this study, the new RCP (Representative concentration path) climate scenarios data of the future climate output under assumed radiative forcing scenario will be used for each catchment in the River basin. Then the data will be used as input to the hydrological model (HEC-HMS 3.5) to simulate the effect of climate on the hydrological regimes of the catchments.

In this study the upper Blue Nile River Basin (Gilgelabay, Gumera, Ribb, and Megech) catchments will be modelled and the impacts of the climate change using (RCP 4.5 and RCP 8.5 emission scenarios) climate data will be used to analyse the future water availability in the region.

Description of the Study Area

The upper Blue Nile River basin which is located in the Ethiopian Highlands. The Blue Nile River runs from its origin, Lake Tana, to the Sudanese border and eventually meets the White Nile River at Khartoum, Sudan. The Lake Tana basin is located in north-western Ethiopia (latitude 10.95° and 12.78°N, and longitude 36.89° and 38.25°E) with a drainage area of about 15,000 km² [11]. The Lake Tana, the largest lake in Ethiopia and the third largest in the Nile Basin, is located in this basin. The major rivers feeding the Lake Tana are Gilgel Abay, Gumera, Ribb, and Megech. These rivers contribute more than 93% of the flow to the Lake Tana [12] (Figure 1).

Hydro Climatic Nature of the Blue Nile River Basin

The climate of Upper Blue Nile River basin (Tana basin), is dominated by highland tropical monsoon. Even if the basin is located near to equator the climate is comparatively mild due to high elevation. Most of the rainfall (70-90% total rainfall) occurs from June to September, [13]. April and March are intermediate seasons with some rainfall. The mean annual rainfall of the area is about 1465 mm with significant spatial variation (Figure 2). The mean annual maximum temperature is 25.5 °C and mean annual minimum temperature is 10.8 °C (1988-2005).

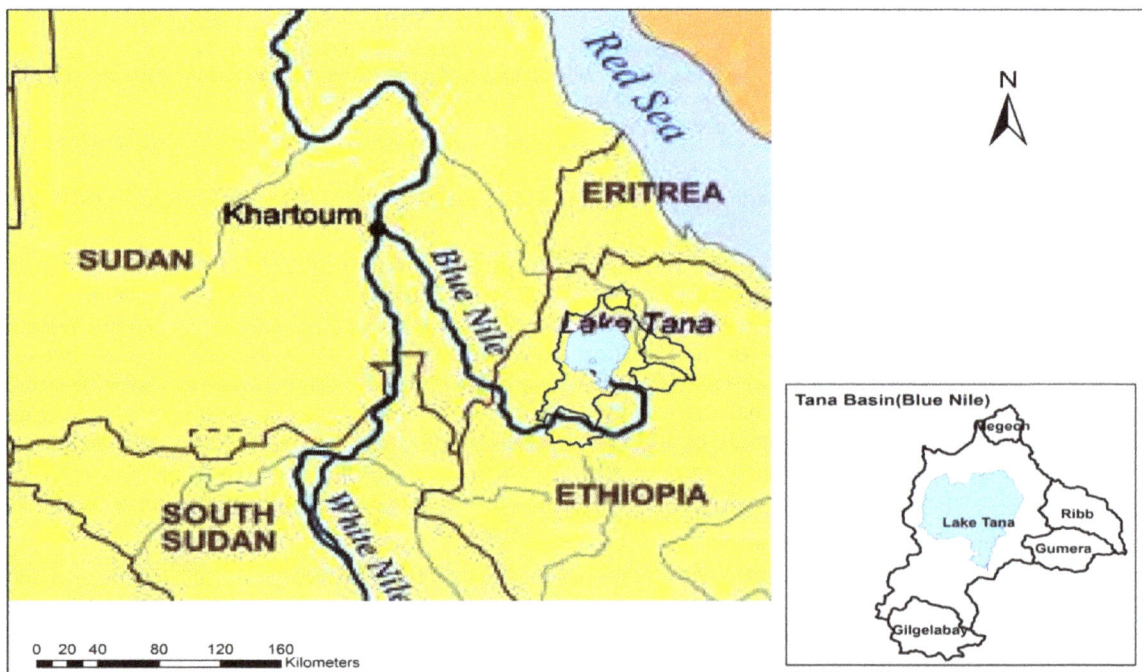

Figure 1: Location of map of Upper Blue Nile River basin.

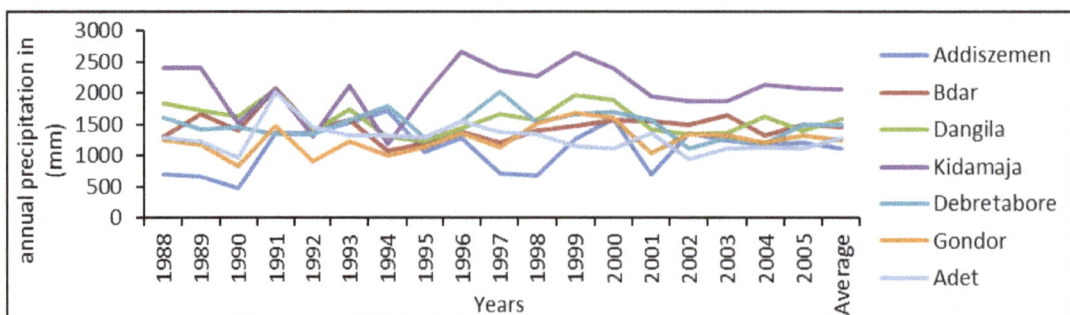

Figure 2: Average annual precipitation for six meteorological stations located in upper Blue Nile River basin from 1988 until 2005.

Hydrology of the Basin

Lake Tana has more than forty tributaries, but the major rivers feeding the Lake are Gilgelabay (the largest river from the south direction), Gumera, and Ribb from the east and Megech from the north, these four main rivers accounts about 93% of inflow. The only river flowing out of the Lake Tana is the Blue Nile River (Abay River) as shown in Figure 3. The Blue Nile flow approximately reaches annually about 4 billion cubic metric at the out let of the lake Tana. From the Lake Tana, the Blue Nile travels around 35 Kms and reaches to a fountain place so called Tisesat which is 50 meter high, then flows in gorges towards the Sudan border. The Blue Nile flow at the Ethio-Sudan border annually reaches about 50 billion cubic meters. In the mean while major tributary rivers joins the Blue Nile, like Beles,Didessa,Fincha, Guder,Muger,Wenchit,Jemma,Beshilo and Temcha. The Blue Nile contributes two third of the Nile River Basin flow [11].

Materials and Methods

General frame work of the study

Data sources and availability

GIS-data SRTM 90 m DEM data was used as an input data for Arc GIS software for catchment delineation and estimation of catchment

characteristic. Hydrological data; stream flow of upper Blue Nile River catchments were used for the calibration and validation the model. Soil, land use, and geological data also used for better understanding of the catchments. All these data were collected from the Ethiopia Ministry of Energy and Water Resources (MoEWR). Meteorological data also collected from National Metrological Station Agency (NMSA), in Addis Ababa and Bahirdar (Figure 4).

Catchment size and weather stations

For each catchment areal precipitation was prepared using Thiessen polygon techniques. The number of observed weather station which contributes for each catchment determined and their areal contribution calculated (Table 1).

Climate scenario data

Five GCMs 0.5 degree by 0.5 degree resolution of RCP 4.5 and RCP 8.5 emission scenarios collected from (Wageningen University, ESS group). These future projections generated based on the Coupled Model Intercomparison Project5 (CMIP5). The inter comparison project started few years ago under the international climate scientist agreed for AR5 (Fifth assessment report). The projections rely on the bases of the new green house gas concentration emission of representative concentration pathways (RCP) stated in [14]. The five different climate models are MPI-ESM-LR, IPSL-CM5A-LR, HadGEM2-ES, EC-

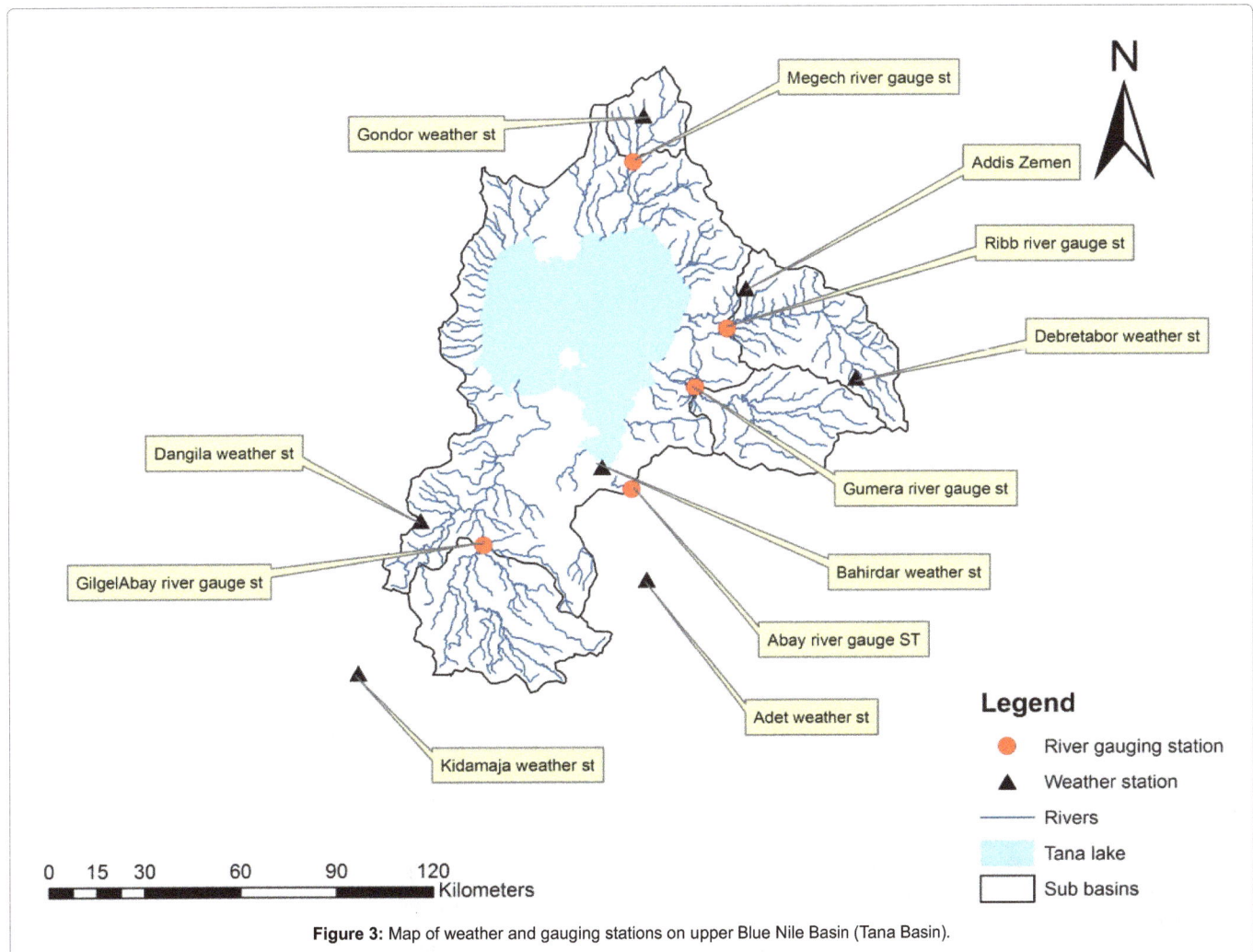

Figure 3: Map of weather and gauging stations on upper Blue Nile Basin (Tana Basin).

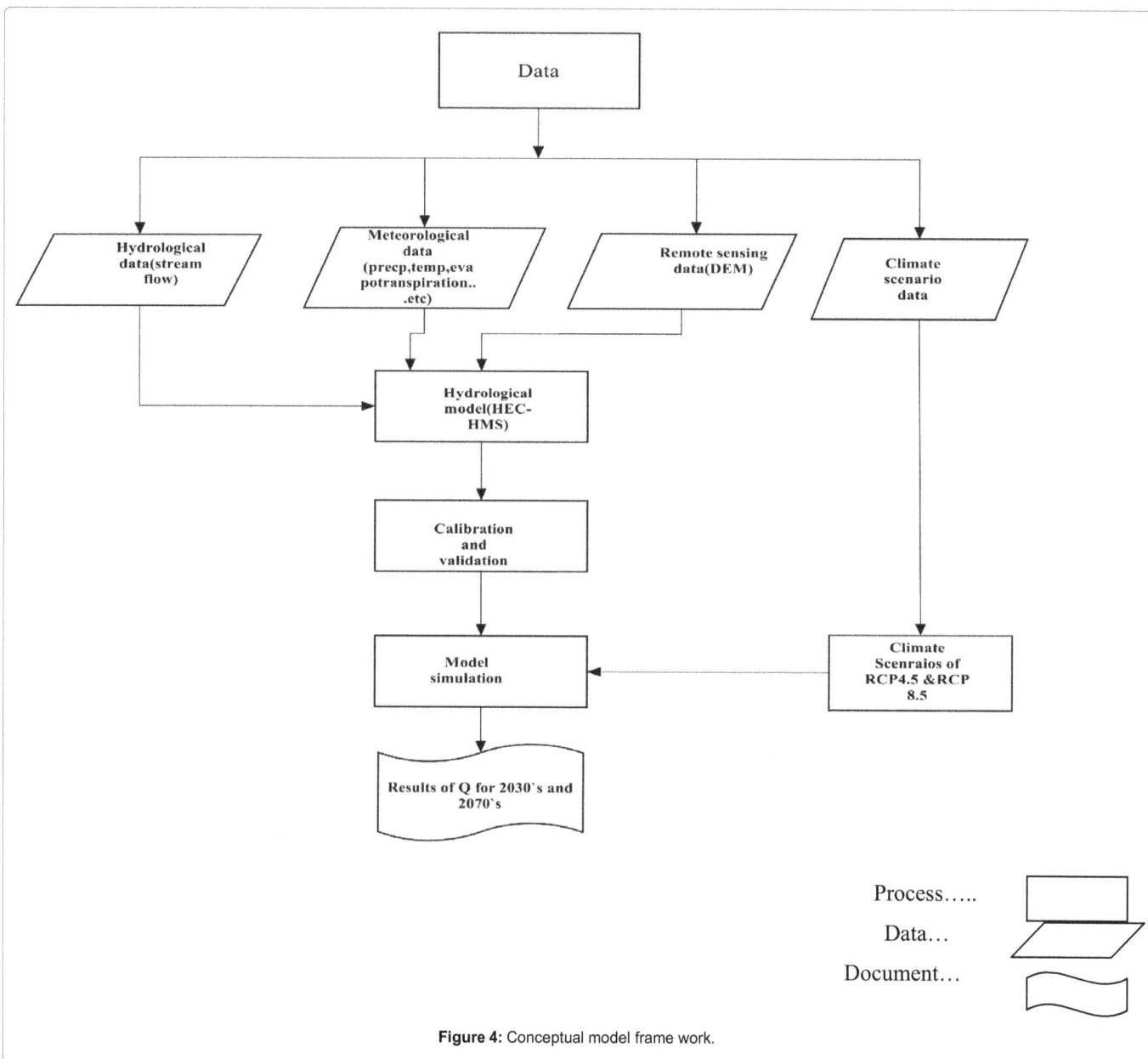

Figure 4: Conceptual model frame work.

Names	Area size(Km²)	Weather stations inside the catchment
GilgelAbay	1664	Kidamaja,Adet and Dangila
Gumera	1335	Addis Zemen,Debre Tabore and Bahirdar
Ribb	1595	Addis Zemen and Debre Tabore
Megech	531	Gondor

Table 1: Area size and weather stations of four catchments in Upper Blue Nile River basin.

EARTH-DMI, and CNRM-CM5. In this study we used RCP 4.5 and RCP 8.5 emission scenarios output of the earth`s system model. The meteorological variable includes; maximum temperature, minimum temperature, mean temperature, wind speed, surface down welling short wave radiation, surface down welling long wave radiation, snow fall and precipitation. The climate data ranges from 1st January 1960 until 31st December 2005 as historical period and for RCP 4.5 and RCP 8.5 comprised data from 1st January 2006 until 31st December 2100 except for HadGEM2-ES which ended at 31st December 2099. The temperature, precipitation, and snow fall data were bias corrected

using method developed by [15]. The radiation and wind speed series data bias corrected used by [16]. Both the bias correction methods used WATCH forcing data series (1960-1999) as reference [17] (Table 2).

Areal precipitation for grid based GCM data

Areal precipitation also prepared for the 50 Kms by 50 Kms grid based GCM data for upper Blue Nile River basin (Figure 5).

Potential Evpotranspiration (PET)

There are a number of methods to estimate potential

No	GCMs	Modelling Group,orgin of country	IPCC model ID	References
1	MPI	Max Planck Institute	MPI-esm	[18]
2	IPSL	Institut Pierre Simon Laplace, France	IPSL-CM5	[19]
3	HadCM	Hadley Center for Climate and UKMO-Prediction and Research, UK	Hadgem$_2$-es	[20]
4	Ecearth	European Insititute and ECMWF	Ec-Earth	[21]
5	CNRM	Centre National de Recherches Meteoroliques,France	CNRM-cm5	[22]

Table 2: GCMs description and sources.

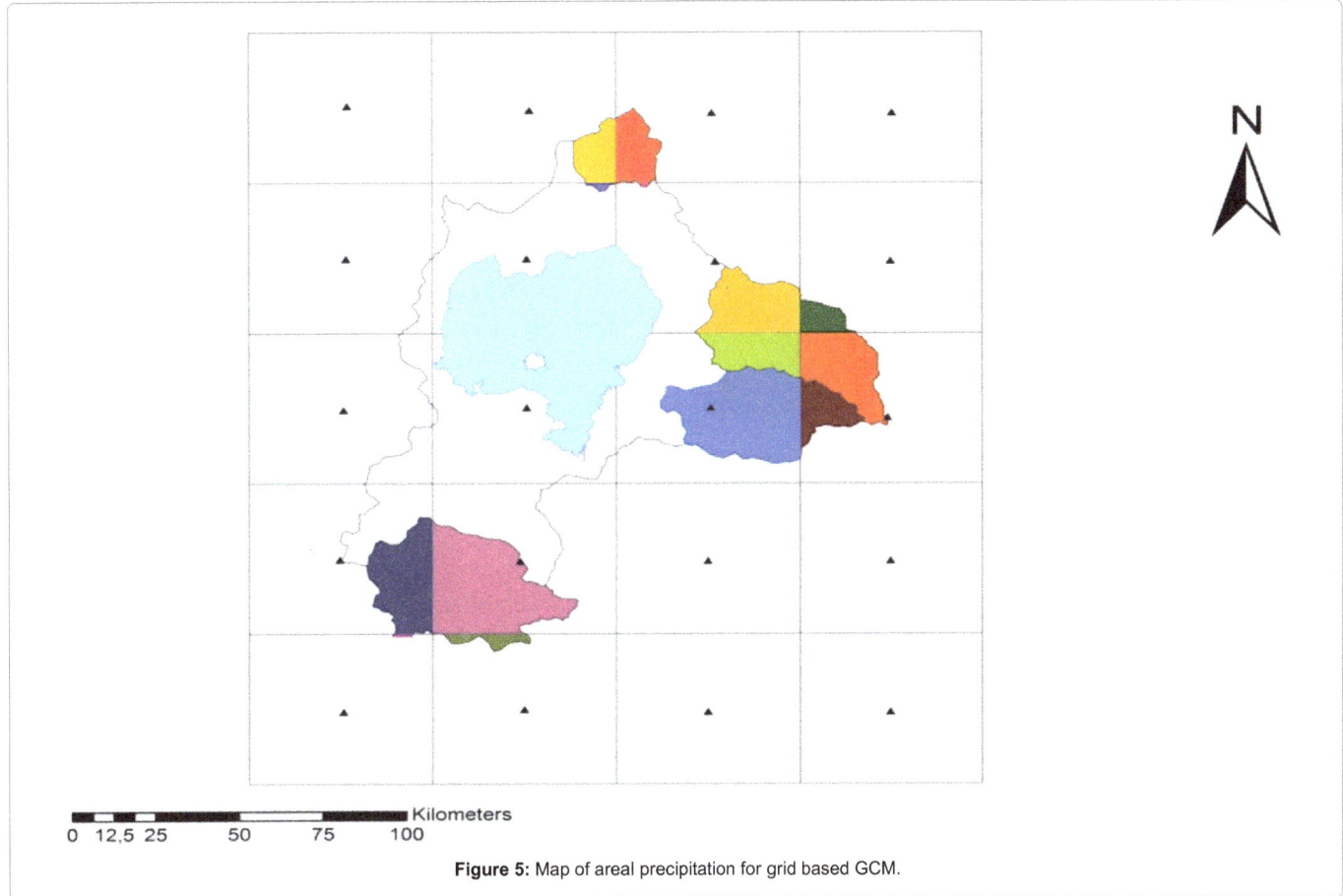

Figure 5: Map of areal precipitation for grid based GCM.

evapotranspiration. However, the methods vary based on climatic variables required for calculation. The temperature based method uses only temperature and day length; the radiation based method uses net radiation and air temperature and some other formula like, Penman requires a combination of the above net radiation, air temperature, wind speed, and relative humidity.

The FAO Penman_Monteith method is recommended as the sole ET_0 method for determining reference evapotranspiration when the standard meteorological variables including air temperature, relative humidity, and sunshine hours are available [23].

$$ET_0 = \frac{0.408\Delta[R_n - G]}{\Delta + \gamma(1 + 0.34u_2)} + \frac{\gamma}{\Delta + \gamma(1 + 0.34u_2) \times} \frac{900U_2(e_s - e_a)}{(T + 273)} \quad ....(1)$$

Where,

ET_0 = Reference Evapotranspiration, mm/day,

Rn = Net radiation, MJm^{-2} day^{-1}

G= Soil heat flux, MJm^{-2} day^{-1}

e_s= Saturated vapor pressure, KPa,

e_a= Actual vapor pressure, KPa,

e_s_e_a= Saturated vapor pressure deficit, KPa,

Δ = Slope of the saturation vapor pressure temperature relationship, KPa°C^{-}

γ = Psychometric constant, KPaC^{-1}

U_2= Wind speed, ms^{-1} and

T = Mean daily air temperature at 2m height (°C)

In this study the potential evapotranspiration for the study area was computed by FAO Penman-Monteith method for each weather station and for grid based GCM data which falls on the study area.

Methods

Arc GIS 10.2 was used to delineate the catchment area. The watershed and sub basins delineation was carried out based on an automatic delineation procedure using a Digital Elevation Model (DEM) and digitized stream networks.

Potential evapotranspiration

Potential Evapotranspiration (PET) is determined primarily by net radiation and temperature but also by the moisture-holding capacity of the air and other factors (e.g. wind speed). Increased temperature will lead to more evaporation, although the effect is complicated by the above factors [24]. Estimates of potential Evapotranspiration (is calculated using the penman- montheith equation with temperature, wind speed, solar hour and humidity as input). Air temperature data are used for calculations of snow accumulation and melt, or to calculate potential evaporation. If none of these last options are used, temperature can be omitted in snow free areas. The calculated Long term average monthly potential evopotranspiration for each observed weather station from 1988-2005.

HEC-HMS hydrological model

HEC-HMS-3.5 software used to simulate future runoff model. The hydrological model was calibrated and verified using observed stream flow data of the basin. The model was selected because of its simplicity, availability, and widely acceptance. HEC-HMS Model setup consists of four main model components: basin model, meteorological model, control specifications, and input data (time series, paired data, and gridded data) [25].

HEC-HMS calibration and validation

The deficit and constant loss method used to model infiltration loss. For the transformation of excess precipitation into direct surface runoff, synder unit hydrograph method was used, and for the base flow recession method was employed. These methods selected based on checking up of every methods for the best fit options.

Therefore, automated calibration in conjunction with manual calibration was used to determine a practical range of the parameter values preserving the hydrograph shape and minimum error in volume. The model calibrated from (1988-2000) and validated from (2001-2005).

HEC-HMS model performance criteria

Model performance was evaluated for both calibration and validation in different ways including coefficient of determination (R^2) and Nash-Sutcliffe efficiency [26]. R^2 and E_{NS} [27] are often used to assess hydrological model performance. In addition to Nash-Sutcliffe efficiencies (E_{NS}) and coefficient of determination criteria, the simulated and observed hydrograph visually inspected and checked.

Nash-Sutcliffe efficiencies (E_{NS})

$$E_{NS} = 1 - \frac{\sum(Q_{obs} - Q_{sim})2}{\sum(Q_{obs} - \overline{Q}_{obs})} *100 \quad\text{.....(2)}$$

Where:

Q_{obs}=observed discharge

Q_{sim}= simulated discharge

$\overline{Q}_{obs}$= mean of observed discharge

$\overline{Q}_{sim}$ = mean of simulated discharge

Nash-Sutcliffe efficiencies can range from -∞ to 1.

An efficiency of E_{NS} = 1 corresponds to a perfect match of modelled discharge to the observed data. An efficiency of E_{NS} = 0 indicates that the model predictions are as accurate as the mean of the observed data, whereas an efficiency less than zero (-∞<E_{NS} <0) occurs when the observed mean is a better predictor than the model. The closer the model efficiency is to 1, the more accurate the model is [28].

Coefficient of correlation (R^2)

$$R^2 = \left(\frac{\sum(Q_{obs} - \overline{Q}_{obs})^2 - \sum(Q_{Sim} - \overline{Q}_{sim})^2}{\sum(Q_{obs} - \overline{Q}_{obs})^2} \right) \quad\text{.....(3)}$$

Where:

Q_{obs}=observed discharge

Q_{sim}= simulated discharge

$\overline{Q}_{obs}$ = mean of observed discharge

$\overline{Q}_{sim}$ = mean of simulated discharge

R^2 is indicates how the simulated data correlates to the observed values of data. The range of R^2 is extends from 0 (unacceptable) to 1(best) [29].

Results

In this study, 0.5 degree by 0.5 degree grid resolution of five different bias corrected GCM model outputs based on RCP 4.5 and RCP 8.5 emission scenarios for upper Blue Nile Basin used for analysis. Period from 1971-2005 taken as a base period and two future periods considered for impacts investigation of 2030`s (2035-2064) and 2070`s (2071-2100).The following GCMs model used, MPI, IPSL, Hadgem-es, Ecearth and CNRM-cm5.Only for Hadgem-es of future horizon of 2070`s ranges from 2071-2099.Inorder to check the exactness replication of the multimodal prediction for the basin. Each historical climate data output compared against observation data for each catchments. The mean monthly precipitation, maximum temperature, minimum temperature and potential evapotranspiration of observed (1988-2005) and GCMs (1971-2005) compared for four catchments (Gilgelabay, Gumera, Ribb, and Megech) for Upper Blue Nile Basin.

Historical GCMs output comparison with observed data for blue nile river basin

The raw GCM of long term mean monthly precipitation, temperature, and evapotranspiration indicated that there is good agreement in trend and pattern with the observed data as shown in Figure 6. In each case the coefficient of determination (R^2) has resulted more than 0.96, which proofs that the GCM simulated the reality of the observation of climate data over the basin. However the mean monthly precipitation has shown a very slight discrepancy during the major rainy season. Relative annual percentage change comparison technique also used such kind of performance measurement metrics also has been done by [30]. The assessment result shows a slight under estimation of the GCM models prediction relative to the observed precipitation (Table 3).

Generally, all the GCMs output prediction of precipitation, maximum temperature, minimum temperature, and evapotranspiration resembled in producing the observed data for base period. Therefore, it is plausible to use the GCMs data output for future prediction for the basin.

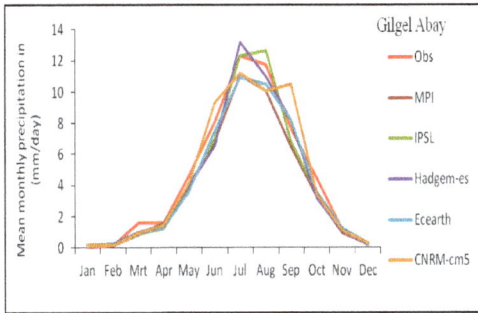

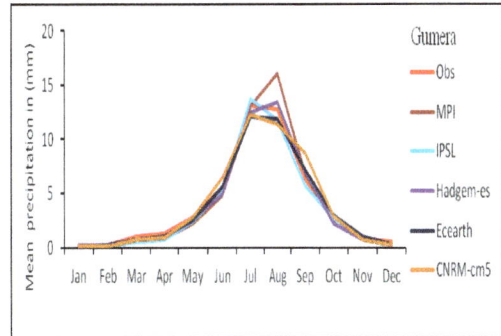

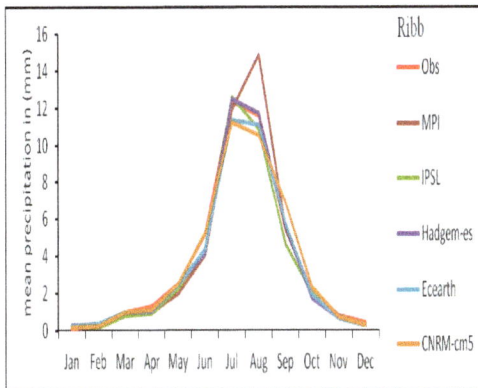

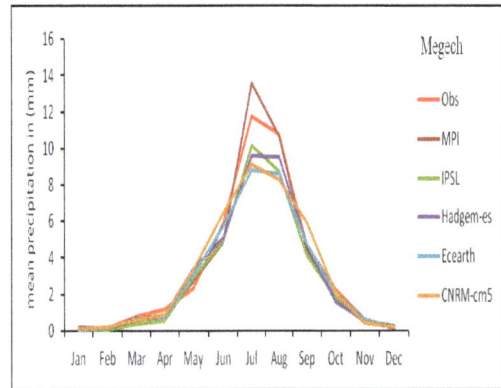

A.

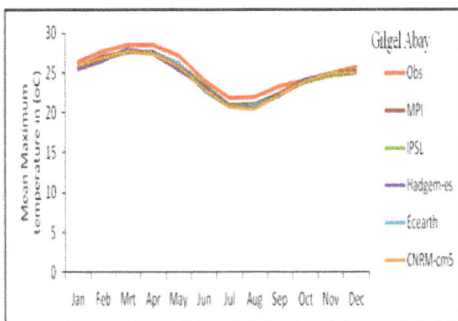

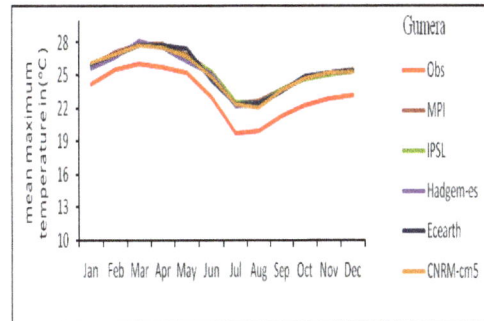

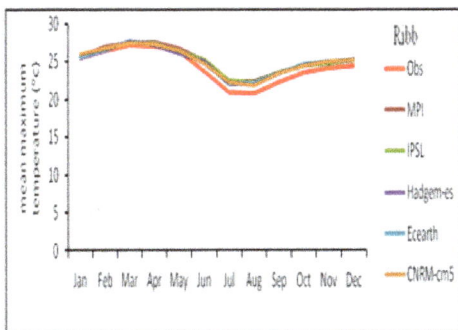

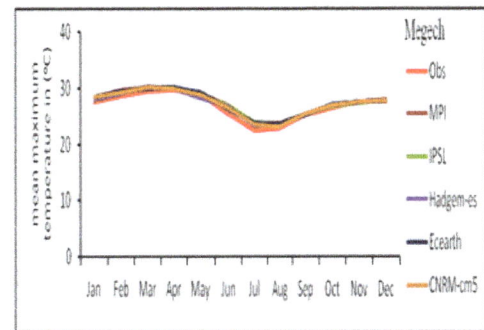

B.

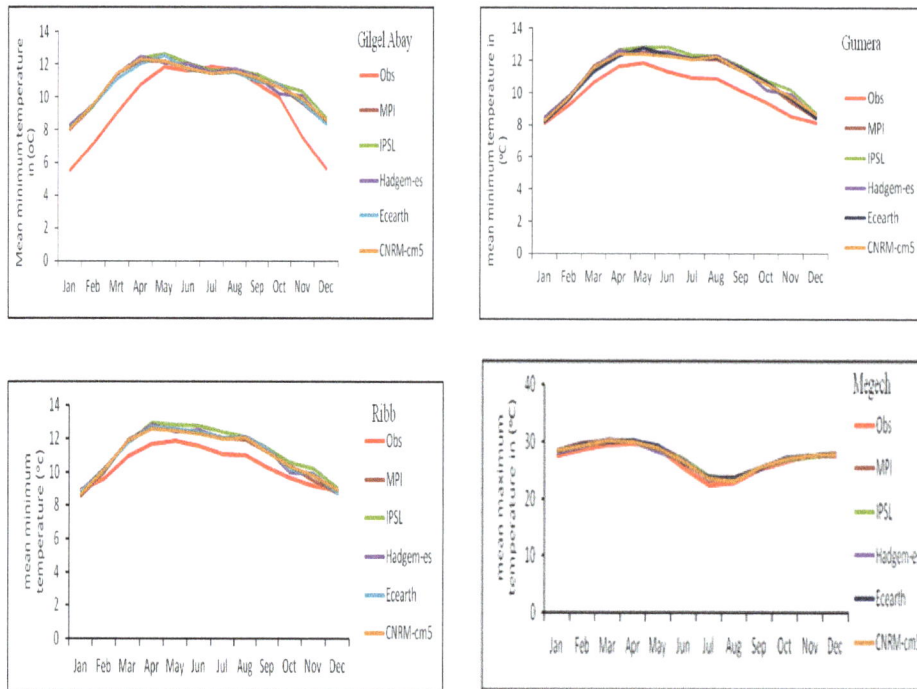

C.

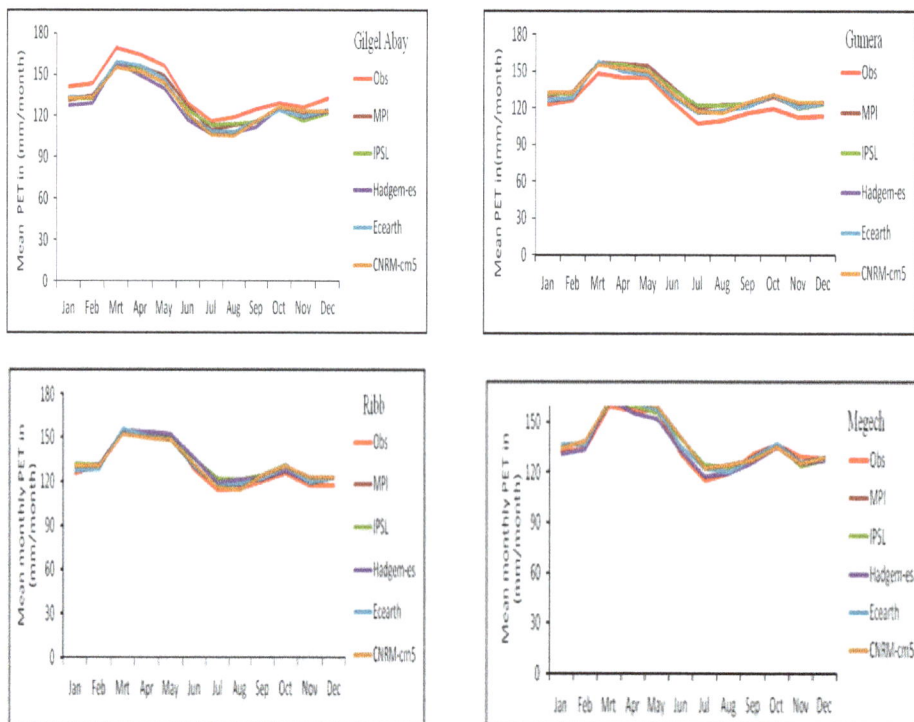

D.

Figure 6: Comparison of mean monthly precipitation, maximum temperature, minimum temperature, and evapotranspiration of observed (1988-2005) and the five selected GCMs (1971-2005) of historical periods over the basin. A) Precipitation B) Maximum temperature C) Minimum temperature and D) Evapotranspiration.

Catchment name	Factor	Observed	MPI	IPSL	Hadgem2-ES	Ecearth	CNRM-cm5
Gilgel Abay	Average Annual Rainfall in mm	1630.3	1388.7	1511	1528.3	1472.4	1605.1
	R^2		0.99	0.98	0.98	0.99	0.93
	Percentage change in (%)		-14.8	-7.3	-6.3	-9.7	-1.5
Gumera	Average Annual Rainfall in mm	1442.1	1467.7	1341.5	1392.7	1386.8	1438
	R^2		0.97	0.99	0.99	0.99	0.96
	Percentage change in (%)		1.8	-7.0	-3.4	-3.8	-0.3
Ribb	Average Annual Rainfall in mm	1306.4	1325.8	1210.2	1269.4	1228.5	1283.4
	R^2		0.96	0.99	0.99	0.99	0.98
	Percentage change in (%)		1.5	-7.4	-2.8	-6.0	-1.8
Megech	Average Annual Rainfall in mm	1240.8	1229.5	1051.2	1112.1	1078.9	1149.9
	R^2		0.98	0.99	0.98	0.97	0.94
	Percentage change in (%)		-0.9	-15.3	-10.4	-13.1	-7.3

Table 3: Comparsion of observed precipitation with five slected GCMs for the Blue Nile Basin.

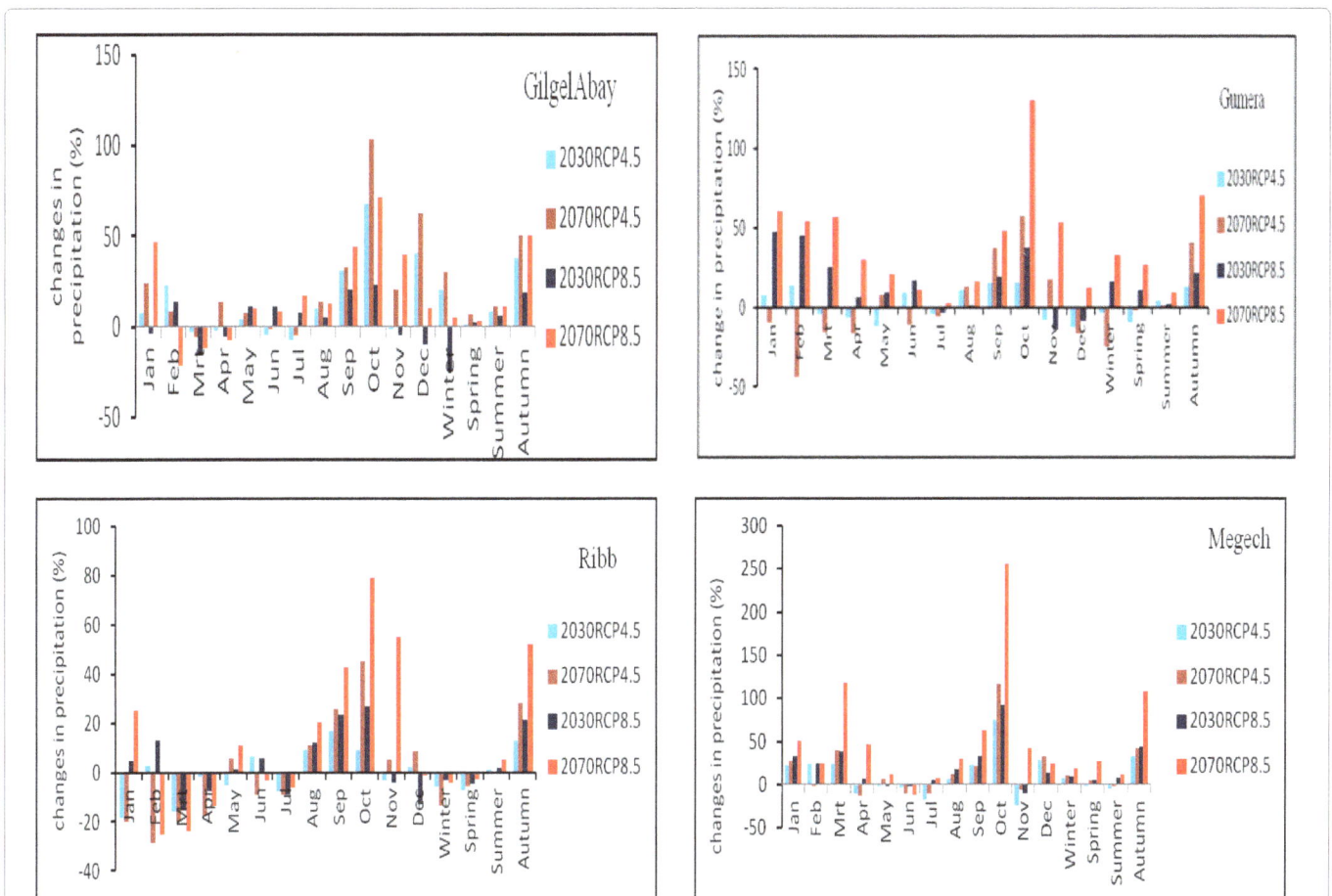

Figure 7: Relative percentage change of multimodal average monthly and seasonal precipitation for 2030`s and 2070`s as compared to the base line period of the Blue Nile River basin.

Future projection of climate impacts on precipitation, maximum temperature, minimum temperature, and potential evpotranspiration of the upper blue nile river basin.

Precipitation: The multimodal average monthly and seasonal precipitation result showed that in the future precipitation generally increases over the basin. In all four catchments, the mean monthly precipitation increased in a positive direction particularly in August, September, and October under both future periods. At 2070`s maximum precipitation change projected during the month of October ranging from +71.4% to +255.2% as shown in Figure 7. In the future, mean seasonal precipitation generally increase in all catchments. Particularly, during spring (Belg season-mild rainy season) and summer (Kiremt-main rainy season), precipitation will increase, this will help rainy fed agriculture dependent farmers to produce more crops (Figure 8).

Long term average annual precipitation change showed that, precipitation significantly increases under both future periods and RCPs scenarios. At 2070`s of RCP 8.5, all the GCM models projected that precipitation will increases in the future. Particularly, IPSL GCM model predicted maximum change in average precipitation than the other GCM models on the basin.HadGEM2-ES GCM model projected

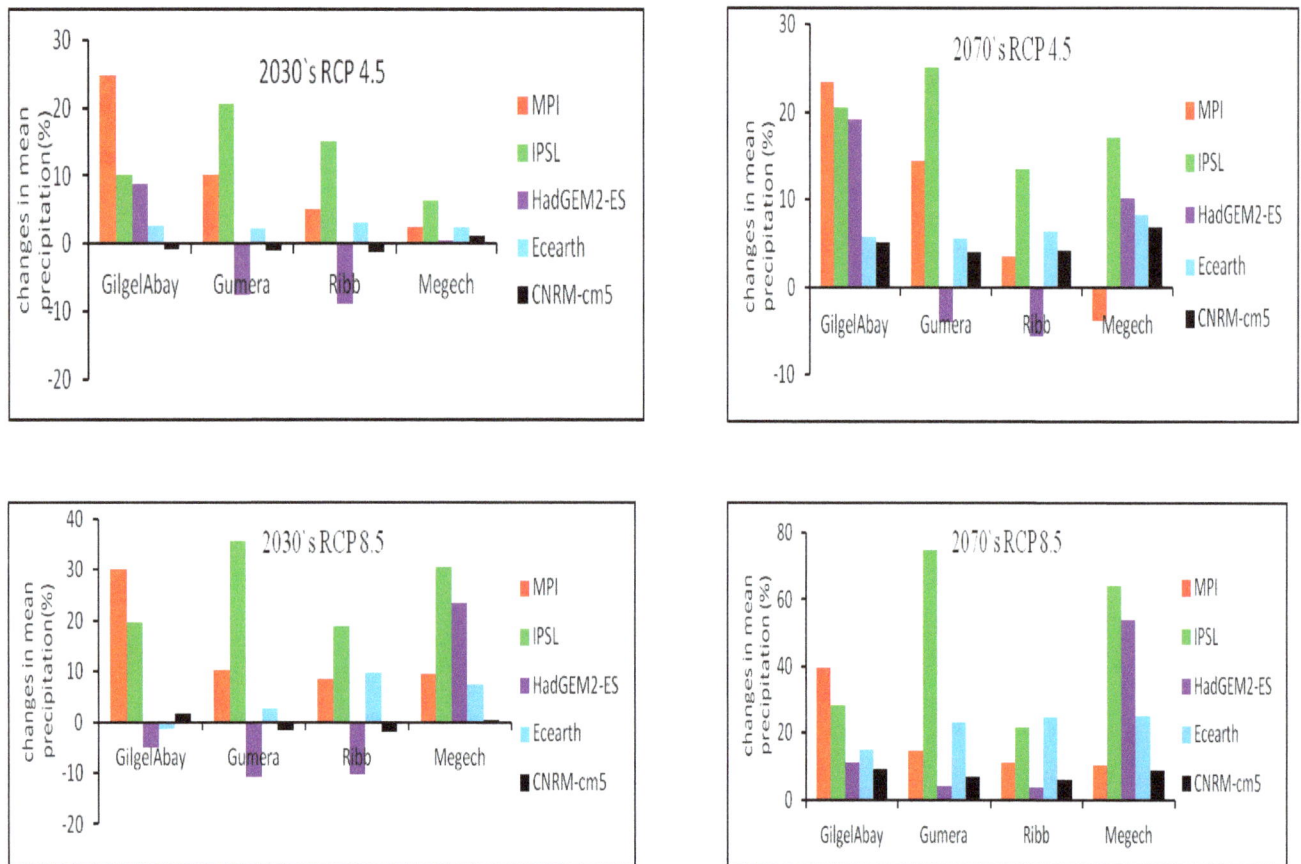

Figure 8: Changes in average annual precipitation on different catchments based on five GCMs of RCP 4.5 and RCP8.5 for 2030`s and 2070`s as compared to the base period 1971-2005.

relatively a decrease in average annual precipitation over the basin. CNRM-cm5 GCM model also projected minimum change for the basin. The relative change in average annual precipitation in RCP 8.5 is more pronounced than RCP 4.5. Moreover the change at 2070`s is relatively more than at 2030`s over the basin. For 2030`s average annual precipitation change projected between (-8.9% and +24.6%) for RCP 4.5 and between (-10.8% and +35.4%) for RCP 8.5. At 2070`s average annual precipitation change projected between (-5.6% and +25.2%) for RCP 4.5 and between (+3.8% and +74.7%) for RCP 8.5 emission scenario.

Maximum temperature, minimum temperature, and potential evapotranspiration

Average annual maximum and minimum temperature significantly increases in both future periods and RCPs scenarios. At 2070`s of RCP 4.5 and RCP 8.5, the change in average maximum and minimum temperature is magnificent compared to 2030`s of RCP 4.5 and RCP 8.5 for the Blue Nile basin. At 2030`s and 2070`s, average annual maximum temperature increases up to +3.9°C for RCP 4.5 and up to +7.1°C for RCP 8.5 emission scenario, respectively. The mean annual minimum temperature may increases up to +4.5°C for 2030`s and up to +8.4°c for 2070`s. Average annual evapotranspiration also increases significantly especially for RCP 8.5 emission scenario. HadGEM2-ES GCM projected the maximum relative change under all the variables for the basin (Figure 9).

Over all, it is clearly observed that individual GCM projected

different variation for the basin in precipitation and temperature. Maximum temperature, minimum temperature, and potential evapotranspiration show fairly a consistence patterns, but there is much less consistency and instability of certainty about future precipitation pattern because of the low convergence in climate model projections in the region of upper Blue Nile River basin. On the basis of the result from seven GCMs model experiments, Conway [31] indicated that there is large inter-model difference in the detail of rainfall changes over Ethiopia. Also in IPCC (2008a) [32] indicates that there is a considerable variation in various models projection. A study done by [33] shows that, the observed precipitation of the last century resulted that precipitation decreases over Africa, even if many climate model projected in increase in precipitation. Hulme et al. [34] and Giannini et al. [35] indicated in their studies, particularly precipitation changes predicted by GCMs in much of Africa involve considerable uncertainty because of the lack of capability of climate model predictions to account for the influence of land use changes on future climate and the relatively weak representation in many of climate models of the important aspect of climate variability that are crucial for Africa (e.g., ENSO).

Over all, we found in our analysis that temperature and potential evapotranspiration significantly increases in the future period in all five GCM models, high increment in RCP 8.5 than RCP 4.5 scenario indicated, the reason is because due to high radiation concentration projection in RCP 8.5 emission scenario. Even if, Precipitation prediction varies over the basin, but precipitation normally increases.

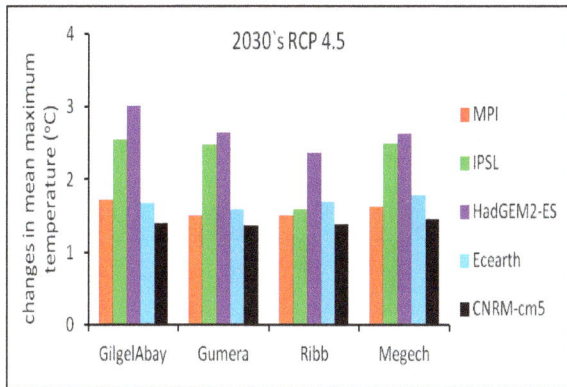

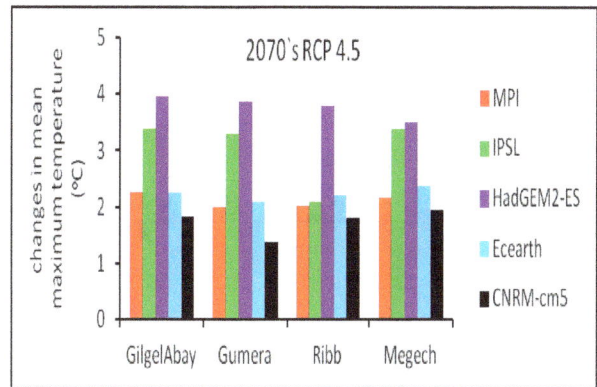

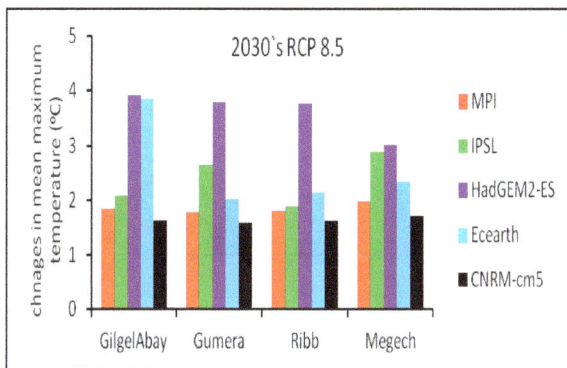

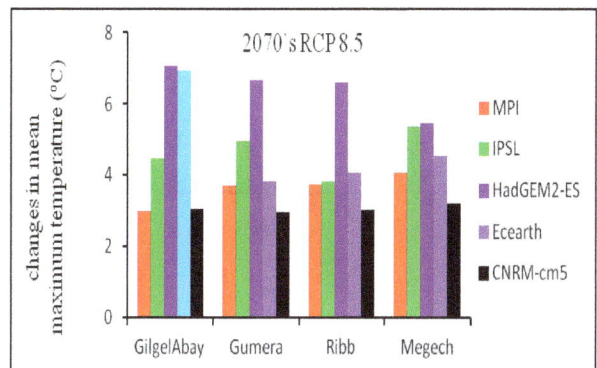

A.

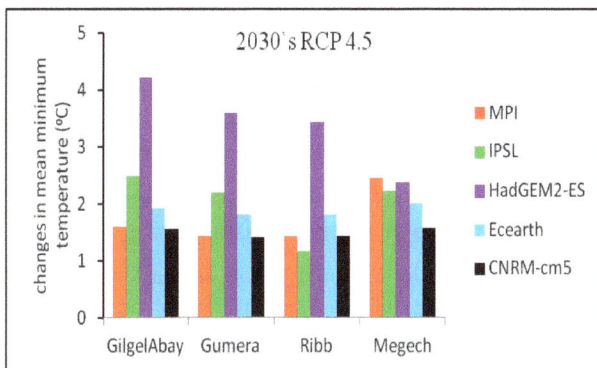

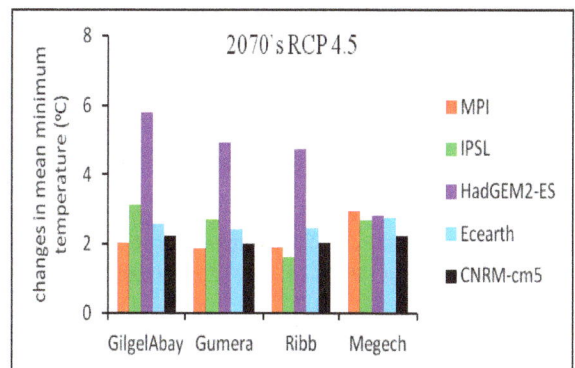

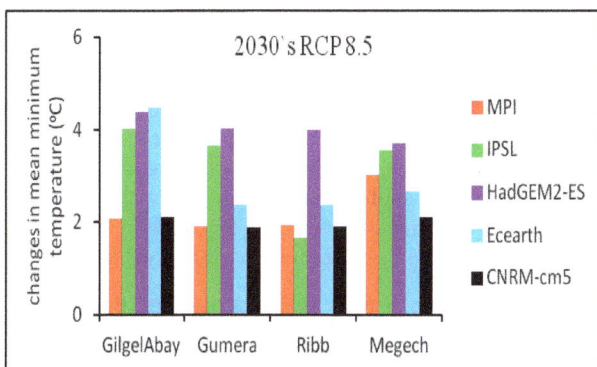

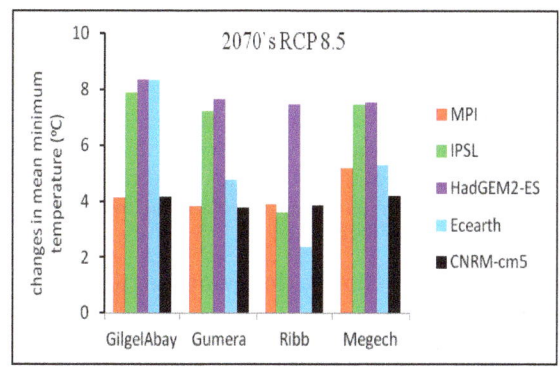

B.

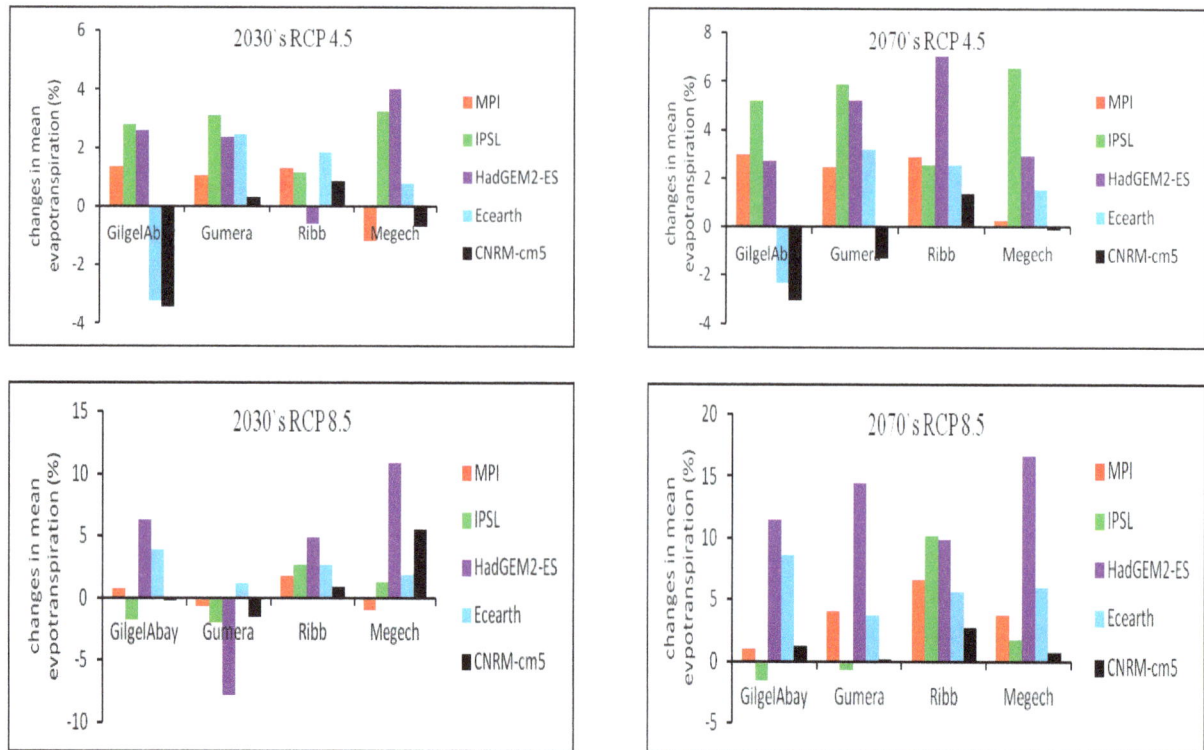

C.

Figure 9: Relative changes in average annual maximum temperature, minimum temperature and evapotranspiration on Blue Nile River basin based on five GCMs of RCP 4.5 and RCP 8.5 for 2030`s and 2070`s as compared to the base period 1971-2005.A) maximum temperature, B) minimum temperature, C) evapotranspiration.

The prediction is in line with the previous studies carried out for the basin, though there are some differences in magnitude of change that is due to the difference in GCM models and emission scenarios.

Hydrological modelling of catchments

A semi-distributed hydrological modelling technique applied for GilgelAbay, Gumera, and Ribb catchments in order to increase the performance of the model. However a lumped system applied for Megech catchment due to its small area size. The catchments are classified into sub basins and each sub basin parameters manually adjusted by trial and error method and automatically optimised to get the best fit. The model performance was checked using E_{NS} and R^2, the values during calibration period (1988-2000), for Gilgelabay (0.71, 0.73), Gumera (0.52, 0.72), Ribb (0.55, 0.72) and Megech (0.50, 0.51) respectively. During validation period (2001-2005), the values of E_{NS} and R_2, for Gilgel Abay(0.77,0.78), Gumera (0.57,0.76), Ribb (0.53,0.78) and Megech(0.5,0.51) respectivly,the results obtained are satisfactory and acceptable to simulate the basin runoff for future projection. The closer the model efficiency is to 1, the more accurate the model is. E_{NS} values greater than 0.5 are fairly acceptable to use the hydrological modele for simulation [28].

Future impacts of climate change on runoff on upper blue nile river basin

The impact of climate change on stream flow predicted on upper Blue Nile based on the changes in temperature and precipitation projected in the five different GCMs under RCP 4.5 and RCP 8.5 emission scenarios. The future stream flow prediction was analysed

taking stream flow from 1988-2005 as a base line against future runoff projection of 2030`s (2035-2064) and 2070`s (2071-2100). In this study ,namely Gilgelabay,Gumera,Ribb and Megech catchments were considered and analysed and presented in the following sections.

Mean monthly and seasonal average multimodal GCM runoff of the upper blue nile river basin

Generally, mean monthly and seasonal runoff of multimodal GCM projection indicated that, average runoff increases under both future periods and RCPs emission scenarios as compared to the base line period (1988-2005). Runoff change in the mid of the century is relatively small in both RCP 4.5 and RCP 8.5 emission scenarios. At 2070`s (2071-2100), average monthly and seasonal runoff by far will increase over the basin than 2030`s (2035-2064). At 2030`s average monthly runoff change projected in between -25% and +84.5%, while at 2070`s average monthly runoff change ranges between -14.9% and +127.9%. Relatively small change of average seasonal runoff projected during mild and main rain seasons (spring and summer) compared to autumn and winter. During rainy season, average seasonal runoff change reach up to +46.2% and up to +90.7% in the mid of the century and at the end of the 21st century respectively (Figure 10).

Comparison of future impacts of climate change on average runoff of the upper blue nile river basin based on five GCMs model projections

In this study analysis (Figure 11), the relative long term average runoff projection of the different GCMs showed that runoff increases

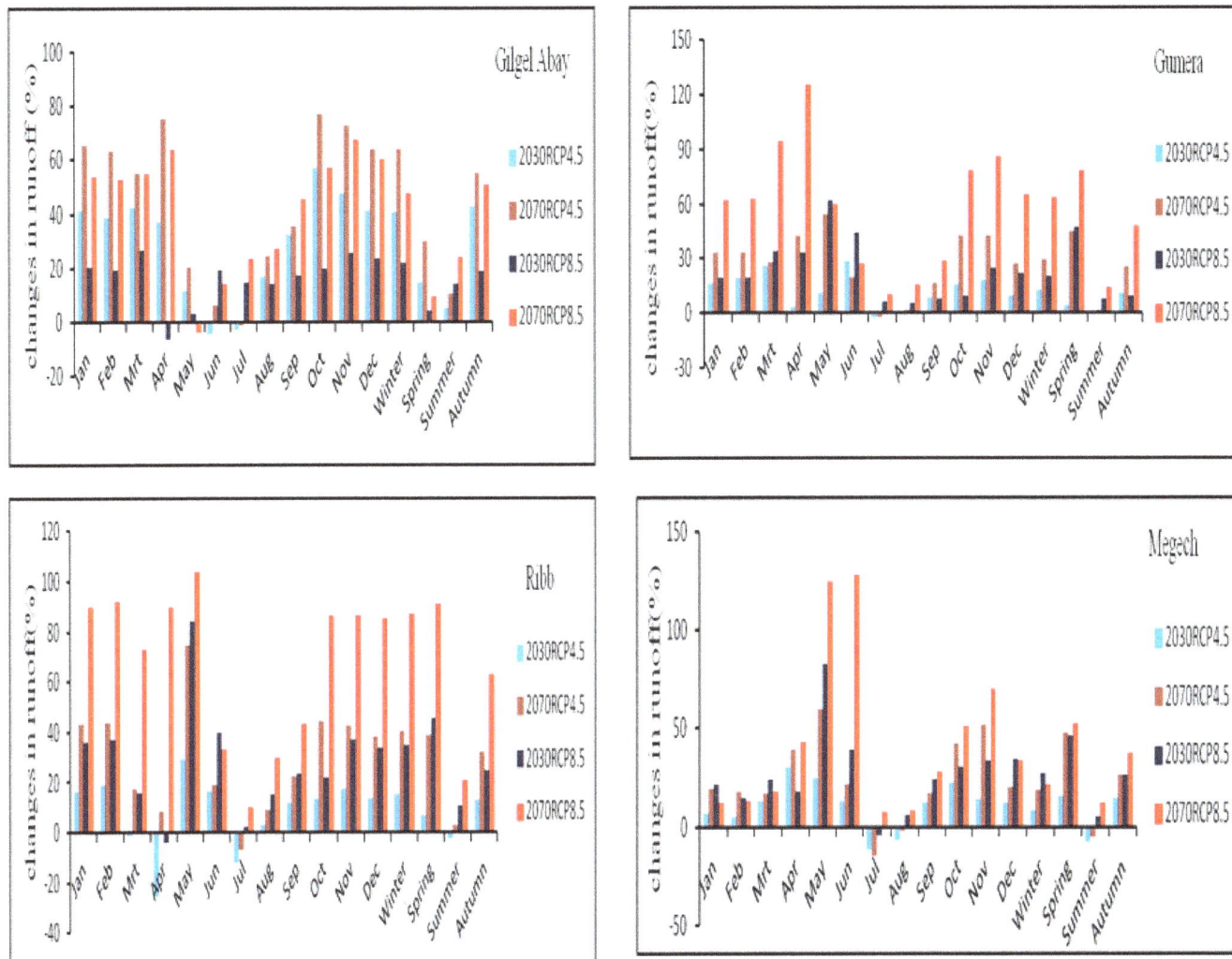

Figure 10: Relative percentage change in mean monthly and seasonal runoff of multimodal GCM projection for 2030`s (2035-2064) and 2070`s (2071-2100) under RCP 4.5 and RCP 8.5 scenarios as compared to the baseline period (1988-2005) for the Blue Nile basin.

on the upper Blue Nile River basin under both RCP 4.5 and RCP 8.5 emission scenarios. Future projection change at 2070`s of RCP 8.5 is more than at 2030`s of RCP 4.5 emission scenario, due to high increase in precipitation and potential evapotranspiration over the basin. At 2030`s, average annual runoff changes in between (-13.5% and +67.7%) for RCP 4.5 and in between (-16% and +74.8%) for RCP 8.5 scenario. At 2070`s, average annual runoff changes in between (-12.2% and +73.5%) for RCP 4.5 ,and between (-19.6% and +127.4%) for RCP 8.5 emission scenario under five GCMs out puts as compared to the base line period.

In this study, the different GCMs model resulted different projection response to climate change over the basin. Ecearth and IPSL GCM projected more or less increase in runoff change where as HadGEM2-ES projected decrease in average runoff change for the different of the catchments of the Blue Nile basin.

In previous study conducted by [36] estimated that in the mid-century the mean annual flow change ranges from −72% to 75% using VHM and from −81% to 68% using NAM model. The result obtained in our study is quite within the estimated range of change in mean annual flow of the upper Blue Nile River basin. Generally in our study almost

4 out of 5 models projected increase in mean annual flow. According to IPCC study, 18 models out of 21models projected that there will be a great robust of precipitation increase in core of East Africa [1].

Generally, in this study almost all of the five GCMs model result projected that, precipitation increases in the future over the basin, this suggests that the increase in runoff may be due to the increase of precipitation.

Discussion and Conclusion

Climate impact has potential impacts on future hydrological and meteorological variables due to increased green house emissions which is associated in increasing temperature of the globe. The future impact of climate change on hydro- meteorological characteristics of the basin has been studied like: precipitation, Temperature (Maximum and Minimum) and Potential evpotranspiration (PET) for 2030`s (2035-2064) and 2070`s (2071-2100) using CMIP5 projection output. HEC-HMS 3.5 hydrological model was used to study impacts of climate change on runoff.

In the future, mean monthly and seasonal precipitation will increase

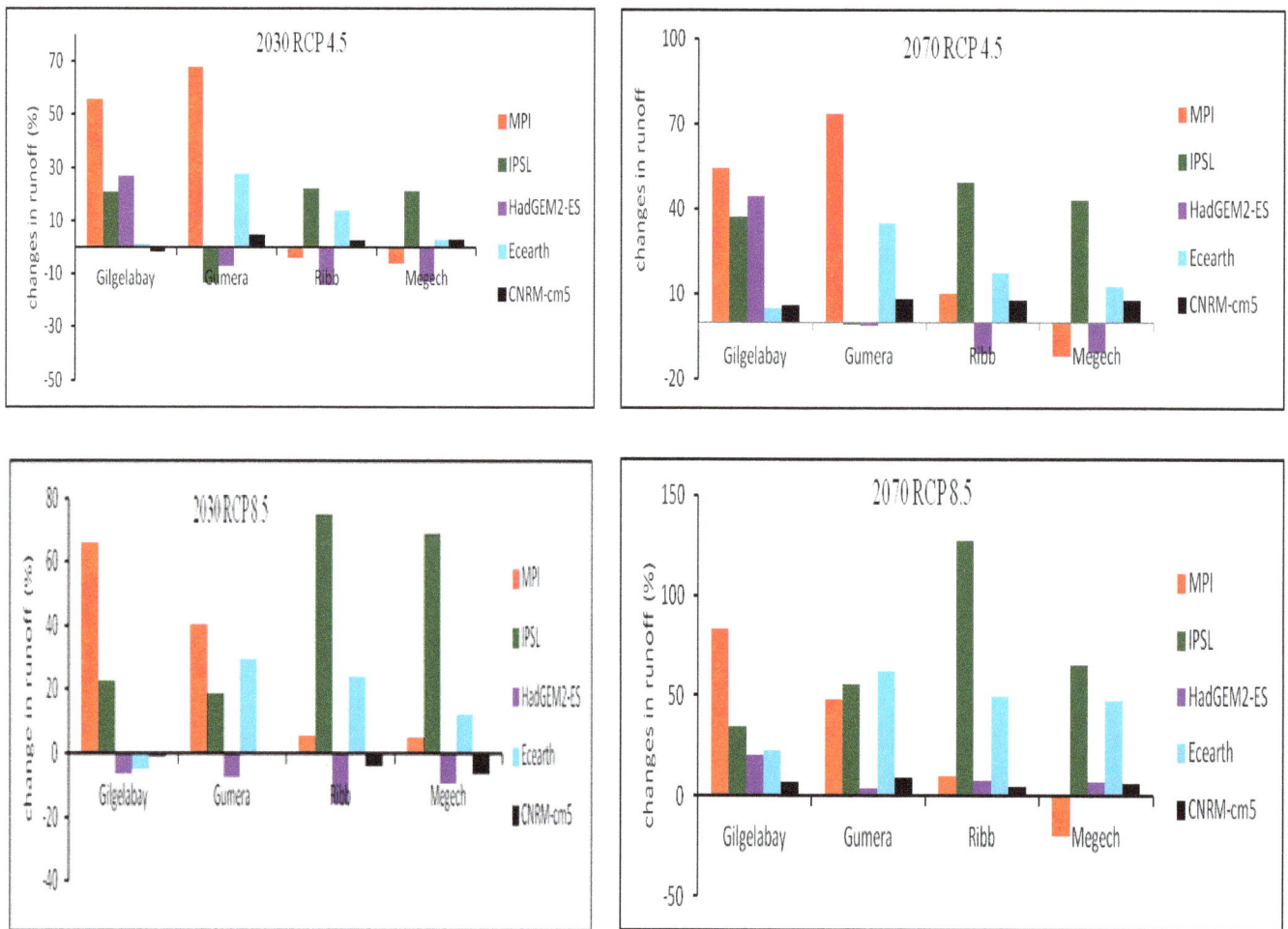

Figure 11: Relative percentage change in mean annual runoff of five GCM projection for 2030`s (2035-2064) and 2070`s (2071-2100) under RCP 4.5 and RCP 8.5 scenarios as compared to the baseline period (1988-2005) for the Blue Nile basin.

over the basin. At 2030`s and 2070`s, average annual precipitation may increases by +35.4% and by +74.7%, respectively. The average annual potentialevpotranspiration, maximum and minimum temperature projection results showed that temperature will increase in both future horizon periods. High maximum change predicted at the end of 21st century for RCP 8.5 emission scenario. The direction and magnitude changes of projection results are in line to the global change projection by IPCC, 2007.Moreover, the results shown in this report are inconsistent with previous studies conducted in the basin by Kim et al. [8] and Setegn et al [37]. They both had used CMIP3 results while in this study we used CMIP5 results.

HEC-HMS hydrological model calibrated and validated for each catchments. The daily Nash and Sutcliffe efficiency (E_{NS}) and coefficient of determination (R^2) of model performance criterion used to evaluate the model applicability for different catchments. The results obtained are satisfactory and acceptable. Therefore, we assured in this study, HEC-HMS model can be used for modelling and projection of future impacts of climate changes on runoff for upper Blue Nile river basin.

The impacts of climate change in precipitation and temperature has produced a significant change on runoff in the basin. Conway et al. [38] Clearly indicated that the combined effects of precipitation and temperature changes would have profound effects on the stream flow

regime of the Blue Nile River. According to our study of multimodal average GCM shows that, mean monthly and seasonal runoff will increases in all months and seasons for both future periods of RCP 4.5 and RCP 8.5 emission scenarios. At 2030`s and 2070`s, average annual runoff volume may increases up to +74.8% and +127.4%, respectively. The increase in runoff on the future may highly associated to the increase in precipitation over the basin. The average runoff change in magnitude and direction report results are similar compared to other studies report results indicated by Beyene et al. [9] and Abdo et al. [39].

In this study, we used different GCM model out puts and hydrological model, different models gives different results, particularly there is high biases in rainfall in climate models. Using of bias corrected data also adds its own uncertainty to our results [40].

This study result confirmed that, in the future average annual runoff increases due to climate change at the out let of each catchment that feeding the great Lake Tana on upper Blue Nile River. The increases in water availability will play significant benefits for small and large scale farmers for agricultural activities more over for water resources development projects. The climate change may contribute in a positive direction for crop water availability, if and only if farmers are adopted themselves to cropping schedule. However, precautionary notion has to be taken to control flooding on the flood plains.

Acknowledgment

The authors gratefully appreciate Dr.Iwan Supit (Wageningen University, ESS department) for providing NETCDF GCMs Climate data. Our gratitude also extends to Ethiopian Ministry of Water and Energy (MoWE), and Ethiopian National Meteorology Service Agency (NMSA) for providing hydro meteorological data.

References

1. SD Solomon (2007) Climate change: The physical science basis. IPCC, Contribution of working group I to the fourth assessment report of the intergovernmental panel on climate change, Cambridge university press, UK, New York, NY, USA.

2. Luqman Atique, Irfan Mahmood and Farman Atique (2014) Disturbances in atmospheric radiative balance due to anthropogenic activities and its implications for climate change. American-Eurasian J Agric & Environ Sci 14: 73-84.

3. IPCC (2007b) Towards new scenarios for analysis of emissions, climate change, impacts, and response strategies. Technical summary, Noordwijkerhout, Netherlands.

4. Alerts JCH, Droogers P (2004) Climate Change in contrasting river basins: adaptation strategies for water, food and environment. Walling: CABI.

5. Frederick Kenneth D, David C Major, Eugene Z Stakhiv (1997) Water Resources Planning Principles and Evaluation Criteria for Climate Change. Climatic Change 36

6. Yates DN, Strzepek KM (1996) Modelling economy-wide climate change impacts on Egypt: a case for an integrated approach. Environ Model Assess 1: 119-135.

7. Conway D (2005) From headwater tributaries to international river: Observing and adapting to climate variability and change in the Nilebasin. Global Environmental Change 15: 99-114.

8. Kim U, Kaluarachchi J (2009) Climate Change Impacts on Water Resources in the Upper Blue Nile River Basin, Ethiopia. JAWRA J Am Water Resource Assoc 45: 1361-1378.

9. Beyene T, Lettenmaier DP, Kabat P (2010) Hydrologic impacts of climate change on the Nile River basin: Implications of the 2007 IPCC scenarios. Climatic Change 100: 433-461.

10. Andrew Wiltshire et al. (2010) Implications of climate change for agricultural productivity in the early twenty-first century. Philos Trans R Soc Lond B Biol Sci 365: 2973-2989.

11. BCEOM (1998) Abbay River Basin Integrated Development Master Plan Project Phase 2, Data Collection and – Site Investigation Survey and Analysis, Section II, - Sectoral Studies, Volume XIV – Demography and Sociology. The Federal Democratic Republic of Ethiopia, Ministry of Water Resources.

12. Setegn SG, Srinivasan R, Dargahi B, Melesse AM (2009) Spatial delination of soil erosion vulnerability in the Lake Tana Basin, Ethiopia. Hydrol Process 23: 3738-3750.

13. Conway D, Schipper EL (2011) Adaptation to climate change in Africa: Challenges and opportunities identified from Ethiopia. Glob Environ Chang 21: 227-237.

14. Moss RH, Edmonds JA, Hibbard KA, Manning MR, Rose SK et al. (2010) The next generation of scenarios for climate change research and assessment. Nature 463: 747-756.

15. Piani C, Haerter J, Coppola E (2010) Statistical bias correction for daily precipitation in regional climate models over Europe. Theoretical and Applied Climatology 99: 187-192.

16. Haddeland I, Heinke J, Voß F, Eisner S, Chen C et al. (2012) Effects of climate model radiation, humidity, and wind estimates on hydrological simulations. Hydrol Earth Syst Sci 16: 305-318.

17. Ludwig F, Supit I, Franssen W, Biemans H (2013) Climate Change Impacts on Hydrological and Meteorological Extremes. Excutive summery of COMBINE project.

18. Block K, Mauritsen T (2013) Forcing and feedback in the MPI-ESM-LR coupled model under abruptly quadrupled CO2. Journal of Advances in Modeling Earth Systems 5: 676-691.

19. Dufresen J et al. (2012) Climate change projections using the IPSL-CM5 Earth system Model: From CMIP3 to CMIP5. Clim dyn.

20. Liddicoat et al. (2013) CO_2 emissions determined by hadgem2-es to be compatible with the representative concentration pathway scenarios and their extensions. J Climate 26: 4381-4397.

21. Van Noije TPC, Le Sager P, Segers AJ et al. (2014) Simulation of tropospheric chemistry and aerosols with the climate model EC-Earth. Geosci Model Dev Discuss 7: 1933-2006.

22. Voldoire A et al. (2012) The CNRM-CM5.1 global climate model: description and basic evaluation. Climate Dynamics 40: 2091-2121.

23. Allen RG, Pereira LS, Raes D, Smith M (1998) Crop Evaporation: Guidelines for computing crop water requirements. FAO Irrigation and Drainage paper 56: 300.

24. Sælthun NR (1996) The "Nordic" HBV Model. Description and documentation of the model version developed for the project Climate Change and Energy Production. NVE Publication 7, Norwegian Water Resources and Energy Administration, Oslo

25. HEC (2006) Hydrological Modelling (HEC-HMS) user's Manual, US Army Corps of Engineers.

26. Nash JE, Sutcliffe JV (1970) River flow forecasting through conceptual models part I-A discussion of principles. Journal of hydrology 10: 282-290.

27. Kalin L, Hantush M (2006) Hydrologic Modeling of an Eastern Pennsylvania Watershed with NEXRAD and Rain Gauge Data. J Hydrol Eng 11: 555-569.

28. Rientjes THM, Perera BUJ, Haile AT, Reggiani P, Muthuwatta LP (2011) Regionalisation for lake level simulation-the case of Lake Tana in the Upper Blue Nile, Ethiopia. Hydrol Earth Syst Sci 15: 1167-1183.

29. Krause P, Boyle DP, Base F (2005) Comparison of different efficiency criteria for hydrological model assessment. Advances in Geosciences 5: 89-97.

30. McMahon TA, MC Peel, DJ Karoly (2014) Uncertainty in runoff based on Global Climate Model precipitation and temperature data – Part 1: Assessment of Global Climate Models. Hydrol Earth Syst Sci Discuss 11: 4531-4578.

31. Conway D (2005) From headwater tributaries to international river: Observing and adapting to climate variability and change in the Nilebasin. Global Environmental Change 15: 99–114.

32. IPCC (2008a) Special report on Climate Change and Water. A Special Report of Working Group II of the Intergovernmental Panel for Climate Change. Cambridge University press, Cambridge.

33. IPCC (2013) Climate change 2013 of the physical science basis. Working group I, fifth assessment report of intergovernmental panel for climate change, summary for policy makers. IPCC Switzerland.

34. Hulme M, Doherty R, Ngara T, New M, Lister D (2001) African climate change: 1900-2100. Clim Res 17: 145-168.

35. Giannini A, Biasutti M, Held IM, Sobel AH (2008) A global perspective on African climate. Clim Change 90: 359-383.

36. Taye M, Ntegeka V, Willems P (2011) Assessment of climate change impact on hydrological extremes in two source regions of the Nile River Basin. Hydrol Earth Syst Sci 15: 209-222.

37. Setegn SG, Rayner D, Melesse AM, Dargahi B, Srinivasan R (2011) Impact of Climate Change on the Hydroclimatology of Lake Tana Basin, Ethiopia. Water Resources Research 47:W04511.

38. Conway D, Hulme M (1993) Recent fluctuations in precipitation and runoff over the Nile sub-basins and their impact on main Nile discharge. Clim Change 25:127-151.

39. Abdo KS, Fiseha BM, Rientjes THM, Gieske ASM, Haile AT (2009) Assessment of climate change impacts on the hydrology of Gilgel Abay catchment in Lake Tana basin, Ethiopia. Hydrol Process 23: 3661-3669.

40. Ehret U, Zehe E, Wulfmeyer V, Warrach-Sagi K, Liebert J (2012) HESS Opinions Should we apply bias correction to global and regional climate model data?. Hydrol Earth Syst Sci 16: 3391-404.

6

Impact of Extreme Climate Events on Water Resources and Agriculture and biodiversity in Morocco

Mohamed Aoubouazza[1]*, Rachid Rajel[2] and Rachid Essafi[2]

[1]Centre de la Recherche Forestière, BP. 763 Agdal-Rabat, Morocco
[2]Direction de la Recherche et de la Planification de l'Eau, Agdal-Rabat, Morocco

Abstract

This work is devoted to examine trends and extreme climate variability observed over the past five decades in Morocco, and future projections of these trends for the coming five decades. It also aims to assess the potential impact of these trends on water resources as well as threats to agricultural production. Five weather stations representing the type different types of climate in Morocco who has been the subject of statistical analysis of their extreme temperature indices (minimum and maximum) and their daily rainfall, during the time span 1960 to 2004. These indices of extreme temperatures (TX90[eme], TN90[eme], TX10[eme], TN10[eme], IVF, IFC) and precipitation (R10: number of days or total precipitation above 10 mm), were recorded through the four seasons of the year. Overall, the curves of change of temperature indices show a significant upward trend of hot days and a significant downward trend in the case of cold days, reflecting a slight warming of the climate of Morocco. Regarding the precipitation, in general, the curves of rainfall indices showed a downward trend but not to a significant extent, except for the "Oujda" station, where a significant declining tendency was noted. In order to understand future trends and forecast future projections, software MAGICC/ CENGEN was applied to average emission scenario P50 powered variable of average temperatures and precipitation for the period 1981 to 2004. The projection of these meteorological parameters to 2050 confirms a clear trend toward warming (1.7°C and 2.6 °C) for the 1st parameter while a significant reduction is registered in the case of the 2nd parameter (15-22% in the northeast, 9 and 16% in the south), reflecting a transition of climate to the semi-arid in the north of Morocco. Due to climate warming and decreased precipitation, the overall intake between 1940 and 2005 for all dams in Morocco recorded a deficit of 20%. This deficit is estimated for 2050 to be between 13.6 to 21.3 % compared to the present situation in the simulated results of climate models used in this study for three large dams. It is forecasted that the northern region of Morocco would be most affected by climate change resulting into a significant reduction in irrigated area (4200 to 6500 ha) and, consequently, a substantial drop in agricultural production, thus threatening food security for the inhabitants of these regions.

Keywords: Extreme weather; Temperature; Precipitation; Future projection; Water resources; Agriculture, Morocco.

Introduction

Around the world, climate change can be observed more and more. Record temperatures were recorded from year to year in major cities. Heat waves and important recurring aunts floods have occurred in recent decades. These disorders weather hitting the planet causing dramatic affect humans and agriculture, so the economies touched. Given this situation, the concern is to predict what climate changes can we expect to be able to build a socio-economic adapted to this climate. Morocco is not immune, either, extreme weather events. Covering an area approximately 710,850 km², it has nearly 3,500 km of coastline along its Atlantic and Mediterranean coasts. The four mountain ranges in the High Atlas, Anti Atlas, Middle Atlas and Rif give it a diverse geography and a subdivision into at least four climatic zones: the Mediterranean coastal area, the Atlas Mountains and steppe or Saharian climate zone south of the line of the watershed. The regions of the Atlantic and Mediterranean coasts receive most of their rainfall between November and March. Because of the terrain, they vary between 300 mm on the Mediterranean coast a little drier and more than 700 mm in the north-west of Morocco. In winter there is often above 2500 mm, a snow cover that persists more than six months a lot. The foothills of the Sahara receive less than 200 mm per year, mostly in autumn and spring. This study aims to examine trends in extreme temperatures and rainfall recorded in various weather stations over the past five decades and future projections of these trends over the next five decades. It also aims to assess the potential impact of these trends on water resources and threats to agricultural production and biodiversity [1].

Methodology

Climate data were used, namely daily values of the precipitation and maximum temperatures (TX) and minimum (TN), spread over the period 1960-2004. The data from five weather stations (Agadir, Ifrane, Marrakech, Oujda and Tangier) and are representative of regional and climatic contrasts of Morocco. Used for the calculation of trends, the slope estimator based on Kendall's criterion, since this method is robust and does not impose any particular distribution for residues. The estimated slope is admitted only if it is strictly less than the error [5%]. The trend is considered significant at 95% if the probability is less than 0.05. The indices of temperature and precipitation used in this study appear in Table 1. Percentiles of a statistical series are the 99 values (P1, P2,. ..., P99) of character who share the population into 100 parts of equal strength (Figure 1): -10% of the population, the character value is less than P10; in these southern - 90% of the population, the character value is less than P90, $F(Pk)=k/100$, $1<k<99$, F: the distribution function of random variable.

---

*Corresponding author: Mohamed Aoubouazza, Centre de la Recherche Forestière, BP. 763 Agdal-Rabat, Maroc, E-mail : abouazzayna@gmail.com

Indexes	Defining
TX90e : Tmax > 90e percentile	Number of days where maximum temperature is above the 90th percentile.
TX 10e : Tmax < 10e percentile	Number of days where maximum temperature is below the 10th percentile
TXS : Tmax > Threshold	Number of days where the temperature exceeds a threshold.
IVC: Tmax> 90th percentile for at least 3 consecutive days	Number of "waves" or times during the summer season for three consecutive days, the maximum temperature was above the 90th percentile.
TN 10e Tmin < 10e percentile	Number of days with minimum temperature is below the 10th percentile
TN 90e : Tmin > 90e percentile	Number of days with minimum temperature is above the 90th percentile
IVF: Tmin <10th percentile for at least 3 consecutive days	number of "waves" or times during the winter season for three consecutive days where the minimum temperature was below the 10th percentile.
PRCPTOT	Cumul annuel des précipitations.
R 10 mm	Number of days per year where total precipitation> = 10 mm

Table 1: Indices of temperature and precipitation used

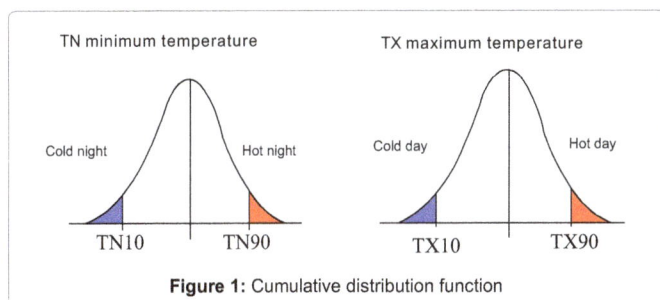

Figure 1: Cumulative distribution function

Result and Discussion

Evolution and trends of climate change indices

Temperatures: To more easily compare results from one city to another and from one season to another, the trends have been expressed in number of days per 44 years. Thus, a positive trend reflected in an increase in frequency, and a negative trend to decrease. The trend values thus calculated are shown in Table 2.

Overall, we find, for the five stations and the four seasons of the year, a downward trend in the trend for the indices and TN10 TX10 while in the case of indices TN90 TX90 and the evolution of the trend is increasing. The indices of cold wave (ICW), the downward trend in the trend occurs in most cities of Morocco. Conversely, the heat wave indices (IHW) have seen their tendency to move upward rather in the northern cities of Morocco.

Analysis of indices of extreme cold (TN10, TX10): For the cold nights or days where the temperature is below the 10th percentile, except for Tangier, located in north-western Morocco, where a slight upward trend was held in the fall (+0.4 days/44 years) and winter (+3.2 days/44 years), there were four other cities for the frequency indices of the extreme cold (TN10) has shown a significant trend downward. This decrease varies between -0.96 days/44 years, in Ifrane, and fall to the lowest values, and -13.2 days/44 years in Agadir, in the summer, for values larger, indicating a significant reduction in the number of the cold nights especially in the southern Morocco (Figure 2a). For

the cold days or days when the maximum temperature is below the 10th percentile, regardless of the season and for all cities, changing the frequency of extreme cold indices (TX10) have also shown a downward trend but significantly lower compared against the cold nights. This decrease varies between -1.21 days/44 years in Oujda, in the fall, for the lowest values, and -11.6 days/44 years in Agadir, in the summer, for larger values meaning, again, a sharp reduction in the number of the cold days especially in the south of Morocco (Figure 2b). We find, again, the same observation made for the cold nights: "These are cold days that are reduced further".

Thus, for the cold nights and for the cold days, the sharp declines in trends, particularly in the Moroccan south areas, reduce the duration of the cold episodes, and promote, thereby warming the climate regions.

Analysis of indices of extreme heat (TN90, TX90): Contrary to what was previously seen for the indices of extreme cold, to warm nights or number of days when the minimum temperature is above the 90th percentile (TN90), excluding the city of Tangier, located in the north-west of Morocco, where there was a slight downward trend in spring (-0.14 days/44 years), the other four remaining cities shows a significant increase in trends. This increase varies between +1.43 days/44 years, in Ifrane, a mountain station with the Atlas in the winter, for the lowest values and +11.64 days/44 years also for the Ifrane higher values, but during the summer, meaning an increase in the number of the hot nights in particular in the northern of Morocco (Figure 3a). For the hot days or days when the maximum temperature is above the 90th percentile (TX90), except Marrakech, located in the central of Morocco, where a slight downward trend took place twice a year : during the summer (-0.22 days/44 years) and spring (-0.03 days/44 years), but also Agadir in the southern of Morocco, which has experienced a decline trend in

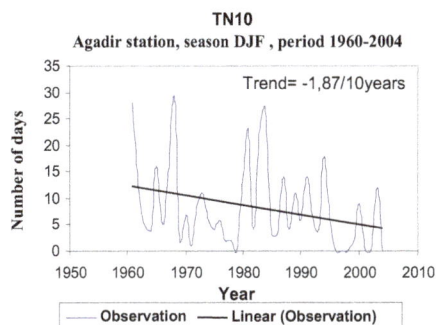

Figure 2a: Number of times in winter when Tmin < TN10 for the Agadir station

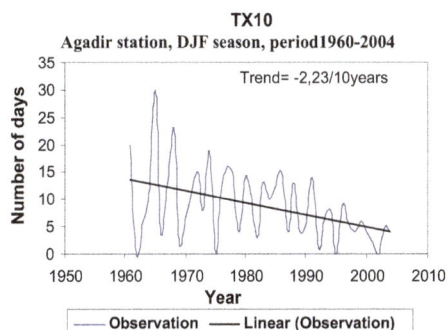

Figure 2b: Number of times in winter when Tmax <TX10 for the Agadir station.

the fall (-1.02 days/44years), other towns in the northern Morocco, they, have trended upward significantly. It varies between +0.24 days/44 years, to Ifrane in the spring to the lowest values, and +12.85 days/44 years, again in Ifrane, in the summer, for values larger, indicating an increase in the number of hot days especially in the north of Morocco (Figure 4a). We find, again, the same observation made earlier for warm nights, "it is the warm days that are growing more".

Thus, for the warm nights like the hot days, the sharp increases in trends, particularly in the Moroccan northern areas, increase the duration of extreme heat and promote, thereby warming the climate in these northern regions.

Analysis of indices of cold waves (ICW): For ICW or number of times in the winter, where for three consecutive days, the minimum temperature was below the 10th percentile, regardless of the season and for all cities, the indices of the cold wave lagging in general, a slight downward trend (Figure 4a) with the exception of Tangier, located in the northwest of Morocco, where a slight increase in trend (+0.88 days/44 years), took place then in the winter. For the other four cities, Oujda, in the lead, with -0.58 days/44 years, followed by Agadir (-0.562 days/44 years), Ifrane (-0.29 days/44 years) and Marrakech (-0.008 days/44 years), meaning thereby a small decrease in the number of the cold waves.

Analysis of indices of heat waves (IHW): Unlike IVF, or IVC for the number of times or for three consecutive days during the summer, the maximum temperatures were above the 90th percentile of the five cities studied, only three of them showed a trend up, to proved

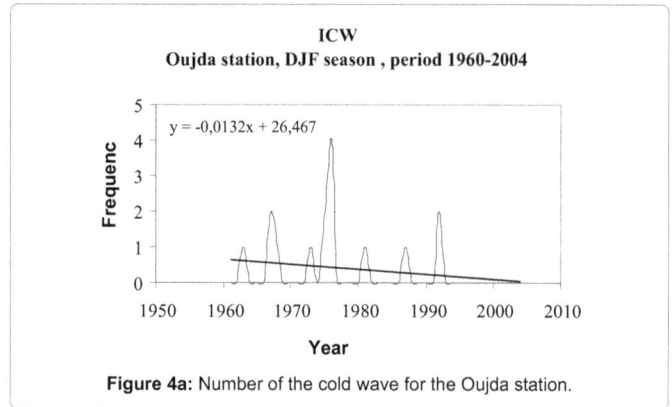

Figure 4a: Number of the cold wave for the Oujda station.

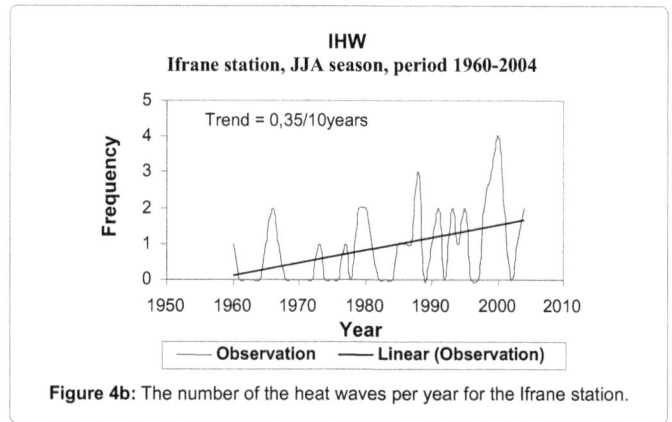

Figure 4b: The number of the heat waves per year for the Ifrane station.

significant : Ifrane and Oujda in the north-east and the Middle Atlas with respectively +1.52 and +1.28 days/44 years and the third largest city, Tangiers, located in north-west of Morocco, with an increase of +1.025 days/44 years but insignificant. (Figures placed in Appendix I). By contrast, cities in the south of Morocco, they, show a slight decline in non significant trend as in Agadir (-0.14 days/44 years), and Marrakech (-0.34 days / 44 years) (Figure 4).

Thus, as indices of extreme cold and extreme heat, the downward trend even insignificant, registered in the southern regions of Morocco, for the ICW, and that the rise in the northern regions of Morocco, in the case of the IHW, reflect a possible migration of the warm southern regions to the north of Morocco. This increase in temperature, we have seen previously, was particularly pronounced in the recent years. But these upheavals are they due to chance or do we witnessing the early signs of climate change announced? It is clear that the number of extremely hot days will tend to increase with rising average temperatures. Indeed, the mathematical behaviour of the frequency distributions showed a typical distribution of the average summer temperature measurements (Figure 5).

It's a classic curve, where values near the mean are more frequent and less frequent than extreme values. Under a warmer climate where the temperatures vary regularly, the curve keeps the same shape but is shifted to the right with considerable consequences. The number of summer with the high average rates temperatures between the two graphs. In addition, where there was a low value of temperature on the left curve, there is a high value on the right. In other words, an extremely hot summer could be repeated frequently in a warmer climate. It is a theoretical illustration, but that shows how a small change in mean temperature causes significant changes in extremes temperature [2].

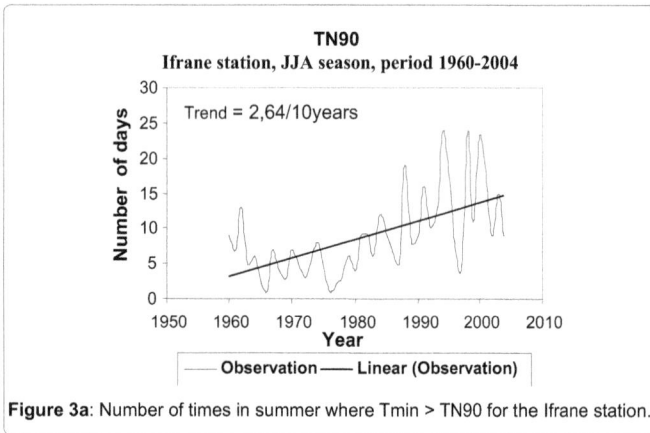

Figure 3a: Number of times in summer where Tmin > TN90 for the Ifrane station.

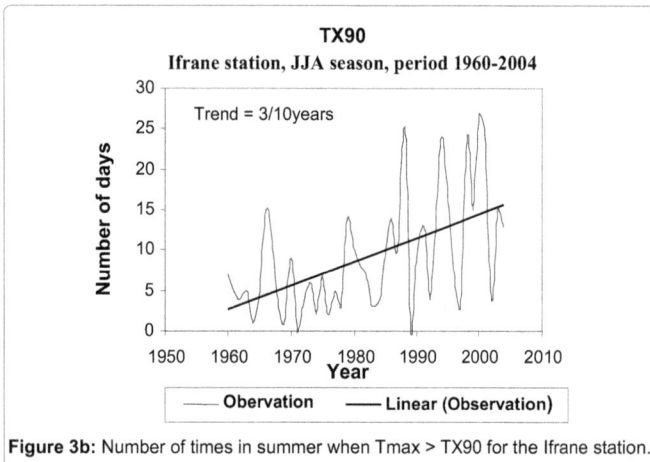

Figure 3b: Number of times in summer when Tmax > TX90 for the Ifrane station.

Thus, the observations made on extreme temperatures are illustrative of climate change gaining ground.

Conclusion

The study of variations and trends in extreme temperature conducted over the period 1960-2004, at five weather stations, subject to climate indices, selected according to the seasons, shows, in general, the frequency of cold days has significantly decreased and that of hot days has increased significantly, indicating a slight warming of the climate in Morocco. This warming leads on the one hand, a decrease in the number of cold days, especially in areas south of Morocco, on the other hand, by increasing the number of hot days, especially in areas north of Morocco, indicating thereby, a possible migration of the arid regions of southern Morocco to the north of Morocco.

Precipitation

If the analysis of the precipitation indices, calculated according RCliDex, has found no significance of trends for the all stations, the approach of the precipitation indices measured "and PRCTOT R10mm" used in this study, however, shows a downward trend but not significant for most cities exception for the Oujda city where a significant reduction at 95% is noted. Annual changes over the period 1960-2004, differences in the average precipitation, are shown in Figure 6 for the Oujda city.

We note in Oujda city, the majority of histograms are directed upwards from 1960 until 1982, and beyond that date, a reversal in the direction of the histogram occurs. In Figure 7 it has been shown, for the Oujda city, the annual variations in the number of days with precipitation totals above10 mm. It is seen that the trend line, which expresses a good approximation of the evolution of deviations of the average rainfall experienced a reversal trend: 12 days at the beginning of the period and 6 days in the end, a decrease by half during the past five decades, reflecting a net reduction of rainfall and the north-eastern of Morocco.

For temperature, this study of trends in foreign exchange and precipitation in Morocco was on the following period 1960-2004. The

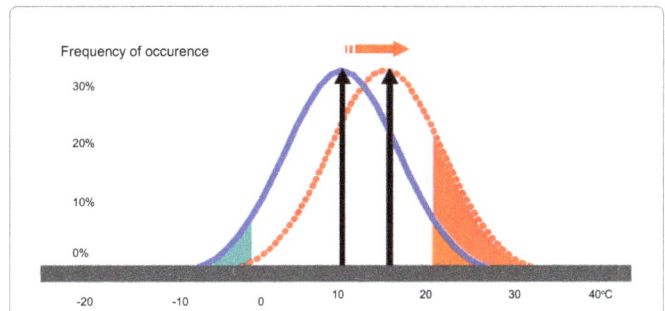

Figure 5: Changes in the frequency of extreme temperatures with the changes in average temperature

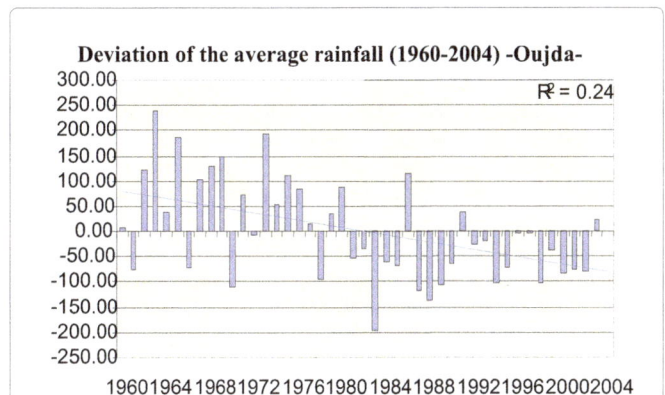

Figure 6: Deviation from average rainfall over the period 1960-2004 for the Oujda city

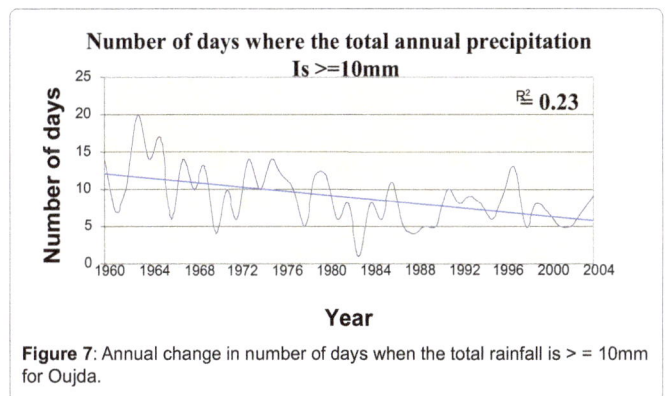

Figure 7: Annual change in number of days when the total rainfall is > = 10mm for Oujda.

Four Seasons of the Year, shows that the trends of simulated rainfall indices are not significant for all the stations, while the quantity rainfall, are not either significant for the majority of the cases with an exception in Oujda station where trend down seems to take place.

Forecasting future climate in Morocco

Future projections of climate in Morocco, our study is based were developed The Software MAGICC / SCENGEN.

Presentation Templates

The MAGICC model

MAGICC (Model for the Assessment of Greenhouse-Gas Induced

	Indices	Trend during the period 1960-2004 (days/44ans)				
		Agadir	Oujda	Marrak.	Tanger	Ifrane
JJA	TX90	2.4	8.3*	-0.22	10.77*	12.85*
	TN90	9.2*	10.8*	3.33	2.9	11.64*
	TX10	-11.6*	-8.2*	-6.76*	-7.85*	-8.98*
	TN10	-13.2*	-8.7*	-10.85*	-3.22	-7.17
	IVC	-0.14	1.28*	-0.34	1.025	1.52*
SON	TX90	-1.02	3.7	0	2	2.38
	TN90	6.2*	2.84	4.15	1.49	3.41
	TX10	-2.5	-1.21	-3.81	-3.22	-5.09*
	TN10	-2.84	-5.17	-4.15	0.4	-0.96
DJF	TX90	7.53*	2.17	7.38*	6.51*	10.3*
	TN90	5.25*	2.37	6.67*	3.29	1.43
	TX10	-9.81*	-4.22*	-5.41*	-4.51*	-5.79*
	TN10	-8.25*	-2.64	-5.11*	3.2	-2.09
	IVF	-0.562	-0.58	-0.008	0.88	-0.29
MAM	TX90	4.18	2.26	-0.03	3.36	0.24
	TN90	8.51*	5.62*	4.29*	-0.14	3.82
	TX10	-7.99*	-6.99*	-6.33*	-8.38*	-6.77*
	TN10	-11*	-4.14	-11.8*	-4.7	-4.92*
	TXS	-0.198	16.5*	13.5*	13.5*	15.5*

*Significant trend at 5%

Table 2: Trends in extremes of temperature on the period 1960-2004 and for the four seasons of the year

Climate Change), consists to set of coupled cycle greenhouse gas emissions, climate and ice melt in one integrated package. It allows the use to determine the changes in the atmospheric concentration of the carbon dioxide (CO_2), the average temperature of the air at the surface and the mean sea level between 1990 and 2100. These are the result of anthropogenic emissions of CO_2, methane (CH_4), nitrous oxide (N_2O), halogenated hydrocarbons (eg., HCFCs, HFCs, PFCs) and sulphur dioxide (SO_2). The years 1990 and 2100 are respectively the beginning and the end of such changes. This software is designed to: Compare, within the same application, the consequences on the global climate scenarios of two separate programmes. MAGICC appoint one of these scenarios "references", which corresponds to the high values of the parameters and the other "political", which corresponds to the low values (optimistic). It is possible to evaluate two different emissions scenarios.

Determine the sensitivity of the results of the different emission scenarios to changes in the parameters of the model. The basic interval of uncertainty is calculated implicitly. But it is possible to compare the results of a given emission scenario for the parameters of the model specified by the user and the results produced by the implied values parameter. The software offers 48 scenarios of the future greenhouse of gases and aerosols. For our study, we chose the P50 scenario, because it is the median of all the greenhouse gas emissions scenarios. The parameters of the model can be changed; it cites the model of carbon cycle, climate feedbacks on the carbon cycle, aerosol forcing, climate sensitivity, the thermohaline circulation, the vertical diffusivity and the melting snow. Thus, the nature of the output calculated and displayed by the MAGICC including one year "reference" for the outputs of the model on the basis of which changes in global mean temperature and mean sea level are calculated. A year "start" for the model outputs for which information on the diagnosis of the MAGICC is listed in the report files viewable from the main menu. And the year "end" to the model output, which informs on the MAGICC year, stops the calculations and reporting. The year 2100 is implied, but the user can opt for one year before or after (until 2500). However, the MAGICC only calculates the times, which correspond to the files of gas emissions, if these files go to 2100.

The SCENGEN model

SCENGEN is a scenario generator, global or regional climate change. This software consists of a database containing the results of many experiments of General Circulation Models (GCMs) and also observed climate data on a global scale. He manipulates these various data fields using the information on the speed and magnitude of the global warming that provides MAGICC and depending on the choice made by the user of the main characteristics of the climate scenarios. This model can be operated in conjunction with MAGICC, but it can also be used alone with a more limited number of functions. When running, the software from MAGICC, window or control panel is available to users at least five parameters: the first concerns the choice of variables to be displayed (average temperature or precipitation), the 2nd, concerns the choice of the scale "linear or exponential" and choose the period (monthly, seasonal or annual), the third, selection of a model among the 17 models available from GCM (Global Circulation Model), the fourth , refers to the choice of the place where you can view the results, the 5th, the projection year or date for which the calculations were made, dates are an average of 30 years around that date and scenario displays both scenarios (Reference and Strategy) used for the calculations and selected software and MAGICC, analysis of change in relation to the current situation defined as an average over the period 1961-1990 for temperature and for the period 1981-2000 precipitation.

Projection of the future climate

The first part to develop a climate change scenario process (CCS) deals MAGICC and the selection of the emission scenarios and the settings MAGICC. The medium scenario P50 was used, without changing the default settings of the MAGICC. The second step is to use the emission scenario B1 A1F and to define a wide range of change.

Projection that the average scenario P50

Under this scenario, by 2050, compared against the current situation, the average global temperature could rise from a low of 0.9°C and a maximum of 1.9°C, with an average of 1.4°C, CO_2 concentrations would be associated with these projections, between 490 and 571 ppmv, with an average of 516 ppmv, and rising sea level (ENM) will be between 5 and 27 cm, the average would be15.5 cm. In 2100, the corresponding increases in temperature are respectively 1.9°C, 4.28°C and 2.9°C, corresponding ENM are respectively16.8 cm, 67.5 cm and 38.5 cm.

Projection scenarios as A1F and B1

In order to define a wide range of possible future global climate, given the uncertainty associated with future emissions of greenhouse gases, we also used emissions scenarios A1F and B1 to define an interval changes in mean temperature and concentrations of CO_2 worldwide. The range of warming projected for these three choices, in 2050, evolves from 0.7 to 2.3°C, CO_2 concentrations between 472 and 595 (ppmv), and ENM, for each scenario vary between 6.3 (B1 scenario) and 27.6 cm (scenario A1F). This gap will be wide in 2100, to oscillate between 1.2 and 5.8°C, CO_2 concentrations between 503 and 1032 ppmv.

Climate change at regional level

Once MAGICC was executed, is performed after the development of the regional climate change scenarios for Morocco in 2050 by combining the software SCENGEN. For the profile selection warming, it was used in the development of an SCC is using several GCMs, or a scenario CSC with an average profile of GCMs. Both options have been successively used to examine the effects of CSC products for different regions of Morocco. Like all the observed climate data and available in SCENGEN is a resolution of 5° latitude / longitude, Morocco was divided into seven dials (Figure 8) which correspond to the climatic regionalization of the Kingdom.

Projections of Future Climate Models by MAGICC/SCENGEN

To represent the range, the future regional climate change was used, we have seen above, several models. Then proceeded to the determination of GCMs SCENGEN proposed, which exhibit extreme changes in temperature and precipitation in each region of Morocco. Table 3 summarizes the extreme variations in temperature and precipitation in 2050, according to the emission scenario P50 and in a climate sensitivity of 2.6°C for each dial.

Analysis of results has yielded the following observations: Global warming called on all models but with low temperature dispersion within each region and widely dispersed across the regions

- A strong dispersion between the models for the simulation of the precipitation changes, both quantitative and qualitative.

- A difference of presentation, models used, extreme changes in the temperatures; to better to frame projections, in 2050, the future climate in Morocco next month, seasons and years, we opted for a composite profile of 17 profiles of GCMs.

Monthly cycle

To detect the changes in the likely changes following the months of the year, we plotted the histogram curves illustrating both the annual cycles of temperature over the period 1961 to 1990 and estimated the rainfall on the other from 1981 to 2000. The changes to the dial north-eastern of Morocco, the annual cycles of the temperature and precipitation averages, were represented in the Figures (placed in Appendix I) Figure 9. It notes that in 2050 the overall trend of the global warming will be widespread across Morocco, as the P50 scenario, with a magnitude greater in the northern regions of Morocco compared to southern regions of Morocco. The annual trend of warming in the north of Morocco, shows that the maximum of the temperature will be reached between the months of May and July, the minimum, it would appear between January and February. By cons, in southern Morocco, the maximum of the temperature is reached, from October to November but with lower amplitude. This decrease in the annual change of the temperature from west to east and from north to south of Morocco is due mainly to the ocean and latitudinal effect. As for the precipitation, large disparities both qualitative (drought and humidity)

Regions	Variation of temperature		Variation of precipitations	
	Minimum °C	Maximum °C	Minimum (%)	Maximum
NW	+1.5	+2.8	+7.6	-31.1
NE	+1.7	+3,0	-1.2	-34.9
W	+1.5	+2.9	-2.4	-47.9
E	+1.8	+3.1	-6.1	-48.8
SW	+1.4	+2.4	+15.6	-31.3
SE	+1.8	+3.2	+29	-56.7
S	+0.6	+2.1	+30.7	-33.9

Table 3: Changes in 2050, extremes of temperature and precipitation for each quadrant according to the emissions scenario P50.

and quantitative (magnitude of change) will be perceived, however, the downward trends remain the most likely.

Seasonal cycle

The seasonal distribution of potential changes in the temperature and precipitation arouse particular interest to the agricultural sector. To illustrate these changes, we used the composite profile as the P50 scenario. The evolution of changes in the temperature and precipitation compared to observed values, respectively, during the years 1961-1990 and 1981-2000, was represented in the Figure 10 (placed in Appendix I).

It can be seen here again in 2050, the overall trend in the global warming, will be on all Morocco. In general, the increase in Δtemperature would be very substantial during the summer in the northern of Morocco (Temperature=2°C), where the ocean and latitudinal effects reappear. But the maximum increase characterized mainly the eastern of Morocco (Temperature = +3.1°C). As for the southern Morocco (Temperature =1.7°C), the increase will occur in the fall (the reductions of precipitation, consistency does is demonstrated or the spatial scale or temporal scale. Indeed, the strong decrease during the winter in western of Morocco, substantial falls in the north-west of Morocco, and would be more intense in the east than in the west of Morocco. Thus, despite the fact that the maximal decrease in the precipitation would be recorded outside the rainy season, even small decreases in rainfall, though they had been projected during the winter, should have had an influence on the potential of water resources particularly in the north of Morocco.

Annual cycle

It is observed for the temperature, the increase was estimated between 1.7 ° C in the south of Morocco and 2.6°C in the north-east. As for precipitation, the reductions are estimated between 15 and 22% in northern Morocco and between 9 and 16 % in its southern part. The effect of latitude and the ocean is net in the north 30th parallel (Figure 11). The temperature elevation and instability resulting from lower precipitation contributes to this situation.

Sectorial Impact of Extreme Weather Events

Water resources

Surface water: Were Represented in Figure 12 Overall contributions Recorded at all dams. Observed reductions in water yield at all the Existing dams Have Reached 20% When Comparing the Period To The 1970-2005 period 1940-2005. This reduction is estimated at 35% when compared To the Same Period The Period 1940-1970.

Groundwater: In Figure 13, have been shown some piezometric levels. The later recorded an alarming decline in most quifers: 40 in the Souss area, 30 m in the Haouz and 60 m in the Saïss area reflecting

Figure 8: SCENGEN domain and area

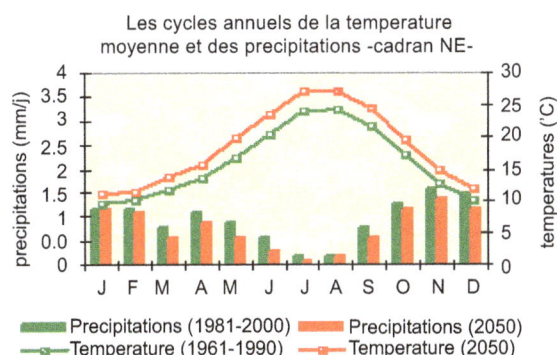

Figure 9: Annual cycles in mean temperature and precipitation - Dial NE Morocco.

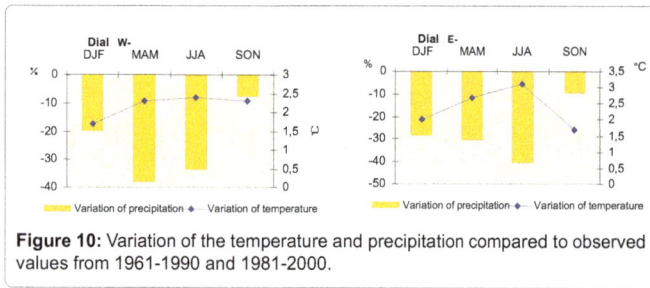

Figure 10: Variation of the temperature and precipitation compared to observed values from 1961-1990 and 1981-2000.

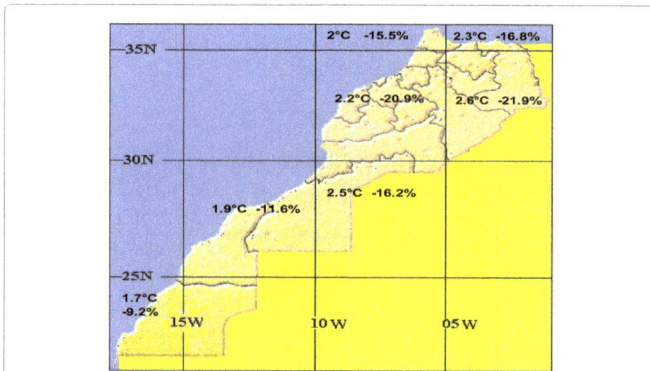

Figure 11: Variation of temperature (ΔT / T) and precipitation (ΔP / P) compared with observed values respectively over periods 1961-1990 and 1981-2000

lower recharge. This reduction would result from the imbalance between water withdrawals and groundwater recharge of that aquifer.

Future projection: The correlative analysis between rainfall recorded in three basins (Moulouya Sebu and Errachidia) and flow rates recorded at their outlets controlled by three major dams (Mohamed V and Hassan Idriss 1st Eddakhil) was used to estimate a reduction water supplies in 2050 from 14% to 20% for dams in question.

Agriculture: The study of the impact of climate change on agriculture has been the subject of two national communications [3]. The first, dated early 2001, involved essentially two key sectors of the Moroccan economy: agriculture and water resources for 2020. The second, dated late 2009, affected, in addition to agriculture and water, other sectors such as health, forests and climate in 2030. These impacts can be summarized as follows:

Crops grown-in will suffer the most from the effects of drought, seeing their grain production vacillate between 100 million quintals a year of good rainfall to 18 million quintals for dry year and yields between 17 quintals hectare and 4 quintals per hectare. This is especially the period of growth that would be affected and there is no doubt that the risk of dry periods will increase. - Reduction of the crop cycles; - The gap and reducing the growing period; - Increased risk of dry spells in the beginning, middle and end of the cycle of annual crops; - The disappearance of some crops such as canary grass and some trees as Argan; - The emergence of new diseases. - The impact on livestock is associated with the impact on agriculture; livestock production in Morocco is inseparable from crop production system.

Forest

The observable climate change has visible consequences on the forests with Moroccan kills standing trees, and sometimes a complete lack of germination of seeds of cedar. According to LEPOUTRE &

PUJOS (1964), the germination can occur only when the temperature reaches a value of about 9°C to +10 °C for a period of at least 9 to 10 days for temperatures minimum of not less than -5 °C, until April or May, as is the case throughout the Middle Atlas. If one stands in the climatic conditions of the Middle Atlas cedar forests, especially in Ifrane, where the number of cold days has accused the spring, a decrease of -6.77 days/44 years (Table 2), we can estimate that the natural regeneration of this species will disappear in 65 years, hence the idea of finding an alternative solution to this problem. Flora fauna follows regressive in this dynamic and also shows his vulnerability.

Health and risk of reactivation of disease outbreaks

The health vulnerability related to climate change in Morocco, can be explained by the presence of endemic foci of disease may be exacerbated by climate change, including malaria, bilharzia, typhoid and cholera [4]. Although efforts in the fight against these diseases are deployed, the risk of reactivation as a result of climate change is always possible.

Coastal

The impact of CC on the coastal environment has been very little studied. A study conducted by BAZAIRI Synthesis [5] showed that the two environments that suffer the most vulnerable are: the first is low-lying coastal plains and beaches, but also coastal wetlands such as estuaries where the slightest rise sea level (ENM) has a direct impact on erosion and / or flooding of coastal wetlands (lagoon of Nador Lagoon Smir), mouthpieces (Moulouya) and the beaches of the Mediterranean Morocco (Tetouan Mdiq , Restinga-Smir, Al Hoceima, Cala Iris, Nador and Saidia) causing losses of large areas of farmland causing, thereby, reduced agricultural productivity. It also causes sodification soil and groundwater salinization, leading to a shortage of drinking water especially in the central and eastern Mediterranean coast of Morocco, threatens the port infrastructure (Tangier, Nador, Al Hoceima) and resorts adjacent to the coast (Bay of Tangier, a seaside resort of Saidia), and is responsible for the proliferation of phytoplankton and the emergence of new diseases, particularly in aquaculture sites located offshore Mdiq. Added to indirect impacts that affect the socio-economic crises including declines in activities in the fisheries sector where declines in catches are considered in relation to the CC, and services continued to decline attractiveness of coastal areas that would be affected by heat waves as is the case of resorts Saidia and Tangier. The second, where the impact of climate change on ecosystem vulnerability manifests itself differently in different settings: ENM only especially for coastal habitats of Mediterranean Monk Seal by loss of small Moroccan beaches inside the caves used as breeding; across the Mediterranean as in the case of species of flora and fauna of the southern areas, where

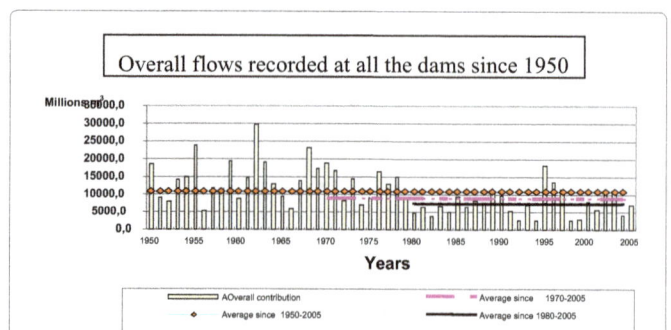

Figure 12: Contributions to global water recorded at all the dams over the period 1950-2005

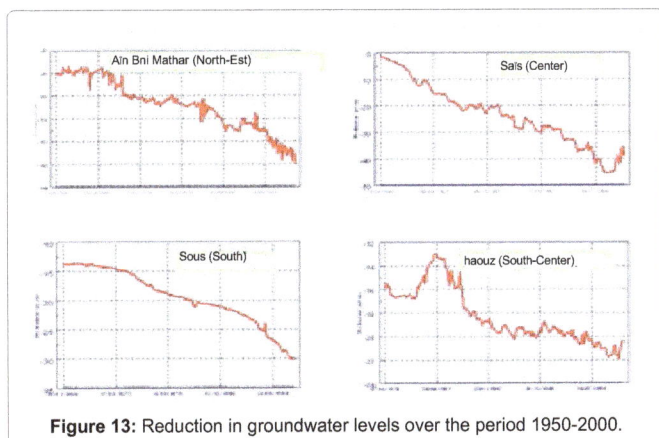

Figure 13: Reduction in groundwater levels over the period 1950-2000.

the conditions necessary for their move up the shore does not arise. So, for example, the giant limpet (Patella ferruginea) and vermetid (Dendropoma paetreum) and changes the geographical distribution of marine invertebrate species of coral (Anthipathella Wallaston) in the region of Ceuta (Dendrophyllia Laborel) in the region of Cabo Negro respectively Macaronesian and limited distribution between Mauritania and Ghana including the Canary Islands, have been encountered in the Mediterranean Sea, too, the red coral (Corallium rubrum), the mollusk bivalve (Lithophaga lithophaga) and cnidarian (Paramuricea clavata) are sensitive to climate; Same in marine seagrass beds (Zostera noltii, Cymodocea nodosa) are relatively more abundant on the Mediterranean coast of Morocco in relation to seagrass Posidonia oceanica, which them are known only around Chafarinas Islands; Biological invasions of ecosystems hosting a significant biodiversity on the Mediterranean coast (the lagoon of Nador, the mouth of Moulouya and coastal areas of Al Hoceima and Tetouan) although many species and habitats are protected and identified as "hot spots" of pollution; significant change in phenology of avian migration of wetlands.

Conclusion

In this study, we examined the evolution of the climate of Morocco on the basis of extreme temperatures (maximum and minimum) through the hot and cold extreme (TX90, TN90, TX10, TN10, IVF,

IVC) and precipitation through indices and R 10mm PRCTOT the four seasons of the year on the period 1960-2004 in five cities. The two parameters examined, the temperature is the one with the clearest effect. Indeed, for the temperature, and for the five study sites for the four seasons of the year, the frequency of hot days per year have increased in recent years and the frequency of cold days decreased, reflecting a slight warming Morocco. For precipitation, if no clear conclusion has been reported by this study about their changing trends for all stations surveyed, the approach of deviations from the mean rainfall, revealed a statistically significant rainfall in the only station in Oujda. Climate projections, in 2050, at the regional level were conducted using the new software version MAGICC / SCENGEN seventeen coupled general circulation models (GCMs), as the average emission scenario P50, with a sensitivity average climate (2.6°C) and a mesh SCENGEN. The projection results for the whole country showed a marked warming trend of about 1.7°C to 2.6°C, whereas for precipitation, the trend is down about 15 to 22% in the northern part of Morocco and between 9 and 16% in the south. For cons, the reduction in annual rainfall is more important: it varies between 6 and 13% in the north and between 0.7 and 8% in the south. Moving to the north of the arid zone; we recommend the following studies: Looking for a possible correlation between the indices of extreme temperature and sea surface; the use of regional models for projecting future climate of Morocco.

References

1. EL Rherari E (2005) Impact of climate change on water ressources. Work report at the end of study. EHTP/CERHY.

2. Rajel R (2005) Study of temperature extremes. Report internship study, EHTP/DMNP.

3. Le Poutre B, Et Pujos A (1964) Climatic factors determining the conditions for germination and plant seedlings Cedar. Journal of Forestry Research in Morocco, Tom 7, Forest Research Station Rabat. North African Technical Publications, Rabat.

4. MECHOUARI R (2005) Morocco and climate change. Report internship study, EHTP/MND.

5. Bazairi H (2008) National Overview of vulnerability and impacts climate change on marine and coastal biodiversity in the Moroccan Mediterranean. Contract / RAC/PSA, No. 05-2008 SAPBIO.

7

Detection of Precipitation and Temperature Trend Patterns for Mulanje District, Southern Part of Malawi

Kachaje O[1*], Kasulo V[2] and Chavula G[1]

[1]*University of Malawi-Polytechnic, Chichiri, Blantyre, Malawi*

[2]*Mzuzu University, Mzuzu, Malawi*

[*]**Corresponding author:** Kachaje O, University of Malawi-Polytechnic, P/Bag 303, Chichiri, Blantyre 3, Malawi, E-mail: okachaje@gmail.com

Abstract

This paper reports on the recent trend of climate changes taking place in Mulanje, one of the major tea producing districts in Malawi, with a focus on precipitation and temperature. The study analyzed trends in weather time series (air temperature and rainfall) data from 1980 to 2011 using non-parametric Mann-Kendall test. The MK test gave a mix of positive and negative trends for the monthly and seasonal precipitation. Only the month of December showed a positive significant trend while February, April, May, June September, October and November had a significant negative trend. Cool and wet as well as hot and dry season showed a significant decreasing trend in precipitation. The data also indicated that annual rainfall variation is decreasing with time at the rate of about 4.29 mm per year while temperature is increasing by about 0.04°C every year. The MK test for temperature results showed a significant increasing trend for both mean annual minimum and maximum temperature. The results of the study do not deny climate change is happening, therefore, proper adaptation measures should be employed to reduce the vulnerability to climate change in the area because of its economic significance in tea production.

Keywords: Climate change; Malawi; Precipitation; Temperature; Mann-Kendall

Introduction

There is a worldwide consensus that climate change is a real, rapidly advancing and widespread threat [1]. The change in climate will result in alteration of temperature and rainfall regimes in most regions of the world. While climate change results from activities all over the globe with rather unevenly spread contributions to it, Africa at large and southern Africa in particular will be the most affected part of the world due to lack of adaptation and mitigating measures [2].

Climate change will lead to very different impacts in different countries, depending on local, regional environmental conditions and on differences in vulnerability to climate change [3].

Malawi as a country has experienced extreme weather events over the last few decades, ranging from droughts (1991/92) to floods (1996/97) and flush floods (2000/01). These extreme weather events also affects the country differently since its climate is significantly moderated by the effects of Lake Malawi, high altitudes and proximity to the influence of westerly frontal systems which move eastwards around the South African coast, influencing the temperatures to be relatively cool [4].

The country's climate has two distract seasons, the rainy season from November to April and the dry season from May to October. However, the dry season is subdivided into two parts namely cool and wet (May to August) and hot and dry (September to October).

Detailed rainfall climatology for Malawi by Nicholson et al. [5] revealed that the rainfall regime in 75 weather stations throughout Malawi is not homogeneous. Through correlating the stations from one another, the country was divided into four regions by combining the stations as illustrated in Figure 1. A major implication of the results was that the factors governing inter-annual variability may be quite different for the early and late rainy seasons and global climate change might affect these regions differently.

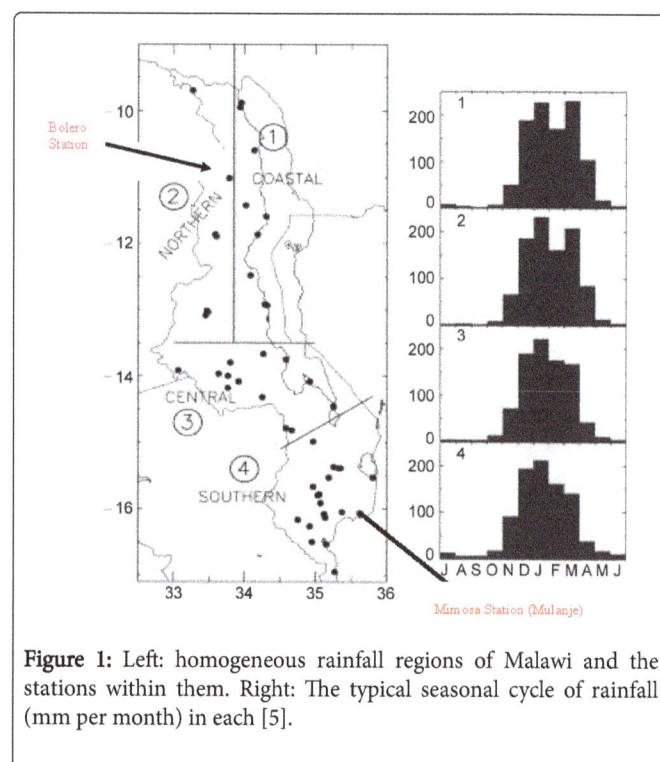

Figure 1: Left: homogeneous rainfall regions of Malawi and the stations within them. Right: The typical seasonal cycle of rainfall (mm per month) in each [5].

In the quest of forecasting climate change pattern for Bolero Agriculture Extension Planning Area in Malawi, Singini et al. [6] concluded that the maximum temperature has increased by 1.6°C from 27.7°C in 1982 and will increase to 29.3°C in 2030. This increase in temperature would have a negative impact on agricultural livelihood options in the area (Bolero). The geographic position of Bolero falls in region 2 (Figure 1), and as attested by Nicholson et al. [5], the regions are not homogeneous and thus the climate for region 2 is different from other regions. Therefore, this calls for reasons to investigate the change in climate for Mulanje district (in region 4) in order to reveal the impact it will have on the people in the district.

Mulanje District is named after the highest Mountain in the South-Central Africa and is located approximately 65 km east of Blantyre district (Figure 2). The district local weather conditions are greatly influenced by the Mulanje Mountain, the southern portion of the eastern Shire Highlands, and the climate is also partly affected by Chiperoni winds causing high rainfall on the windward (South East side of the Mulanje Mountain) while limiting it on the leeward side [7]. The average annual rainfall is 1600 mm and the average minimum annual temperatures ranges from 21°C to 23°C and the maximum temperatures of about 32°C to 35°C [8]. Taulo et al. [7] also note that one most distinctive feature of Mulanje climate is the variation of rainfall over short distances, a characteristic attributed to the influence of Mulanje Mountain.

The aim of this study was to analyse the recent trend of climate changes taking place in the area (Mulanje), with a focus on precipitation and temperature. The district was chosen due to the economic significance it contributes to the growth of the nation. Malawi as a country heavily depends on agricultural products as its main exports. Apart from tobacco, sugar and coffee, tea is one of Malawi's top main exports and accounts for 8% of the export earnings [9]. Mulanje is one of the major districts where the crop is grown because its soils, climatic conditions and sloping ground of the area around Mount Mulanje are well suited to the crop [10]. However, prolonged droughts, increased temperatures, late onset and early cessation of precipitation will affect tea production [11]. Climate change could therefore negatively affect the tea industry of Mulanje district and later on Malawi. The results of the study are to inform practitioners, policy makers and the community on the future projections of climate change in the district, its effects and strategies for adaptation.

Materials and Methods

Data and methods

To assess the recent trend patterns in Mulanje district, we collected the historic temperature and precipitation data for the district. The utilized data in these studies was kindly provided by the Department of Climate Change and Meteorological Services (DCCMS). Pre-quality analyses of the data were done by DCCMS. The mean monthly minimum temperature and mean monthly maximum temperature data as well as the monthly and seasonal precipitation for the past 31 years (1980 to 2011) from Mimosa station was analysed to infer for any changes in climate. The data was logically interpreted along with simple tables, charts and graphs.

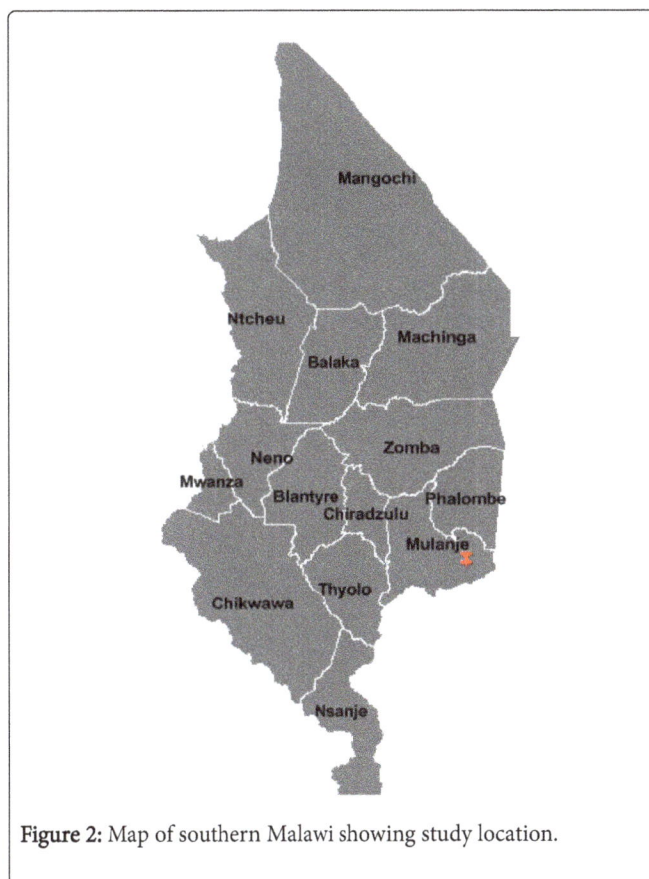

Figure 2: Map of southern Malawi showing study location.

Data analysis

The precipitation and temperature data were used to analyze the changes in trend happening in Mulanje district over the years. XLSTAT 2016 was used for Mann-Kendall (MK) statistical test to detect if any statistically significant trends exist in the data. This test, MK, is a powerful non-parametric method tool for analyzing long time series data such as precipitation, temperature and discharge. Under the null hypothesis (H0), the assumption is that there is no trend in the data and the alternative hypothesis (H1) carries the assumption that there is an increasing or decreasing trend over time. The mathematical computational for the MK test statistics S, Var (S) and the standard test statistic ZS were calculated as follows:

$$S = \sum_{i=1}^{n=1} \sum_{j=i+1}^{n} sign(Tj - Ti) \tag{1}$$

$$sign(Tj = Ti) = \begin{cases} 1 & if\ Tj - Ti > 0 \\ 0 & if\ Tj - Ti = 0 \\ -1 & if\ Tj - Ti < 0 \end{cases} \tag{2}$$

$$\sigma^2 = \frac{n(n-1)(2n+5) - \sum t_i(i)(i-1)(2i+5)}{18} \tag{3}$$

$$Z_S = \begin{cases} \dfrac{s-1}{\sigma} & for\ S > 0 \\ 0 & for\ S = 0 \\ \dfrac{s+1}{\sigma} & for\ S < 0 \end{cases} \qquad (4)$$

where Tj and Ti are the actual time series observation data, n is the period of the time series data, ti denotes the number of ties up to sample i. The test statistic ZS follows normal distribution and was used as a measure of significance of trend. Positive and negative values of ZS signify an increase and decreasing trend respectively. A significance level α is used to test the null hypothesis (increase or decreasing) trend exist. If ZS is greater than Zα/2, the null hypothesis is rejected implying that the trend is statistically significant. The chosen significance level for this study was 1.96 for p-value of 0.05. However, before running a MK trend test, autocorrelation was considered to remove the serial dependence of the time series data that would cause problems in testing of data and interpretation of results, according to the method proposed by Yue and Wang [12].

Results and Discussion

Precipitation

Rainfall distribution: The 31 years' data of monthly and seasonal rainfall from 1980 to 2011 at Mimosa station in Mulanje district was analyzed for changes in rainfall distribution, and results of the analysis are given in Figure 3. The results showed that about 85% of the rainfall occurs during rainy season (November to April), 10% during the cool and wet season (May to August) and finally, 5% during the hot and dry season (September to October).

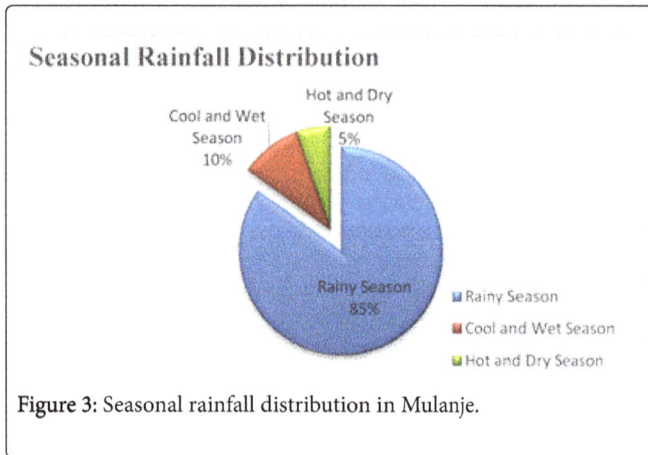

Figure 3: Seasonal rainfall distribution in Mulanje.

The mean daily rainfall distribution in Mulanje district for the same period (1980 to 2011) is shown in Figure 4, and shows that precipitation in Mulanje starts from the 295th day, which is in October, and is highest between about the 341th to 65th days in the year, which is the period during November to March (rainy season in Malawi). These results are in agreement with most of the studies done on Malawi about the onset, duration and end of rainy season, that is, it begins in October or November and continues until March or April [4,5,13-18].

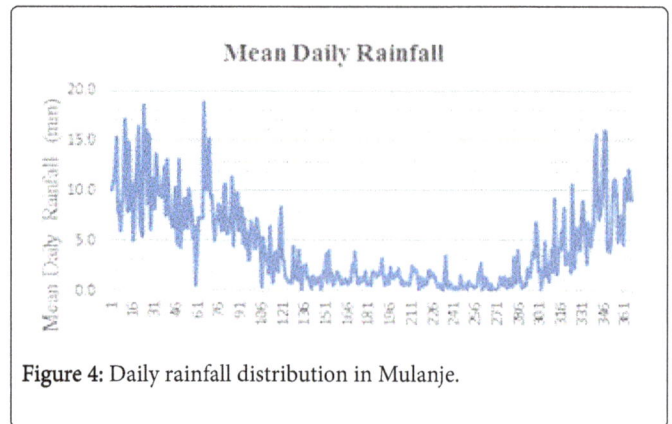

Figure 4: Daily rainfall distribution in Mulanje.

Annual rainfall variation

The average annual rainfall data for period 1980 to 2011 (31 years) used in this study showed that the average annual rainfall of Mulanje is decreasing at a rate of about 4.29 mm per year (Figure 5). This is in agreement with the observation [16] which shows a general decrease in the mean monthly rainfall with time. From this analysis, the maximum occurrence of rainfall occurred in 1988/89 season and the minimum rainfall occurred in 1991/92 season with total annual precipitation of 2356.5 mm and 811.2 mm respectively. It was these years, 1991/92, when Malawi experienced severe drought across all regions of the country. It was also observed from the results that the average rainfall is 1631.1 mm for the 31 year period representing an average of about 135.9 mm of precipitation per month [17].

However, Vincent [18] reports that the amount of rainfall in southern Malawi is between 150 mm and 300 mm per month and hence that for Mulanje is below this range. Bulckens [19] also observed a decrease in rainfall in Mulanje district from an average of 2000 mm in 1960 to about 1500 mm in 2012. This clearly shows that there is a decline in annual rainfall distribution for period 1980 to 2011 and more analysis needs to be done to see whether this trend will continue.

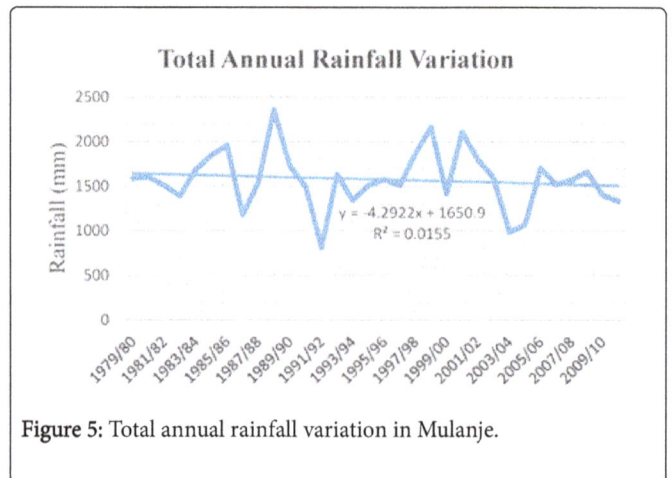

Figure 5: Total annual rainfall variation in Mulanje.

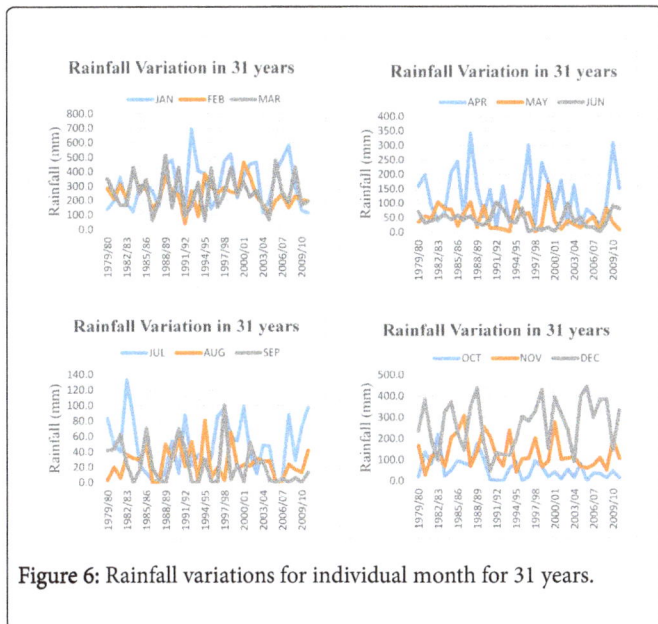

Figure 6: Rainfall variations for individual month for 31 years.

Figure 6 shows the rainfall variation for individual months for the 31 years period. It was observed that the minimum average annual rainfall occurs in the month of September (23.6 mm) followed by August (26.2 mm) and the maximum average rainfall is in January (322 mm) and December (274.3 mm).

In order to detect if the trends in the precipitation were significant, the MK test was applied on a monthly scale for the entire period from 1980 to 2011 (31 years). The summary results of non-parametric MK test for trend analysis of data are presented in Table 1. The results for monthly trend tests showed a mix of positive and negative trends with January, July and December depicting a positive trend and the rest of the months of the year showing negative trend. The increasing trend in precipitation for December was statistically significant (p=0.006). Significant negative trend were detected in the months of February, April, May, June September, October and November; no significant trends were found for the other months. The significant decrease of precipitation in October and November confirms the observation by many studies that the rains are shifting towards December [13,15,17].

Variable	Kendalls tau	S	Var (S)	P-value (Two-tailed)	Alpha	Interpretation	Trend
JUL	0.06	31	792.5	0.287	0.05	Accept	NST
AUG	-0.03	-13	646.9	0.637	0.05	Accept	NST
SEP	-0.22	-104	428.5	<0.0001	0.05	Reject	Decreasing
OCT	-0.29	-145	634.4	<0.0001	0.05	Reject	Decreasing
NOV	-0.16	-78	720.9	0.004	0.05	Reject	Decreasing
DEC	0.15	73	681.6	0.006	0.05	Reject	Increasing
JAN	0.1	52	770.7	0.066	0.05	Accept	NST
FEB	-0.22	-108	692.6	<0.0001	0.05	Accept	Decreasing
MAR	-0.03	-16	137.4	0.201	0.05	Accept	NST
APR	-0.1	-49	299	0.006	0.05	Reject	Decreasing
MAY	-0.21	-104	257.2	<0.0001	0.05	Reject	Decreasing
JUN	-0.15	-75	678.6	0.005	0.05	Reject	Decreasing

Table 1: MK tests results for precipitation in monthly time series. NST: No Significant Trend.

Seasonal rainfall variation

The hot and dry season precedes the rainy season. Rainfall data for about 61 days from September to October was analysed. The hot and dry season average annual rainfall variation together with the non-rainy days in Mulanje results are shown in Figure 7. The observation was that the season annual rainfall is decreasing by about 3.52 mm per year in the district. The UNDP climate change profile on Malawi done by McSweeney et al. [20] also showed that the projections in rainfall tend to decrease in September, October and November. The number of non-rainy days, as depicted in Figure 7, is increasing within the season. Historically, the rains in most parts of the country were beginning in late October [4,16]. This may also explain the changes in the onset of

rainy season, as many studies have attested it is shifting towards December and making agricultural decisions regarding planting more difficult and less reliable [13,15,17].

The average annual rainfall distribution and non-rainy days during the rainy season (November to April) were also analysed as shown in Figure 7. It was observed that the rainy season average annual rainfall showed a mild increase in trend of about 0.63 mm per year. This was also attested by McSweeney et al. [20] where rainfall projections over Malawi showed an increasing trend during the rainy season. The slope of the trend line for number of non-rainy days during the season is not very large in magnitude but it is positive indicating a small increase. Hence, the amount of rainfall being received in the district is almost the same every season of the year but in lesser number of days. The Participatory Rural Assessment (PRA) exercise reported that villagers are experiencing late onset of the rains and an earlier cessation from what they consider "normal rainy seasonal" [13,15,17]. Generally, projections for Malawi as a whole are mainly consistent which suggest a later onset of rainy season and shows an enhancement of rainfall during the months of December through February, followed by an early cessation [4,5,16,17,20].

Similar to hot and dry season, the trend of the cool and wet season (May to August) average seasonal rainfall is decreasing by about 1.4 mm per year, with an increasing number of non-rainy days are also increasing as shown in Figure 7. Rainfall projections for the austral winter months (June, July and August) tend to decrease June to August becoming exceedingly dry over most of Malawi, with seasonal rainfall in the order of 10 mm to 25 mm [5,20]. These trends, therefore, needs to be analyzed further to evaluate if the changes are significant in the district in seasonal time series.

Hence, the MK test was also used to detect trends in seasonal precipitation between 1980 and 2011 period and the results are given in Table 2. Similar to the monthly analysis, results from ZS statistics shows a mix of negate and positive trends. Cool and wet as well as hot and dry season showed a decreasing negative trend in precipitation and trends were statistically significant implying that the precipitation in these two seasons is decreasing over time. However, thought not statistically significant, a positive precipitation trend was detected for rainy season.

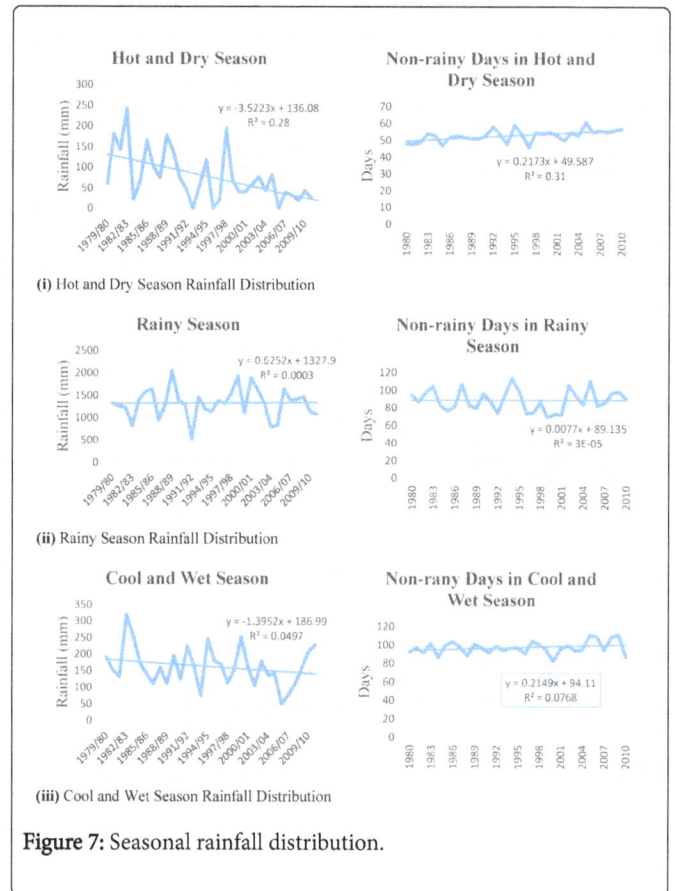

Figure 7: Seasonal rainfall distribution.

Variable	Kendalls tau	S	Var (S)	P-value (Two-tailed)	Alpha	Interpretation	Trend
Rainy	0.03	16	391.2	0.448	0.05	Accept	NST
Cool & Wet	-0.12	-60	465.8	0.006	0.05	Reject	Decreasing
Hot & Dry	-0.3	-150	473.5	<0.0001	0.05	Reject	Decreasing

Table 2: MK tests results for precipitation in seasonal time series. NS: No Significant Trend.

Temperature

The temperature data from 1980 to 2010 was also analyzed. Figure 8 shows that both mean annual minimum temperature and mean annual maximum temperature for Mulanje is increasing by about 0.04°C every year. Phiri and Saka [21] also observed that mean temperatures in the lower Shire (Chikwawa and Nsanje districts) had increased by 2.3% while mean maximum temperatures had increased by 2% between 1970 and 2002. Meanwhile, for the period 1960 to 2006 [20], observed that mean annual temperature increased by 0.9°C, an average of 0.21°C per decade. The observation from the temperature analysis done here shows consistency with the GCMs which is in also in conformity with temperature projections for the whole country. This increase in temperature has also been perceived by villagers in a PRA study [15,17] and the results highlighted warming temperatures as one of the most visible impacts of climate change in Malawi.

On running the MK test on mean monthly and mean maximum temperature data to detect if there was a trend in the 30 years period, the following results in Table 3 were obtained.

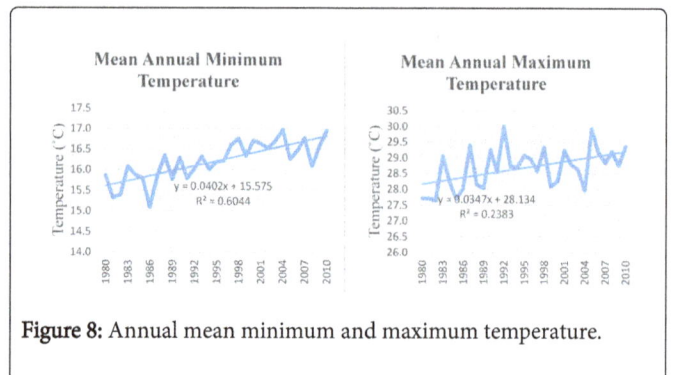

Figure 8: Annual mean minimum and maximum temperature.

The results indicated that there was significant increasing trend for both mean annual minimum and maximum temperature. Therefore, further impacts associated with the increasing temperature in Mulanje district should be looked upon.

Variable	Kendall's tau	S	Var(S)	p-value (Two-tailed)	alpha	Interpretation	Trend
Mean annual minimum temperature	0.58	268	651.9	<0.0001	0.05	Reject	Increasing
Mean annual maximum temperature	0.34	158	912.3	<0.0001	0.05	Reject	Increasing

Table 3: MK tests results for temperature in annual time series.

Conclusion

The evidence from this paper shows that there are some changes in the trend of temperature and precipitation of the district in the 31 years' period analyzed in different months. The data indicated that the annual rainfall variation is decreasing with time at the rate of about 4.29 mm per year. It also showed that the rainy season is shortening; however, the amount of rainfall being received is almost the same every season but in lesser number of days. The MK test gave a mix of positive and negative trends for the monthly and seasonal precipitation. Only the month of December showed a positive significant trend while February, April, May, June September, October and November had a significant negative trend. Cool and wet as well as hot and dry season showed a decreasing negative trend in precipitation and trends were statistically significant. However, thought not statistically significant, a positive precipitation trend was detected for rainy season. Similarly, temperature has been and can be predicted to be rising in the years to come (by about 0.04°C every year) which can lead to high evapotranspiration hence less discharge. On running the MK test for temperature, the results indicate that there is significant increasing trend for both mean monthly and mean maximum temperature. Therefore, the results of the study do not deny climate change is happening and calls for further study on climate change in the district and their impacts on tea production.

Acknowledgement

The work towards this study was financed by the Scottish government through the University of Stathclyde and University of Malawi-Polytechnic.

References

1. Mishra BK, Herath S (2012) An investigation on climate change risk for run-of-river hydropower schemes: a case study of Sunkoshi river basin, Nepal.

2. IPCC (2001) Climate Change 2001, Impacts, adaptation and vulnerability. Contribution of Working Group II, Cambridge, Cambridge University Press for Intergovernmental Panel on Climate Change.

3. United Nations Environmental Program/Earths can (UNEP/Earthscan) (2002) Global Environmental Outlook 3. London: Earthscan.

4. Ministry of Natural Resources and Environmental Affairs (MoNRE) (2002) Initial National Communication to the Conferences of Parties of the United Nation Framework Convention on Climate Change.

5. Nicholson SE, Klotter D, Chavula G (2013) A detailed rainfall climatology for Malawi, Southern Africa. Int J Climatol 34: 315-325.

6. Singini W, Tembo M, Banda C (2015) Forecasting Climate Change Pattern for Bolero Agriculture Extension Planning Area in Malawi. J Climatol Weather Forecasting 3: 145.

7. Taulo JL, Mkandawire RW, Gondwe KJ (2008) Energy policy research baseline for Mulanje and Phalombe districts.

8. Haarstad J, Jumbe CBL, Chinangwa S, Mponela P, Dalfelt A, et al. (2009) Environmental and Socio-Economic Baseline Study - Malawi. Norwegian Agency for Development Cooperation.

9. Fairtrade foundation (2010) Fairtrade Tea: Early Impacts in Malawi.

10. IRIN (2011) Tea tells the future of the climate.

11. Food and Agriculture Organization (FAO) (2014) Intergovernmental Group on Tea, Report of the Working Group on Climate Change-21st Session.

12. Yue S, Wang C (2004) The Mann-Kendall test modified by effective sample size to detect trend in serially correlated hydrological series. Water Resour Manag 18: 201-218.

13. ActionAid (2006) Climate change and smallholder farmers in Malawi: Understanding poor people's experiences in climate change adaptation.

14. Kaunda CS, Mtalo F (2013) Impacts of environmental degradation and climate change on electricity generation in Malawi. Int J Energ Env 4: 481- 496.

15. Magrath J, Sukali E (2009) The winds of change: Climate change, poverty and the environment in Malawi, Oxfam International.

16. Ministry of Natural Resources and Environmental Affairs (MoNRE) (2011) Second National Communication to the Conferences of Parties of the United Nation Framework Convention on Climate Change.

17. USAID (2013) Malawi Climate Change Vulnerability Assessment.

18. Vincent K, Dougill AJ, Mkwambisi DD, Cull T, Stringer LC, et al. (2014) Analysis of Existing Weather and Climate Information for Malawi.

19. Bulckens H (2013) Helping Smallholder Tea Farmers, Malawi. Ethical Tea Partnership.

20. McSweeney C, New M, Lizcano G (2011) UNDP Climate Change Country Profiles: Malawi.

21. Phiri MGI, Saka AR (2004) The impact of changing environmental conditions on vulnerable communities of the Shire valley, southern Malawi.

8

Seasonal Variation in Surface Ozone Concentrations, Meteorology and Primary Pollutants in Coastal Mega City of Mumbai, India

Sagar A. Marathe and Shankar Murthy*

National Institute of Industrial Engineering, Mumbai, India 400 087

Abstract

In the present the study variation in ground level ozone (GLO) concentration with respect to availability of precursor pollutants and meteorological parameters in the tropical climatic conditions in Mumbai, India is discussed. The concentration of various pollutants and meteorological parameters have been monitored at the continuous air quality monitoring station of Maharashtra Pollution Control Board, located in Bandra (Mumbai) for two years. The analysis of the data indicates that the pattern of formation of GLO changes with change in season. The maximum concentration for GLO was experienced in the post monsoon season unlike any other study reported in the country. The study highlights that the post monsoon season experienced the highest number of hourly and eight hourly exceedances in the permissible limits prescribed by the Central Pollution Control Board. These high concentrations can be attributed to favorable climatic conditions and presence of precursor pollutants. This was proved by correlating GLO with precursor pollutants and meteorological parameters.

Keywords: Ground level ozone; Mumbai; Air quality; Correlation of ozone

Introduction

Ozone (O_3) is a colorless and reactive oxidant gas which is a major constituent of atmospheric smog. It is secondary pollutant, which is formed due to photo-chemically active primary pollutants [1-4]. The concentration of Ground Level Ozone (GLO) pollution becoming a major concern because of its increasing concentrations and adverse impacts on plants and animals including human beings, even at low concentrations [5-7].

In order to control the ground level ozone pollution in the country, the Central Pollution Control Board (CPCB) has announced revised air quality standards in the year 2009 which include one hourly and eight hourly permissible limits for GLO for the first time in the country and have also included ozone in the list of criteria pollutants.

Furthermore literature available regarding variation of GLO in India is limited and the studies have different climatic backgrounds from rural areas to urban and sub urban regions [8-11]. Hence, the comparison is difficult. The studies which are available report observations for short duration or for one or two seasons and are mostly conducted with the objectives of assessing the impact of ambient ozone concentrations on various varieties of pollutants [12,13]. There is limited literature available on continuous long term monitoring for ground level ozone concentrations along with the precursor pollutants in urban as well as rural areas. The formation of ground level ozone is non-linear in nature which makes the approach to the study even more difficult. The chemistry of ground level ozone is explained in (Appendix I).

There is a dearth of studies in the Indian context; moreover such extensive studies have not been conducted for the sub-tropical climatic conditions.

After laying the basis for the study, a brief analysis of the literature is presented in section two. Section three documents the chemistry of ozone formation, while section four describes the characteristics of the sampling location. Besides, in the same section, the climatic conditions of the city are described. In addition, the prominent features are also highlighted. The fifth section discusses the methodology of the study.

The results and discussions are presented in the sixth section while the seventh/last section provides conclusions and future scope for the study.

Brief Literature Review

Clinical studies have documented an association between short-term exposure to ground-level ozone (at concentrations of 200-500 µg/m^3) and mild temporary eye and respiratory irritation, cough, throat dryness, thoracic pain, and headache [5,14,15]. Across Europe, O_3 was attributed to 21,400 premature deaths per annum [16]. Against a similar back drop studies by Chameides et al. [6] documented the effects of increased ozone concentrations on crop yield and forests. In this study it was identified that high ozone concentration lead to low productivity and loss of forest cover.

Derwent analysed long-term measurements from Mace Head on the west coast of Ireland and found an increase in background ozone until 1999 and a stabilization or decline thereafter. The reasons are attributed to the rise in emissions from anthropogenic activities. For Europe, Jonson considered the period 1990-2004 and predicted reduction of 5-10 ppb (10-20 µg/m^3) in mean daily maximum summer ozone concentrations (June-August).

In another study for European region, it was observed that the concentration of surface ozone had increased from an estimated pre-industrial value of 10 ppb to 30-50 ppb (Pritchard and Amthor, 2005). The results were found to be applicable for areas lying in the mid-latitudes of the northern hemisphere. Kley D et al. [17] has reported

***Corresponding author:** Shankar Murthy, Associate Professor, National Institute of Industrial Engineering Mumbai, India 400 087
E-mail: murthyshanker@gmail.com

Mesoscale convective systems which contributed to diurnal variation in O_3 concentration; that substantially and abruptly deviated from the diurnal patterns observed during sunny conditions enhancing the boundary layer O_3 concentration by 10-30 ppbv. In the Asian context, study by Min Shao, has discussed the role of NO_x and VOCs in the formation of O_3 at ground level in and around the Pearl River delta (China).

From the above it can be inferred that ozone has dire consequences on not only plant and animal life but also on human existence. The present study aims to evaluate the variation in ground level ozone concentrations with respect to availability of precursor pollutants and change in seasonal climatic conditions.

Methodology

Mumbai is located in tropical zone and is known to have fluctuation in relative humidity in every season, which in turn is bound to affect the ozone formation. Being a mega city, Mumbai has large population density of over 29,000 individuals per sq. km. This high density is likely to result in large scale ozone exposure. Hence a detailed study of formation ozone is necessary for the mega city of Mumbai. The study attains significance as this may be its first of its kind. In addition the monitoring of a secondary pollutant would allow the decision makers to capture the holistic picture of air pollutants.

Meteorology

The site experiences distinct seasons due to the tropical climatic zone. The temperature reaches peak in unison with solar intensity around 13:00 hrs. And drops to minimum during night and early morning hours in the absence of solar radiation around 03:00 hrs. The monthly average temperature recorded at the monitoring site is presented in (Figure 1). The summer season occurs between the months of March to May and is marked with very hot and humid climate with maximum temperature recorded to be 40°C in the month of May.

It is important to note that the relative humidity drops to minimum of 20% in the summer season. The average relative humidity is presented in (Figure 2). The relative humidity and temperature are a function of solar intensity which is also observed to be maximum during the same season. The variation of solar intensity is presented (Figure 3). As a characteristic feature of the tropical site, heavy rainfall for four months is experienced during June to September during the season. The rainfall is the result of north-west monsoon. The wind pattern for the seasons is provided in (Figure 4). Maximum rainfall is experienced in the month of July. The relative humidity in the entire monsoon season is very high. The minimum relative humidity

recorded in this season was 57% while the maximum was 99%. The average relative humidity of the season varied between 78-85%. The solar intensity is observed to be intermittent and weak in the monsoon season. The monsoon season is then followed by small transitional Post-Monsoon season. The season lasts only for two months of October and November, which is distinguished by high temperatures and high humidity. The temperature shoots up in October leading to locally known phenomenon of *"October Heat"* during which the climate is similar to summer. The relative humidity is observed to fluctuate prominently in this season. The minimum relative humidity is recorded

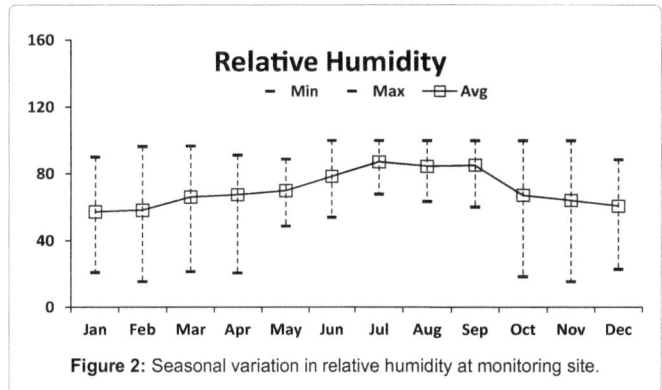

Figure 2: Seasonal variation in relative humidity at monitoring site.

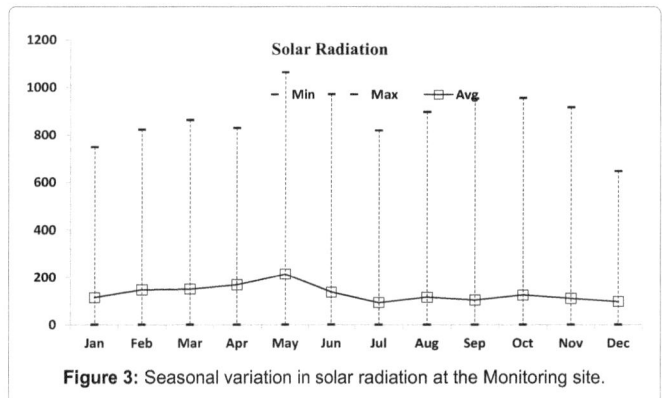

Figure 3: Seasonal variation in solar radiation at the Monitoring site.

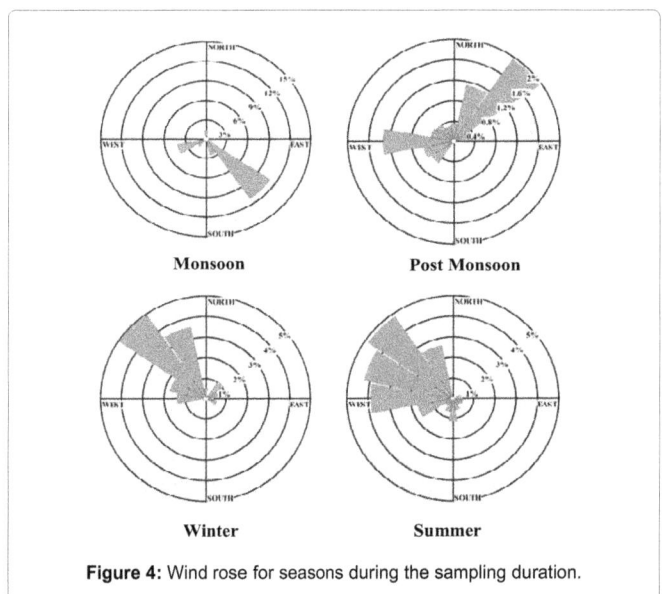

Monsoon   Post Monsoon

Winter   Summer

Figure 4: Wind rose for seasons during the sampling duration.

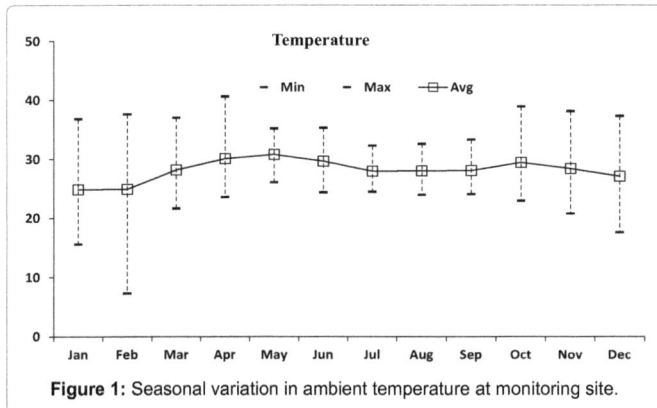

Figure 1: Seasonal variation in ambient temperature at monitoring site.

in this season. The daily average temperature then tends to cool down in November marking the onset of winter (December to February). The temperature in winter occasionally drops below 10°C but that is only for a short period of time.

Site description

The site selected for the present study is Mumbai city which is also known as financial capital of India. It is one of the four Mega cities of the country (Figure 5). The city is located on the western coast of India and exhibits typical urban characteristics such as high population density and enormous vehicular population. The site is surrounded on the western side by Arabian Sea, while the southern side is marked by residential and commercial areas of the city. A small portion on the south-west of the city is covered by a bay (Mahim Bay); whereas the suburban areas cover the adjoining boundaries on the northern and eastern sides of the city. The international airport of the city (circled) is located between the North and North-East side. A major roadway named Western express highway which experiences a heavy traffic during peak hours runs very close to the monitoring site.

The continuous air quality monitoring station is located at Govt. Polytechnic, Bandra (Figure 5). It is located away from the coastline in order to avoid influence of rapid changes in the wind pattern. A major roadway near the monitoring site enables to trap the pollution from vehicular sources. The city also has a small industrial complex at Andheri which is located on the northern side of the monitoring site. Therefore, it is important to study the formation of ground level ozone in the city which has a high population density and dynamic metrological conditions.

Database and analysis

For the present study, hourly average data for a period of two years (from 1st Jan 2008 to 31st Dec 2009) has been considered. This secondary data has been obtained from the Maharashtra Pollution Control Board (MPCB) monitoring station. A continuous monitoring system was installed at the present site in the year 2007, manufactured by Chemtrols and provides data on instantaneous and hourly basis for the pollutants monitored.

The meteorological parameters and primary precursor pollutants which are considered to be responsible for the formation of GLO have been considered. The data for oxides of Nitrogen (NOx), carbon monoxide (CO) and benzene has been obtained. Along with this, the meteorological data for temperature, solar radiation, relative humidity, wind-speed and wind-direction has also been considered. The hourly average data has been grouped into four seasons to study the seasonal variation and was further used to administer the correlation tests. For calculating wind direction mode was preferred over mean values. The

reason to opt the mode value of wind direction is that, if the records would have been averaged the results would show a completely different pattern for wind direction. The percentage of missing values in summer, monsoon, post monsoon and winter accounted for less than 2%. The missing values in the data were replaced with median value for the corresponding pollutant in that particular season. The sampling was not carried out from 9th Jan 2009 to 15th Jan 2009 due to maintenance of instruments. Similarly, the data where in the readings for three or more parameters is not obtained for more than three hours has been deleted from the data set. The hourly data obtained from MPCB was divided into four seasons and was subjected to the four-sigma test in order to remove outliers. It was found that the outliers amounted only 3% of the entire data set combining all the parameters. The data thus obtained was subjected to Pearson's correlation test in SPSS software version 17. The values falling out of the range were replaced with the median value for the respective season. The median value was preferred over mean or mode for replacing the outliers because the mean calculate was arithmetic mean and in case of mode, multiple modes existed in the data sets. To evaluate the effects of meteorological parameters, wind rose was plotted using the software WR PLOT version 7.0.

Results

After applying the outlier test the highest concentration of the parameters during the monitoring period are reported. The descriptive statistics of precursor pollutants and meteorological parameters are described in Table 1.

From the results it can be inferred that the annual hourly average for ozone during the monitoring period was 27.87 μg/m³ and 34.93 μg/m³ in 2008 and 2009 respectively. It was observed that the hourly ozone concentration exceeded the Central Pollution Control Board's norm (180 μg/m³) on 191 occasions which accounts to 1% of the total no.of observations.

Whereas the 8 hourly permissible limit (100 μg/m³) was exceeded 88 times in the monitoring period. It should be noted that ground level ozone concentration remained under the norms during the summer and monsoon seasons while the highest concentration of ozone was observed during post monsoon season (252.4 μg/m³). At the same time, the number of hourly and eight hourly exceedances have more than doubled in 2009 as compared to 2008.

In case of NOx and CO, the annual averages in 2008 and 2009 were computed to be 65.54 μg/m³, 2.06 μg/m³ and 74.26 μg/m³, 2.26 μg/m³ respectively which marginally exceed the annual average standards prescribed by CPCB (NOx: 40 μg/m³ and CO: 2 μg/m³).

The diurnal variations in concentration of ozone concentration are depicted in Figure 6. For the purpose of evaluating the influence

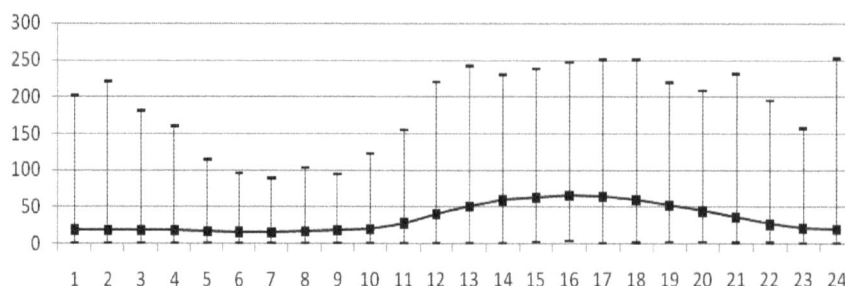

Figure 5: Average Diurnal variation of ground level ozone during the monitoring period at Mumbai with minimum and maximum concentration.

		NOx (µg/m³)	CO (µg/m³)	O₃ (µg/m³)	Benzene (µg/m³)	Temperature (°C)	Solar Radiation (W/M²)	RH (%)
Summer	Min	3.2	0.5	1	0	21.70	0	20.7
	Max	226	6.2	141.5	33	40.60	1063.70	96.5
	Avg.	44.1	2.6	27.6	2.7	29.69	177.08	67.73
	SD	38.1	0.9	23.1	4.3	2.91	264.99	12.16
Monsoon	Min	0.5	0.1	1	0	24.00	0.00	53.9
	Max	188	4.7	73.4	15	35.30	971.50	99.9
	Avg.	40.3	1.3	19.1	1.4	28.43	112.01	86.36
	SD	30.2	0.5	12.9	1.9	2.08	194.05	9.47
Post Monsoon	Min	2.9	0.1	1	0.04	20.80	0.00	15.4
	Max	546	7.6	252	51.4	38.90	956.30	99.9
	Avg.	114	2.3	48.1	3.3	28.94	117.56	65.52
	SD	99	1.3	44.9	4.4	3.44	209.49	18.42
Winter	Min	1.9	0.1	1	0	7.3	0	15.3
	Max	422	7.1	228	46.6	37.6	823.3	96.2
	Avg.	96.2	2.2	41.6	6.1	25.69	118.62	58.67
	SD	80.9	1.2	44.6	6.9	4.13	194.82	14.71

Table 1: Descriptive Statistics of Parameters.

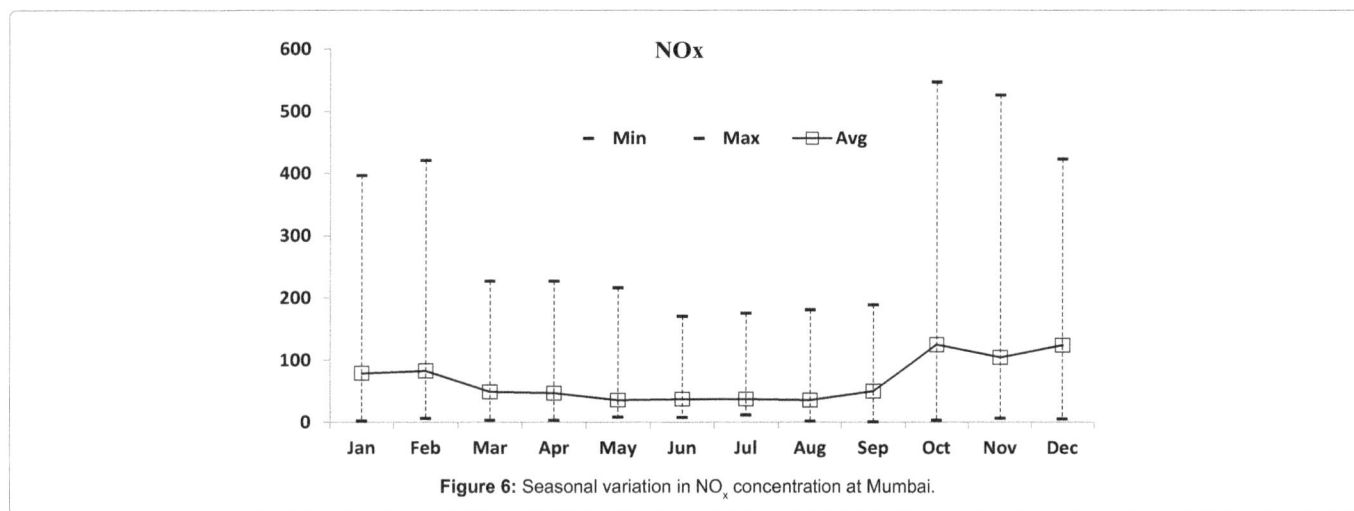

Figure 6: Seasonal variation in NO_x concentration at Mumbai.

of precursor pollutants and metrological parameters on formation of ozone, Pearson's correlation test was administered to the data set. As stated in section 5 the data divided was into four seasons and the resultant diurnal average was considered for the correlation test.

Oxides of nitrogen (NO_x)

The highest one hourly concentration of NO_x was observed in the post monsoon season. The values were observed to be consistently high in both the months (Oct. and Nov.) ranging between 525-547 µg/m³. The daily (24 hr.) average limit of (80 µg/m³) was computed to be exceeding on 77 days in the same season which accounts 63% of the days in the post monsoon season. The minimum concentration of the season was noted be around 2.9 µg/m³.The lowest concentration was observed during 04:00 hrs. in the early morning. The monthly average for both the months in the season remained above 100 µg/m³.

In the winter season, it was experienced that, 60% of the time the concentration of NO_x exceeded the CPCB standard (80 µg/m³). This observation can said to be similar as in the case of post-monsoon season. The average hourly maximum concentration in this season was noted to be 422 µg/m³. This highest hourly average was observed

during 09:00 hrs in the morning same as in case of the post monsoon season, while the lowest value of the season was measured to be 1.9 µg/m³ around 14:00 hr. during afternoon. The concentration of NO_x is observed to decrease drastically in the summer and monsoon seasons, decreasing up to 50% as compared to earlier seasons.

The lowest concentration in the summer season is observed to be 3.2 µg/m³ while the highest concentration was noted to be 226.8 µg/m³. It is noteworthy to mention that the lowest concentration was experienced during the afternoon around 15:00 hr. while the highest concentration was noted around 09:00 hr. in the morning which are considered to have high vehicular activity. The seasonal average value for the season was computed to be 44.2 µg/m³. It is important to note that the instances of the daily limit being exceeded were observed only 3 times in the season which is lowest amongst all of the seasons. The monsoon experienced a low concentration of less than 2 µg/m³ during the afternoon around 14:00hr. The highest concentration of NO_x in this season was noted be around 185 µg/m³ during early morning hours at 07:00 a.m. The average concentration of NO_x (40.31 µg/m³) in this season is calculated to be lowest as compared to rest of the seasons. The daily standard was observed to be exceeding only 8 times in the season.

Carbon monoxide (CO)

As in the case of NO_x the highest concentration of CO was measured in the post monsoon season as provided in Table 1 earlier. The average hourly maximum concentration in the season was noted to be 4.7 µg/m³ at 00:00 during the night. The minimum concentration in the season was noted be less than 1 µg/m³. The maximum average hourly concentration in the winter season reached up to 7.1 µg/m³ while the minimum concentration remained below 1 µg/m³. The highest concentration was observed around 10:00 in the morning and the minimum concentration was experienced during evening around 18:00 hr. in the evening. The maximum hourly average concentration of CO in the summer season was measured to be around 6.2 µg/m³ around 11:00 a.m. in the morning while the concentration was experienced to drop below 1 µg/m³ several times during the season. The highest hourly average concentration of CO in the monsoon season reached up to 4.7 µg/m³ in the night while the lowest concentration was observed to be under 1 µg/m³ at several times.

The above results indicate that the availability of precursor pollutants in the post-monsoon was higher as compared to the rest of the seasons. The reasons for the high concentration are discussed in the later section of the study after presenting the results for variation in the concentration of ground level ozone in different seasons.

Ground Level Ozone

The variations in concentration of ozone in each season are presented as follows

Post monsoon

It was observed that there is a sharp increase in the concentration of NOx in the post monsoon season (Figure 7). The seasonal average concentration of the season was calculated to be 114 µg/m³ which is nearly three times that of summer and monsoon. The peak hourly concentration recorded in the entire monitoring season was observed in this season (546 µg/m³). The peak of hourly average concentration of NOx was observed around 09:00 hrs in the morning. This can be related to high traffic and low solar intensity which negates the chances of photo-dissociation. The concentration of CO was noted to be very low as compared to rest of the pollutants. The seasonal variation in concentration of CO is depicted in Figure 8. The post monsoon season also has the highest number of one hourly (60 times) and eight hourly (25 times) exceedances. The diurnal variations in post monsoon season are depicted in Figure 9. Which shows maximum and minimum concentrations of the time of day in the season.

It is important to note that the highest hourly concentration of 252.1 µg/m³ in the post monsoon season was observed at 00:00 hrs

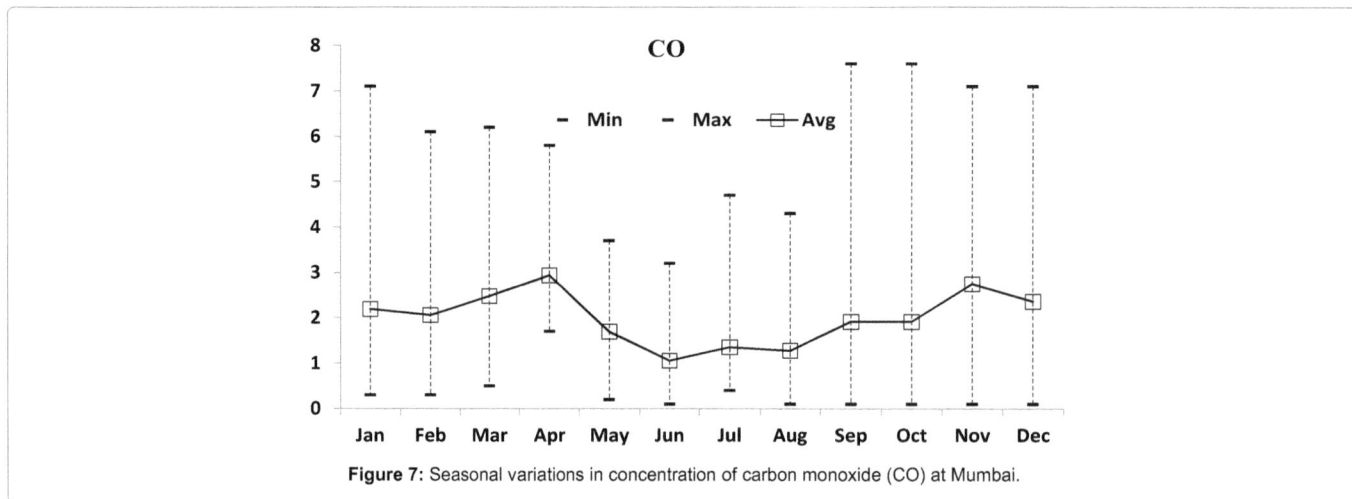

Figure 7: Seasonal variations in concentration of carbon monoxide (CO) at Mumbai.

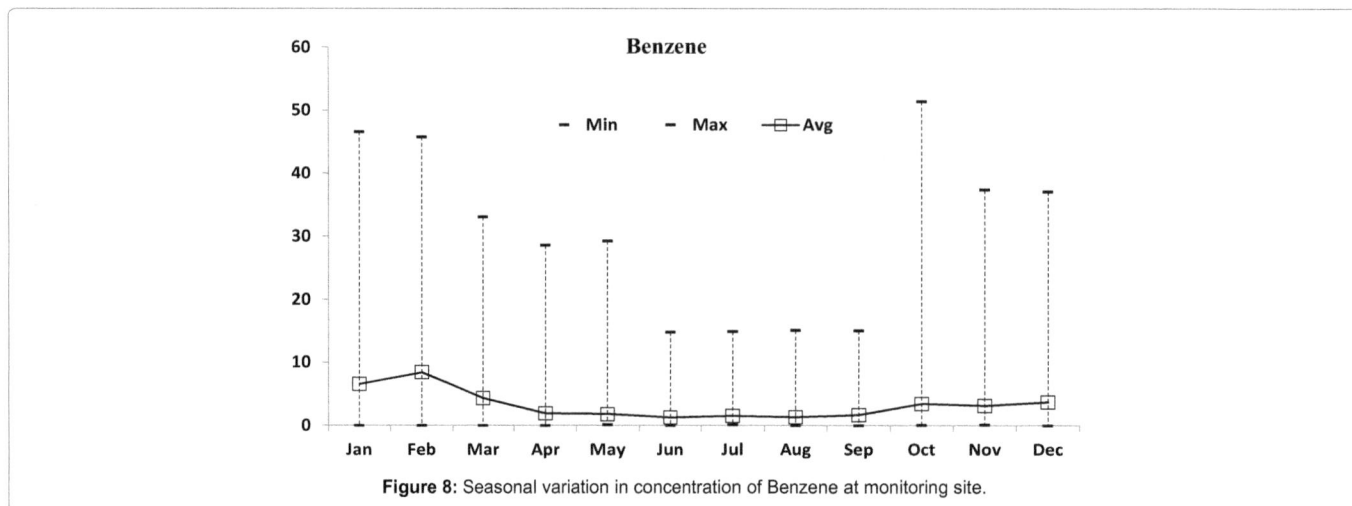

Figure 8: Seasonal variation in concentration of Benzene at monitoring site.

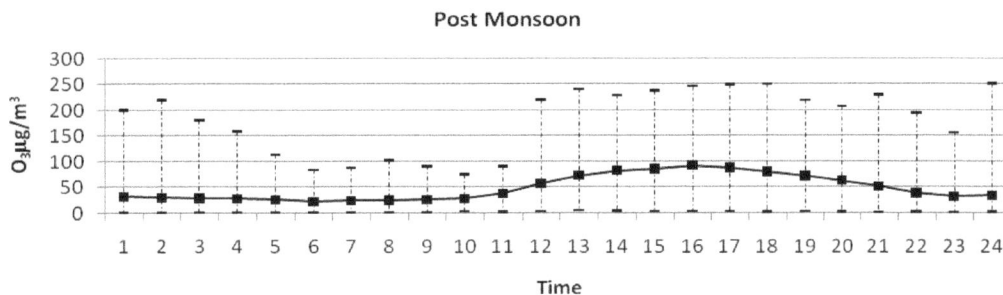

Figure 9: Variations in ground level ozone concentration in the post monsoon season with minimum and maximum concentration.

	Ozone	CO	NOx	Temp	Solar Rad.	RH	WD
Ozone	1	-0.804**	-0.939**	0.814**	0.408*	-0.647**	0.645**
CO		1	0.834**	-0.599**	-0.348	0.503*	-0.569**
NOx			1	-0.643**	-0.206	0.440*	-0.504*
Benz				-0.697**	-0.430*	0.626**	-0.612**
Temp				1	0.821**	-0.963**	0.721**
Solar Rad.					1	-0.923**	0.529**
RH						1	-0.701**
WD							1

Note: Where, CO = Carbon monoxide, NOx = Oxides of Nitrogen, RH = Relative Humidity, Temp = Temperature, Solar Rad. = Solar radiation and WD = Wind Direction respectively.

Table 2: Pearson's Correlation of Ozone with precursor pollutants and meteorological parameters in Post Monsoon Season.

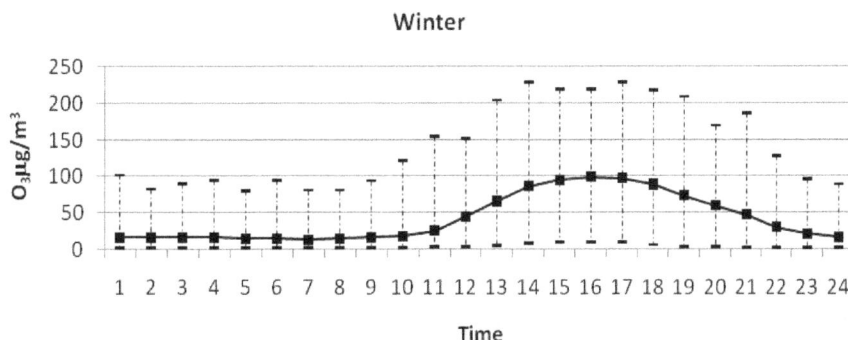

Figure 10: Variations in ground level ozone concentration in the winter season with minimum and maximum concentration.

during the night. The lowest concentration was reported as 1 µg/m³ and it was mostly observed during the night time. The peak of hourly average in the season was recorded at 16:00 hr. and was noted to be 92.02 µg/m³.

The correlation between ozone and selected parameters is shown in Table 2. According to this analysis, a strong negative correlation of ozone with NO_x (-0.939) and CO (-0.804) is observed. This finding indicates that concentration of precursor pollutants not only plays an important role in the formation of ozone but also in its destruction. Amongst the metrological parameters except for relative humidity the correlation of ozone with reset of the parameters considered for the study was found to be positive. Among the metrological parameters the highest correlation was observed with temperature (0.814). The correlation of ozone with solar radiation, relative humidity, wind direction and wind speed was estimated to be 0.408, -0.647, 0.645 and 0.808 respectively.

It may be reckoned that the predominant wind direction for the post monsoon season is north and north east (as provided in the wind rose plot in Figure 4.

Winter

The next highest concentration of ozone was found in the winter season. The average concentrations and minimum and maximum values for ground level ozone are plotted in Figure 10. High ozone concentrations are observed during the afternoon in this season with peak reaching at 14:00 hrs. The season is marked with low temperatures, especially during the night. Since there is limited amount of thermal inversion in this season the pollutants tend to be trapped near the ground level. Therefore this season also experiences high pollution episodes. The highest hourly concentration (228 µg/m³) in this season was experienced close to the maxima of the post monsoon season; while the lowest concentration was observed to be 1 µg/m³.

In the winter season, the correlation of ozone with the precursor pollutants is strongly negative as shown in Table 3 and is above -0.9 suggesting their importance. The correlation of ozone with the

	Ozone	CO	NOx	Temp	Solar Rad.	RH	WD
Ozone	1	-0.938**	-0.917**	0.800**	0.468*	-0.644**	0.922**
CO		1	0.957**	-0.714**	-0.389	0.552**	-0.875**
NOx			1	-0.704**	-0.312	0.517**	-0.834**
Benz				-0.705**	-0.320	0.537**	-0.842**
Temp.				1	.834**	-0.967**	0.735**
Solar Rad.					1	-0.896**	0.449*
RH						1	-0.593**
WD							1

Table 3: Pearson's Correlation of Ozone with precursor pollutants and meteorological parameters in Winter Season.

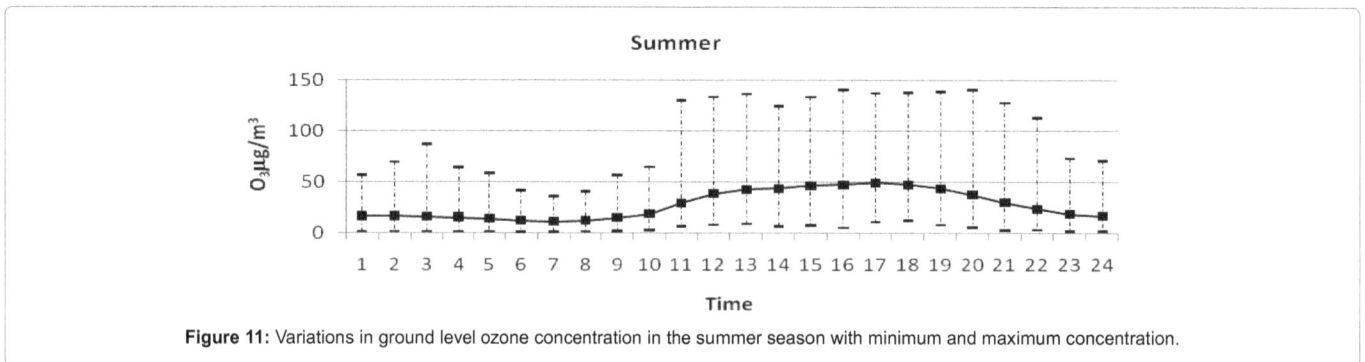

Figure 11: Variations in ground level ozone concentration in the summer season with minimum and maximum concentration.

	Ozone	CO	NOx	Temp.	Solar Rad.	RH	WD
Ozone	1	-0.668**	-0.644**	0.850**	0.619**	-0.818**	0.875**
CO		1	0.922**	-0.404	-0.326	0.319	-0.437*
NOx			1	-0.410*	-0.283	0.328	-0.422*
Benz				-0.140	-0.036	0.064	-0.300
Temp.				1	0.888**	-0.989**	0.855**
Solar Rad.					1	-0.853**	0.647**
RH						1	-0.876**
WD							1

Table 4: Pearson's Correlation of Ozone with precursor pollutants and meteorological parameters in Summer Season.

meteorological parameters is different in this season as compared to that in the monsoon. In winter ozone exhibits a strong positive correlation with wind direction (0.922) clearly indicating the role of wind in transportation of pollutants. It needs to be confirmed whether the precursor pollutants are transported or the ozone formed at some other location is being transported directly to the monitoring site. Moreover the correlation of ozone with wind speed helps in strengthening the above findings.

Summer

The maximum hourly average concentration of ground level ozone was observed to be around141.5 $\mu g/m^3$ at 16:00 hrs. The seasonal variation in the concentration is depicted in Figure 11. The concentration can be observed to have decreased as a result increase in the photochemical breakdown of ground level ozone. The same case may be applicable for the precursor pollutants as the rate of their dissociation may exceed the rate of formation of ground level ozone. In this season the correlation of ozone was negative with the precursor pollutants but not strong as compared to that in the case of post monsoon and winter seasons. The correlation for the summer season is provided in Table 4. The correlation of ozone with wind direction is positive similar to the earlier seasons. It should be noted that the correlation values between wind direction and ozone are higher as compared to that in the post monsoon season but the highest concentration of ozone

was still observed in the post monsoon season. The correlation of ozone with solar radiation also appears to be strongly positive as compared to the rest of the seasons which suggests that higher solar intensity favors the formation of GLO. The correlation of relative humidity with ozone is strongly negative indicating that high relative humidity leads to decrease in ozone concentration. As depicted in Figure 8 it can be observed that the average relative humidity during the daytime in the summer season is higher as compared to the post monsoon and winter season which supports the above findings for destruction of ozone.

Monsoon

The average concentration for the monsoon season is depicted in the Figure 12. During Monsoon the weather is largely cloudy and hence there is minimal availability of solar intensity. Also, the concentration of pollutants is fairly low making it the cleanest season of the year in tropical conditions. The highest concentration of ozone in this season was observed to be 73 $\mu g/m^3$, but it is also important to note that the average concentration never exceeded 30n $\mu gm/^3$. The peak ozone concentrations are observed during the afternoon at 13:00 hrs which is a peculiar feature of the season. The correlation of ozone with selected parameters for the monsoon season is provided in Table 5.

It is important to highlight that the correlation of ozone with NOx (0.1) was found to be weakly positive which is not observed in any of the studies considered so far. The correlation with CO was observed

Seasonal Variation in Surface Ozone Concentrations, Meteorology and Primary Pollutants in Coastal...

69

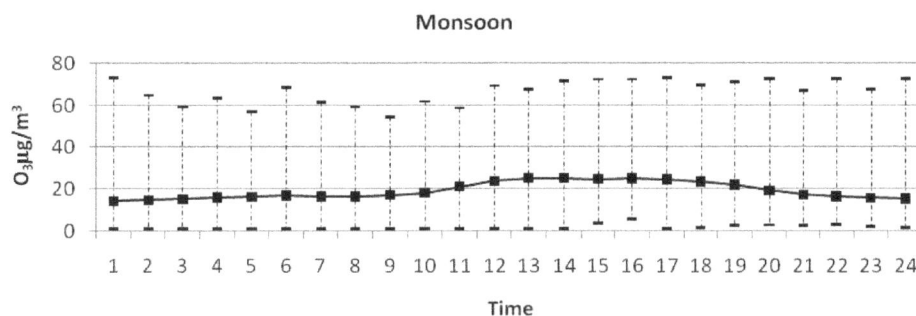

Figure 12: Variations in ground level ozone concentration in the monsoon season with minimum and maximum concentration.

	Ozone	CO	NOx	Temp	Solar Rad.	RH	WD
Ozone	1	-0.250	0.101	0.943**	0.817**	-0.948**	0.758**
CO		1	0.830**	-0.163	0.006	0.251	-0.308
NOx			1	0.208	0.219	-0.170	-0.089
Benz				-0.620**	-0.433*	0.670**	-0.614**
Temp				1	0.906**	-0.988**	0.751**
Solar Rad.					1	-0.838**	0.704**
RH						1	-0.736**
WD							1

Table 5: Pearson's Correlation of Ozone with precursor pollutants and meteorological parameters in Monsoon Season.

to be weakly negative. The correlation of precipitation (rainfall) was observed to be negligible and hence not considered in the study. The correlation of ozone with the meteorological parameters can be observed to increase in this season. Amongst the meteorological parameters, the correlation with Temperature was found to be strongly positive (0.943) while the correlation with relative humidity appeared to be strongly negative having the value of -0.948. The correlation of solar intensity can be observed to be doubled as compared to that in the post monsoon or winter season and is also the strongest. The correlation of ozone with wind speed remains strongly positive but less than in the case of summer or winter season.

Discussion

Since there is no annual average prescribed by the CPCB for GLO it becomes difficult for comparison with other pollutants.

Trend of ground level ozone in India: It is documented that ozone concentration may increase as a result of transport of precursor pollutant mass [7]. In a similar context the study by Naja and Lal [9] report an increase in O_3 concentration by 1-2% per year. This study was conducted for the city of Ahmedabad.

In yet another study for the city of Ahmedabad, elevated level of O_3 concentration up to 110 ppbv were reported [7]. At rural sites it was observed that ozone concentrations were high during the afternoon hours (ibid). According to the study the increased ozone concentration may have occurred due to the trans-boundary carriage of polluted air from the up wind urban areas. Surface O_3 concentration in the city of Pune was reported to be around 46-120 μg/m³; the increase of daytime maximum O_3 concentration being attributed to the increased vehicular traffic density. For the same city during 2001-2005, concentration of ozone was reported to be as high as 180 μg/m³ and the permissible limits were repeatedly breached [18]. In rural areas of western India, the ozone concentrations were reported to be high during 2002-2004 in the summer season. The study attributed this to natural as well

as anthropogenic activities. In the city of Delhi, Singh et al. (1997) reports that ozone concentration during the winter season of 1993were between 68-252 μg/m³ (averaged for 30 min. time interval). It can be said that due to higher the temperatures in summer the ozone and precursor pollutants both undergo photo dissociation at a faster rate than their production or emissions and hence are unable to contribute for higher concentration of ozone. The lowest concentration of ozone is observed in monsoon season which is well in accordance with several other studies from the Indian subcontinent [7,9,10,18].

The studies conducted by Sarkar and Agrawal [13] report the impact of ozone concentration on plants. These studies consider short monitoring duration of one or two seasons which are important for life cycle of plants. The seasonal trend cannot be affirmed from such type of studies as they do not record the data for long term duration. The studies reported so far are conducted in different climatic backgrounds. In addition these studies do not have homogeneity in terms of spatio-temporal land use and emission sources. Other studies which have conducted monitoring for all the seasons have reported highest ozone concentration in summer season (REFS.), while some studies have reported highest concentrations in winter season.

Few studies have reported high ozone concentrations in the month of October (Post Monsoon season) which were conducted at Varanasi [13]. But the study was conducted to assess impacts of ambient ozone levels on productivity and other parameters of two Indian cultivars of rice and the reasons for high concentration were not discussed in detail. Highest GLO concentrations from few cities of the country are reported in Table 6 along with the observations from the present study. From the city of Pune which is situated in the western part of India, the highest concentration for GLO was reported to be around 80-100 μg/m³ during summer whereas the lowest concentration was reported around 8-14 μg/m³ during the monsoon [18]. Other studies conducted in the city of Ahmadabad reported peak concentrations of 93.8 ± 27.6μg/m³ [9]. There is a slight variation in ozone concentration in other cities, Delhi (68-252 μg/m³) in winter and at high altitudinal

Sr. No.	City	Author and Year of Study	Highest conc. (µg/m³)	Season	Lowest conc. µg/m³)	Season
1	Pune	Debaje and Kakade, 2009	80–100	Summer	8–14	Monsoon
2	Ahmadabad	Lal et al., 2000	93.8 ± 27.6	Winter	-	Monsoon
3	Delhi	Singh et al., 1997	68 - 252	Winter	-	Monsoon
4	Mt. Abu	Naja et al., 2003	206	Winter	-	Monsoon
5	Mumbai	Present Study	252.10	Post -Monsoon	1	Monsoon

Note: The studies which have considered data for all the seasons in the year are considered for the comparison.

Table 6: Comparison of ozone concentration in selected cities of India and Mumbai (present study).

zone of Mt. Abu (206 µg/m³) where maximum concentration was also reported in winter season [7].

In the present study the highest concentration of ground level ozone is observed in the post monsoon season unlike studies discussed above. This can be said to be associated with the change in the wind direction and availability of precursor pollutants. These findings are corroborated by the study conducted for the city of Delhi (Chelani, 2009). It can be said that as the site is near to the coast, horizontal mixing of the air can be experienced throughout the day as result of strong wind flow which is experienced due to land breeze and sea breeze effect. It should also be noted that the relative humidity is observed to be very low throughout the season. The presence of relative humidity indicates availability of OH radicals. These radicals are supposed to interfere with ozone formation by scavenging the O_3 molecules. It is obvious that the ozone formation is not possible in the absence of precursor pollutants. There are two possible sources for the emission of precursor pollutants, one of them being the air traffic from the nearby international airport and the other could be due to vehicular traffic from the western express highway running close to the monitoring site.

Conclusion

Based on the above findings it can be concluded that the process of formation and destruction of ozone is dynamic and changes not only seasonally but diurnally. Also it is important to note that ozone formation is influenced by availability of precursor pollutants under favorable meteorological conditions. The present study highlights some important findings which are not reported so far from Indian sub-continent. The highest ozone concentration was observed in the post monsoon season unlike other studies in which the peak concentrations were reported in either winter or summer season. Secondly the correlation of ozone with NOx was found to be weakly positive in the monsoon season which is also a rear phenomenon. Moreover, the concentration of NOx and CO were found to be higher during the early morning and evening during the rush hours of traffic suggesting the influence of high vehicular activity around the monitoring site. The location of the Sahara international airport and the positive correlation of ozone with wind direction in the post monsoon season indicate that the precursor pollutants are being transported to the monitoring site. This can be justified by the fact that highest concentrations for the selected pollutants have been observed in the post monsoon season. Thus it could be concluded the reasons for high ozone concentration in the post monsoon season are the effect of anthropogenic emissions. Moreover the present study highlights the problem of transport of pollutants at local level.

Acknowledgement

We are thankful to MPCB for providing us the required data and NITIE for enabling us with facilities for computational work.

Appendix I

Chemistry of Ground level Ozone:

The available literature suggests that, the process of ozone formation is non-linear and has multiple channels [14,19]. The process of ozone formation can be explained as under:

1. Photolysis of Ozone: In this case the ozone molecules undergo photolysis under UV rays shorter than 320nm to dissociate into active oxygen atom and oxygen molecule [3]. The reactive molecule reacts with either water vapour to form hydroxyl radical or with any inert molecule to produce O_3. The inert molecules are mostly N_2 (ibid).

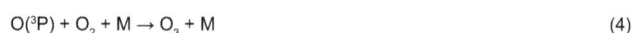

$$O_3 + hu \rightarrow O(^1D) + O_2 \tag{1}$$

$$O(^1D) + H_2OOH + OH \tag{2}$$

$$O(^1D) + M \rightarrow O(^3P) + M \tag{3}$$

$$O(^3P) + O_2 + M \rightarrow O_3 + M \tag{4}$$

2. Hydroxyl radical Channel: The hydroxyl radicals present in the atmosphere prominently react with either photo-chemically active methane or carbon monoxide atoms which leads to formation of ground level ozone [2-4]. It is noteworthy that the low concentration of NO_x in the atmosphere and other precursors can lead to the removal of ozone [3].

3. The NO_x channel: The NO molecules undergo oxidation to form NO_2. This conversion is a result of actions of oxidizing agents like hydroxyl radicals or ozone itself. The photo-dissociation of NO_2 is explained in the equations (5) to (8).

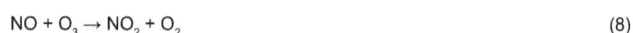

$$NO + OH \rightarrow NO_2 + H \tag{5}$$

$$NO_2 + hu \rightarrow NO + O(^3P) [\lambda \leq 400nm] \tag{6}$$

$$O(^3P) + O_2 \rightarrow O_3 \tag{7}$$

$$NO + O_3 \rightarrow NO_2 + O_2 \tag{8}$$

4. The VOC and nm VOC channel: The reaction of NO with ozone is suppressed by presence of VOCs as the affinity of hydrocarbons towards NO appears to be higher than ozone. This leads to formation of NO_2. At times the hydrocarbons are supposed to compete with ozone leading to destruction of ozone molecules.

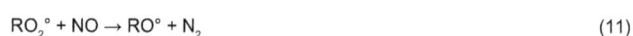

$$RH + OH \rightarrow R° + H_2O \tag{9}$$

$$R° + O_2 \rightarrow RO_2° \tag{10}$$

$$RO_2° + NO \rightarrow RO° + N_2 \tag{11}$$

It may be taken into consideration that the studies also suggest ozone formation has been reported to be water vapour dependent [3]. The water vapour in the atmosphere is usually affected by relative humidity and temperature.

Relative humidity and temperature vary spatially and temporarily making ozone formation a local phenomenon. The studies reported documented for ozone in India do not discuss the role of relative humidity in detail.

References

1. Liu SC, Trainer M, Fersenfeld FC, Parrish DD, Williams EJ, et al. (1987) Ozone production in the rural troposphere and the implications from regional and global ozone distributions, J Geophys Res. 92: 4191-4207.

2. World Bank Group (1998) Handbook of GLO Pollution Prevention and Abatement Handbook World Bank Group Effective.

3. Royal Society (2008) Ground-level ozone in the 21st century: future trends, impacts and policy implications, Science Policy Report.

4. Guttikunda Sarath (2009) Photochemistry of Air pollution in Delhi, India: A monitoring based analysis, SIM-air Working Paper series 25-2009.

5. WHO (2006) Air quality guidelines: global update 2005, particulate matter, ozone, nitrogen dioxide and sulphur-dioxide. WHO Regional Office for Europe: Copenhagen.

6. Chameides WL, Xingsheng Li, Xiaoyan Tang, Xiuji Zhou, Luo Chao, et al. (1999) Is ozone pollution affecting crop yields in China. Geophysical Research Letters 26: 867-870.

7. Naja M, Lal D Chand (2003) Diurnal and seasonal variabilities in surface ozone at a high altitude site Mt Abu in India. Atmospheric Environment 37: 4205-4215.

8. Debaje SB, Jadhav DB (1999) An Eulerian photochemical model for tropospheric ozone over the tropics. Current Science 77: 1537-1541.

9. Lal, Naja,Subbaraya (2000) Seasonal variations in surface ozone and its precursors over an urban site in India. Atmospheric Environment 34: 2713-2724.

10. Debaje SB, Jeya kumar SJ, Ganesan K, Jadhav D, Seetaramayya P (2003) Surface ozone measurements at tropical rural coastal station Tranquebar, India Atmospheric Environment 37: 4911-4916.

11. Salve PR, Satapathy DR, Katpatal YB, Wate SR (2007) Assessing Spatial Occurrence of GLO around Coal Mining Areas of Chandrapur District, Maharashtra, India Environ Monit Assess133: 87-98.

12. David Liji M, Nair Prabha R (2010) Influence of aerosols on near surface ozone mixing ratio at Tropical coastal environment. IASTA 2010 Conference on Aerosols and clouds: Climate change perspectives 28: 155.

13. Sarkar, Agrawal, Chemtrols (2012) Ambient Air Quality Station.

14. WHO (World Health Organization) (1979) "Photochemical Oxidants." Environmental Health Criteria 7 Geneva.

15. WHO (1987) Air Quality Guidelines for Europe. Copenhagen: WHO Regional Office for Europe.

16. EEA (2007) Air Pollution in Europe 1990-2004. EEA Report No2/2007. European Environment Agency: Copenhagen

17. Kley D, Geiss H (1994) Tropospheric ozone at elevated sites and precursor emissions in the United States and Europe. Atmos. Environ 28: 149-158.

18. Debaje, Kakade (2009) Surface ozone variability over western Maharashtra, India. Journal of Hazardous Materials 161: 686-700.

19. Duenas C, Fernandez MC, Canete S, Carretero J, Liger E (2002) Assessment of ozone variations and meteorological effects in an urban area in the Mediterranean Coast The Science of the Total Environment 299: 97-113.

9

Atmospheric Study of the Impact of Cold Surges and Borneo Vortex over Western Indonesia Maritime Continent Area

Paulus AW and Shanas SP

State College Meteorology and Geophysics, Indonesia Meteorological and Geophysical Agency, Jakarta, Indonesia

*Corresponding author: Paulus AW, Regional Centre, State College Meteorology and Geophysics, Indonesia Meteorological and Geophysical Agency, Jakarta, Indonesia, E-mail: pawinarso@gmail.com

Abstract

Cold surges and Borneo Vortex are the synoptic scale disturbances during the Asian Winter Monsoon period (November-March). These disturbances have closely linked with the growth of strong convective clouds and heavy rains in the western part of the Maritime Continent.

This study is subjected to study impact of cold surges and Borneo Vortex to the against atmospheric and rainfall conditions over the western part of the Maritime Continent using compositing technique these parameters for the period of November-March 2004/02-2014/15. The highest frequency of cold surge events occurs in January, while the incidence of Borneo Vortex alone and their interaction occurs in December. The atmospheric parameters of vortices, divergence, and moisture transport indicate that the incidence of cold surge, Borneo Vortex and their interaction having different influences. Region of South China Sea is mostly affected by the most significant of the other regions. Cold surge has the impact of increasing rainfall in all regions except Central Borneo/Kalimantan Island, Borneo Vortex while increasing rainfall throughout the region except Java, and the interaction of both resulted in increased throughout the study area.

This is preliminary study and it would be triggering further study, it was due to the lack study over Indonesia.

Keywords: Cold surge; Borneo vortex; Maritime continent; Atmospheric parameters

Introduction

Indonesia Maritime Continent area comprises from series big and small islands which separates the waters (sea and ocean), this condition mentioned as Maritime Continent Area [1]. As Aldrian [2] study, Indonesia area lies over 2 continents and 2 oceans. So that over this region has interesting weather phenomena to be studied. Indonesia Maritime Continent (IMC) is affected by regional wind of monsoonal wind to be due to the contrast between ocean and continent as well as lad/sea breeze circulation in synoptic scale point of view.

Zakir et al. [3] mentioned the monsoonal wind to be wind circulation pattern which are blowing periodically wind blowing in the consecutive direction and in another period in opposite direction. Tjasyono [4] studied monsoon that reversal on the wind direction on the consecutive monsoon period and differentiation of the seasonal condition in terms cloudy, rainfall and surface temperature. Where Indonesia Maritime Continent (IMC) is affected by Asian and Australian Monsoon.

During the active winter monsoon frequently occurs cold air advection from highland of Siberia of Northern Asian Continent of the so called cold surge. Yihui [5] defined "cold surge" as surging the Asian cold air to the South China Sea area. This Cold surge is the one weather disturbance to cause annual rainfall pattern especially over northwest IMC area the so called formation Monsoonal Asian Winter cycle [6]. Hattori et al. [7] studied that cold surge activity may cause

increasing the rainfall over Java sea, western Borneo and eastern part of the Philippines area. Takahashi et al. [8] study showed that heavy rainfall to be occurred over west Malaysia area last end of December 2006, it was due to the cold surge activity.

Beside cold surge disturbance, during active period of Asia Winter Monsoon, there is another weather disturbance of the so called Borneo Vortex. It has been occurred over Northwest of Borneo Island which has linked with heavy rainfall activities to cause flooding area [9-12]. This weather disturbance is unusual disturbance because one of vortex occurs near the equator.

Asian Winter Monsoon has frequent been linked with occurrences of both weather disturbances of cold surge and Borneo Vortex disturbances. Main factor linked with the interaction of the strong wind occurrence with topographical condition over western Malaysia and IMC areas [13]. Both cold surge and Borneo Vortex strongly affects extreme weather conditions over western and middle IMC region [14].

From these weather disturbances, they are very interesting to be studied where the three aspect will be designed in this paper e.g., Cold Surge activity itself without Borneo Vortex activity, reversal Borneo Vortex without Borneo Vortex and the last both Borneo Vortex and Sold Surge activities. So that three options will be further studied in this paper.

Data and Method

The data might be used in this study to be Reanalysis ERA Interim ECMWF (European Centre for Medium-Range Weather Forecast) data at 00.00 Universal Time Conversion (UTC) in atmospheric layers from 925-300 hecto Pascal (h.Pa.) with spatial resolution $2.5° \times 2.5°$ in terms of wind components of u and v; vorticity field; divergence field and moisture transport. Rainfall study used daily rainfall data from n TRMM (Tropical Rainfall Measuring Mission) and rainfall measurement from Meteorological Stations of Indonesia Meteorological and Geophysical Agency of the so called BMKG.

Used vorticity parameter is subjected to measure vertical microscopic circulation from every study option; meanwhile the divergence parameter to study area convergence/divergence of air mass and moisture transport will be used measure surging/advection the water vapour. The simple formula to use as follows,

$$Bq = \int_0^{300} qV dz \qquad (1)$$

Moisture transport quantity is counted using formula (1). Where level of 300 hecto Pascal will be used as above boundary with small moisture so it obeyed.

Location of the study will be over western IMC area, with 26 points of rainfall station and it will divided into 6 group (Figure 1) as follows,

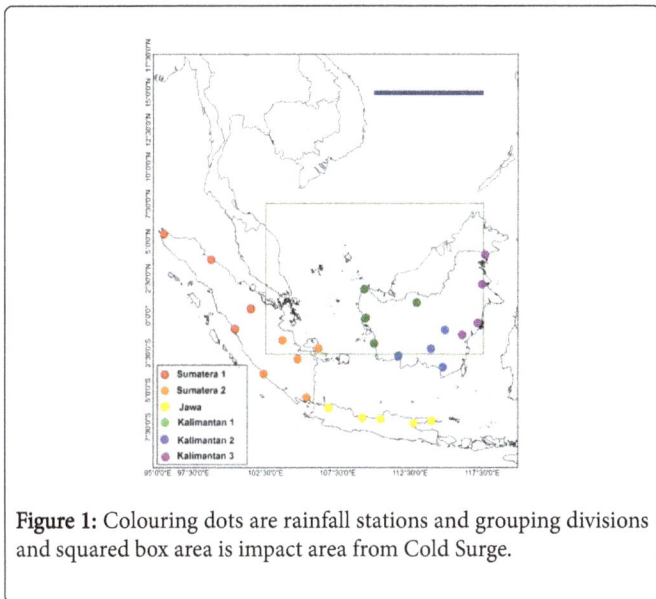

Figure 1: Colouring dots are rainfall stations and grouping divisions and squared box area is impact area from Cold Surge.

Cold Surge Index processing will be computed using Chang's et al. where computations mean meridional wind over 110°E latitude–117.5°E latitude along 15°North longitude. Used Index in this paper having assumption that impact from Cold Surge has not break time more than one day. This study obeys arising Cold Surge intensity.

Occurrence Borneo Vortex will be studied by observing closed counter-clock circulation on the layer 925 h.Pa. at 2.5° South–7.5° North and 102.5° East–117.5° East and at least having one wind component not more than 2 ms^{-1} [10]. Existing monsoonal wind interaction and Borneo topography presented at (Figure 2) such that causing occurrence of Borneo Vortex.

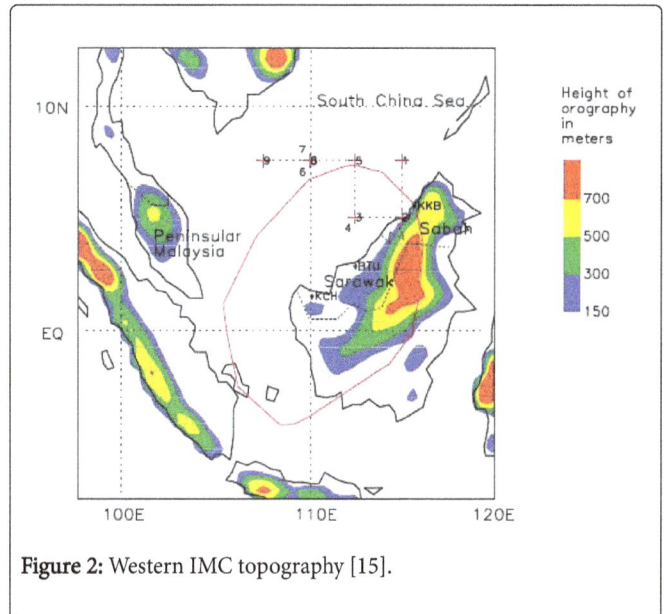

Figure 2: Western IMC topography [15].

After identification of the cold surge and Borneo Vortex occurrences, then they were classified their probability of their occurrences such as normal occurrences (without cold surge and Borneo Vortex occurrences), cold surge occurrence itself without Borneo Vortex, Borneo Vortex occurrence without cold surge, and coincident cold surge and Borneo Vortex occurrences (Figure 3).

Result and Discussion

Identification Borneo vortex

Borneo Vortex occurred; it was due to the shear Westerly wind in the southern area with Northeast Winter Monsoon from North Asia continent, such that those winds interact with Northwest Borneo topography (Figure 3).

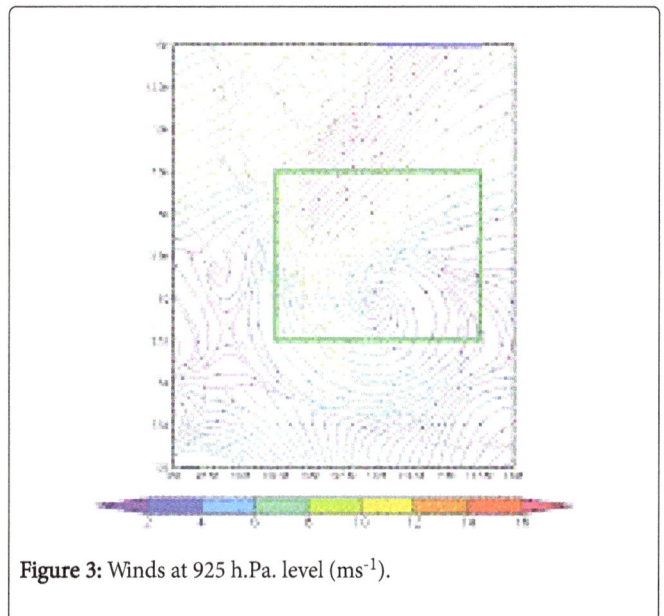

Figure 3: Winds at 925 h.Pa. level (ms^{-1}).

Based on the Figure 4, December is the month having more weather disturbances. If looking from comparison frequency occurrences of Borneo Vortex, it was known that on month of December to have high frequency of occurrences which indicated that this month having more active of Borneo Vortex occurrence [10,11,13,16,17].

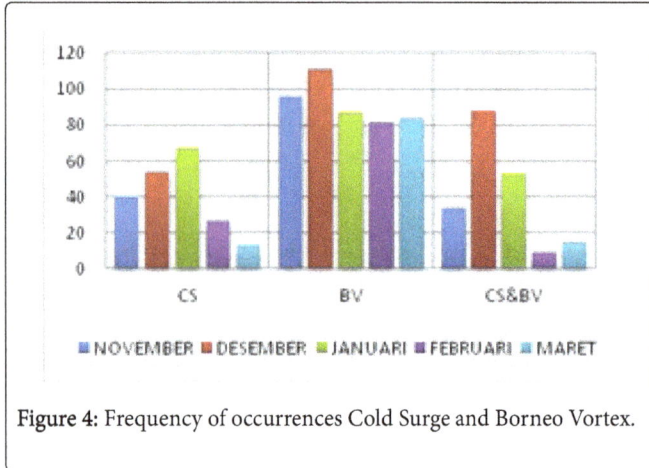

Figure 4: Frequency of occurrences Cold Surge and Borneo Vortex.

Anip and Lupo [10] stated that from November-February vortex centre may shift toward southeast approaching equator line. Vortex has longest life time in December, which indicated that Asian Winter Monsoon was more active during period November–February. More actively of occurrences of Borneo Vortex on December linked with existing strongest southeast trade wind on that month [13]. High frequency of occurrence of Cold Surge itself occurred on January. But, if looking daily occurrence of Cold Surge; so that December is the month more active of this disturbance [6,13].

Cold surge and Borneo Vortex may interact together with highest frequency on December. This occurrence has lowest frequency from another occurrence. Formation this occurrence required wind force from Southeast trade wind to encourage strong wind development from North direction as impact from Cold Surge to form Borneo Vortex. Becoming stronger from the north wind, it was required from strong Southeast trade wind to generate Borneo Vortex.

Impact study

From Compositing of the voracity parameter on the occurrences of the Cold surge, Borneo Vortex, and both Cold surge and Borneo Vortex give the significant result (Figure 5). There was transport high values vorticity advection from South China Sea toward the equator and centralized Northwest of Borneo (Kalimantan) island. It was same study by Wibianto [17] with his study during the period of Cold surge with high vorticity values movement southward. Centralized high vorticity values in Northwest Borneo/Kalimantan coincide with Borneo Vortex occurrence. These conditions may indicate strong counter clockwise circulation during the Borneo vortex occurrence. It indicated that there was increasing Borneo Vortex intensity over Northwest Borneo island. Increasing intensity of Borneo Vortex coincide with the Cold surge. High values of vorticity during Cold surge and Borneo Vortex occurrence reached value of $3 \times 10^{-5-3}$, 5×10^{-5} s^{-1} and this condition might be higher values comparing with others. The strong winds pushed and encourage speed of circulation of Borneo Vortex.

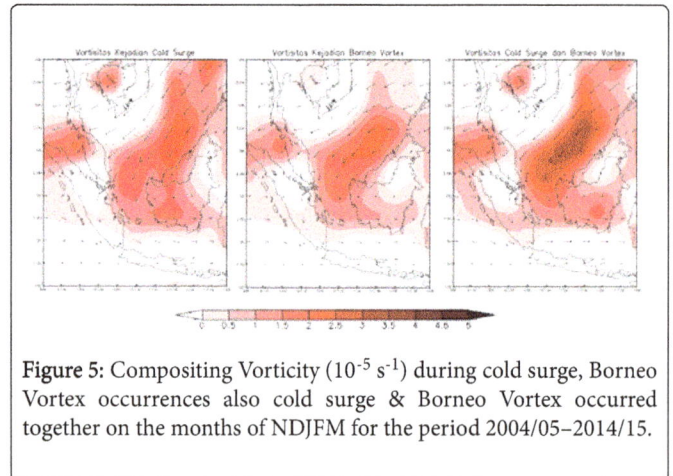

Figure 5: Compositing Vorticity (10^{-5} s^{-1}) during cold surge, Borneo Vortex occurrences also cold surge & Borneo Vortex occurred together on the months of NDJFM for the period 2004/05–2014/15.

Vortex is the main caused of increasing vorticity parameter over the waters Northwest Borneo/Kalimantan. From compositing technique of divergence parameter (Figure 6) produce negative divergence (convergence) in the South China Sea from Equator line reaching the divergence value-6 × 10^{-6} s^{-1} until -7 × 10^{-6} s^{-1}.

It was due to the deceleration of wind speed toward equator, such that there store air mass over South China Sea up to equator line. When Borneo Vortex case found the convergence area over Northwest Borneo/Kalimantan Island with the value 5 × 10^{-6} s^{-1} until 6 × 10^{-6} s^{-1}.

Different occurrence with cold surge activity, convergence area was due to the existing Borneo Vortex with exception of deceleration of air mass motion; it will be cyclonic motion such that converging of the air mass. Converging area occurred when cold surge and Borneo Vortex occurred together, divergence value not more that -10 × 10^{-6}s^{-1} (Figure 7).

Big divergence value was due to the increasing power from Borneo Vortex and cold surge so that they form circulation to be stronger from atmospheric condition without cold surge activity.

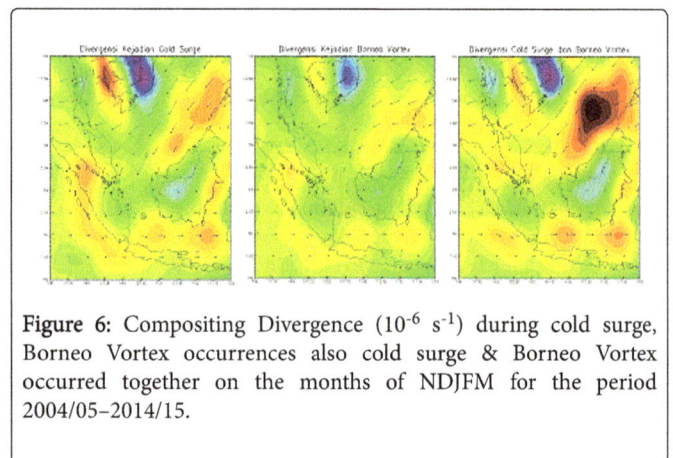

Figure 6: Compositing Divergence (10^{-6} s^{-1}) during cold surge, Borneo Vortex occurrences also cold surge & Borneo Vortex occurred together on the months of NDJFM for the period 2004/05–2014/15.

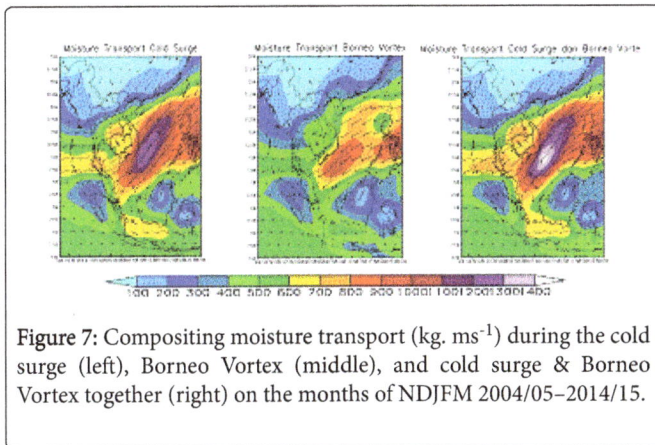

Figure 7: Compositing moisture transport (kg. ms^{-1}) during the cold surge (left), Borneo Vortex (middle), and cold surge & Borneo Vortex together (right) on the months of NDJFM 2004/05–2014/15.

Lowest values (negative) divergence were North Sumatera island with the range values 4×10^{-6} s^{-1}-7 $\times 10^{-6}$ s^{-1}, it was due to the converging air mass from Indian ocean on the western and Asian winter monsoon on the eastern sides. In spite, there was a high negative value over the Java Sea; it was due to the convergence area of Southeast Trade wind on the southern side and Asian and Asian winter on the northern sides. These lowest values of divergence encourage convective cloud development having potency of heavy rainfall occurrences. From compositing study of moisture transport (Figure 8) showed that Cold surge activity having moisture transport values ranging 1200-1300 kg ms^{-1}. These total moisture were due to the strong wind from the Cold surge such that transporting the moisture toward the equator. Different case with Borneo Vortex occurrence, it has moisture transport value ranging from 900-1000 kg ms^{-1}, to lower value comparing during Cold surge occurrence; because the North wind speed was not so strong such as during Cold surge occurrence. High moisture transport values were identified from coincide Cold surge and Borneo Vortex occurrence ranging from 1400-1500 kg ms^{-1}. High moisture transport values were due to pulling impact moisture by the Borneo Vortex such that to store many of the moistures.

In general over Maritime Continent area, incoming moisture transport came from South China Sea, the Pacific Ocean and Indian Ocean [14]. Most of the total moistures were toward the equator having support to the strong potential cloud development of convective cloudy type, if there were support from atmospheric condition. Based upon this finding, the moisture transport study might have important parameter to the developing the weather system over Maritime Continent area.

Rainfall Study using TRMM

Compositing technique of the monthly rainfall on the months NDJFM (November, December, January, February, March) 2004/05–2014/15 from each occurrence (Figure 9) give the different result significantly. During the Cold surge occurrence having high rainfall over South China Sea area ranging 7–15 mm. There was highest rainfall over small area in Western Borneo/Kalimantan and along East coast of North Sumatra. Besides, there were also high rainfall areas (15–21 mm) over the Java Sea. In general, high rainfall occurrences over Western Borneo/Kalimantan, along coastal North Sumatra and Java sea areas, they came from cold advection from Winter Season over North hemisphere toward equator area in the southern part. The movement of cold air mass southward direction may converge with

warm air mass in the equator area in the lower troposphere. These may encourage huge convective cloud development over the tropical area.

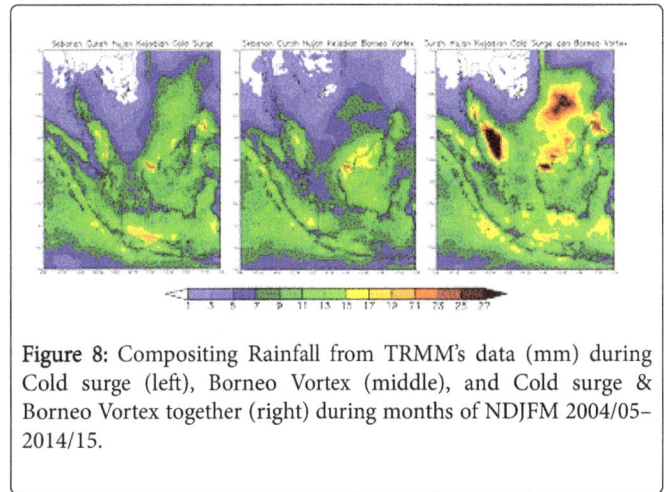

Figure 8: Compositing Rainfall from TRMM's data (mm) during Cold surge (left), Borneo Vortex (middle), and Cold surge & Borneo Vortex together (right) during months of NDJFM 2004/05–2014/15.

Borneo Vortex may encourage the rainfall occurrences over islands of Borneo/Kalimantan (7–23 mm.) and Southern Sumatera (7–17 mm.), but there was decreasing rainfall occurrences over Java island. It was due to deceleration air mass transported by the Asia winter monsoon wind with existing Borneo Vortex such that southern area (Java Island) receiving less of the rainfall.

When cold surge and Borneo Vortex occur coincidently, it might encourage high rainfall occurrences over South China Sea and almost the whole Borneo/Kalimantan Island reaching more than 27 mm. The high rainfall expanded also over some areas Sumatera and Java islands. During Cold surge and Borneo Vortex occurrence (Figure 9) showed that high rainfall occurred over western coast Southern Sumatera Island. This condition was caused by incoming air mass movement from Indian Ocean through Sumatera island. When Borneo Vortex occurred and it was caused from pushing the strong wind came from the Cold surge, so the size of the vortex expanded the area wider area if it occurred without Cold surge occurrence, such that expansion of the high rainfall occurrence over west Maritime Continent area.

Study of the rainfall observation

Processing of the rainfall observation from the meteorological station was done with the compositing technique from every occurrence. Computation of the rainfall was done using classification rainfall observing station followed rainfall distribution areas (Figure 8). From this figure areas were classified into 6 areas/regions (Figure 1).

From the result, there were increasing and decreasing areas of the rainfall over some part areas. Figure 2 presented that impact Cold surge and Borneo Vortex may cause increasing numbers of the rainfall over most the research areas except over area Borneo 2. It was due that area Borneo 2 was in the back of the mountainous area in Borneo/Kalimantan Island such that the developing of the cloud was less. Only Cold Surge occurrence itself may cause the decreasing the rainfall over Borneo 2 area around 6.9% and then additional number of rainfall occurred in area Sumatera 1 (2.7%), Sumatera 2 (1.1%), Borneo 1 (13%), Borneo 3 (25.8%) and area Java (21.9%). Increasing rainfall was caused cold advection process from cold air mass transported by the cold surge converged with warm air mass lying over the equator area. Serious impact the cold surge was over area Borneo 3 (25.8%). Because area Borneo 3 was over font area of the mountainous region of incoming strong wind flow coincides with cold surge activity, these areas were initial area when surge arrived.

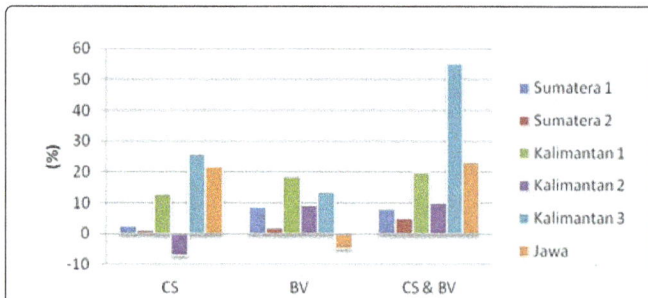

Figure 9: Rainfall Anomaly (mm) during occurrence of Cold surge & Borneo Vortex occurrences, and their interaction from compositing rainfall percentages on the months of NDJFM period 2004/05–2014/15.

Only Borneo Vortex occurrence might cause increasing number of the total rainfall overall impact locations, except over Java Island. When active Borneo Vortex occurrence had trend the air motion centralized in the center of the vortex. Their impact may be increasing rainfall over locations of Sumatera 1 (8.6%), Sumatera 2 (1.9%), Borneo 1 (18.5%), Borneo 2 (9.3%) and Borneo 3 (13.4%). It was reversal with rainfall occurrence over Java Island to decrease of the amount 4.8%. Most of the increasing total number of rainfall was over area nearest the central vortex (Locations of Sumatera 1 and over Borneo). Increasing rainfall over location Sumatera 2 where was away from central vortex, it was due to the wind convergence of Asia Winter Monsoonal wind and Southeast trade wind. This convergence zone extended from Indian Ocean, location Sumatera 2 and Java Sea. Decreasing of the rainfall in Java Island was due to blocking air mass in central vortex, such that it caused over some part Java Island not receiving these moist air mass to support the convective cloud development. Increasing the total highest rainfall was over Borneo location because this location was closed with central vortex. Beside the location Borneo 3 received the highest number rainfall, this location was convergence zone before the wet air mass centralized to form cyclonic pattern (Figure 7).

When Cold surge and Borneo Vortex occurred together, it might cause the increasing of the rainfall overall the study areas. It was due to the mixing process of the mechanical and thermal processes. Increasing of rainfall percentages as follows: Sumatera 1 area, 8.3%; Sumatera 2 area 4.9%; Borneo 1 area 19.9%; Borneo 2 area 10%, Borneo 3 area 55.2%, and Java area 23.3%. During Cold surge and Borneo Vortex occurred together, moisture transport didn't fully store in the vortex center. There was part of moisture transport during Cold Surge occurrence expanded southward direction. Increasing rainfall over Java Island was higher than rainfall causing from only Cold Surge occurrence itself. It was due the migrating convergence zone southward direction because it came from strong pushing cold surge activity from the Northern hemisphere. Meanwhile, over Sumatera 1 area there was increasing rainfall higher than rainfall occurrence over Sumatera 2 area, because position Sumatera 1 area was closed with vortex center. Observation over Borneo Island indicates the same condition when it was closed with vortex center, so that more closely with this center, they caused increasing their rainfall. But it might be different with Borneo 3 area which had own local topography to support increasing the highest rainfall than others areas.

Summary and Conclusion

From this research study can be summarized as follows:

A highest frequency occurrence of the only cold surge was in January, meanwhile only Borneo Vortex in December.

There was a significant parameters change such as vortices, divergence, moisture transport and rainfall; it was due to the Cold surge and Borneo Vortex activities. The areas were having significant changes to be over South China Sea and most Island over Western Indonesia area.

Most of the Cold surge activities to support the development of Borneo Vortex.

Impact of the twin activities of Cold surge and Borneo Vortex might cause increasing rainfall overall observation areas, meanwhile Cold surge activity causing increasing rainfall overall station except Borneo 2 area and only Borneo Vortex activity causing rainfall overall areas except over Java Island.

The highest increasing of the rainfall occurred over Borneo 1 area, it came from only Borneo Vortex activity; increasing rainfall over Borneo 3, it came from both only Cold surge activity and both twin Cold surge and Borneo Vortex activities.

References

1. Ramage (1971) Role of a tropical maritime continent in the atmospheric circulation. Mont Weat Rev 96: 365-370.

2. Aldrian E (2014) Pemahaman Dinamika Iklim Di Negara Kepulauan Indonesia Sebagai Modalitas Ketahanan Bangsa.

3. Zakir A, Sulistya W, dan Khotimah MK (2010) Perspektif Operasional Cuaca Tropis, Pusat Penelitian dan Pengembangan BMKG, Jakarta.

4. Tjasyono B (2008) Sains Atmosfer, Pusat Penelitian dan Pengembangan, Badan Meteorologi dan Geofisika, Jakarta.

5. Yihui D (1991) Advanced Synoptic Meteorology, China Meteorological Press, China pp: 717-751.

6. Aldrian E, Utama GSA (2007) Identifikasi dan Karakteristik Seruak Dingin (Cold Surge) Tahun 1995-2003, Jurnal Sains Dirgantara 4: 107-127.

7. Hattori M, Mori S, Matsumoto J (2011) The Cross-Equatorial Notherly surge Over The Maritime Continent and Its Relationship to Precipitation Patterns. J Meteor Soc Japn 89: 27-47.

8. Takahashi HG, Fukutomi Y, dan Matsumoto J (2011) The Impact of Long-lasting Northerly Surges of the East Asian Winter Monsoon on Tropical Cyclogenesis and its Seasonal March. J Meteor Soc Japan 89: 181-200.

9. Tangang FT, Juneng L, Salimun E, Vinayachandran PN, Seng YK, et al. (2008) On the Roles of the Northeast Cold Surge, the Borneo Vortex, the Madden-Julian Oscillation, and the Indian Ocean Dipole Mode during the Extreme 2006/2007 Flood in Southern Peninsular Malaysia. Geophys Res Lett 35: L14S07.

10. Anip MHM, dan Lupo A (2012) Interannual and Interdecadal Variabillity of the Borneo Vortex During Boreal Winter Monsoon, University of Missouri-Columbia, USA.

11. Ardianto R (2014) Kajian Dampak Borneo Vorteks Terhadap Curah Hujan Di wilayah Borneo Barat, Skripsi Sekolah Tinggi Meteorologi Klimatologi dan Geofisika, Jakarta.

12. Prakoso A (2015) Weather Disruption Event Study On Heavy Rain in Batam. Studi Kasus Tanggal pp: 18-19.

13. Chang CP, Harr PA, dan Chen HJ (2005) Synoptic Disturbances over the Equatorial South China Sea and Western maritime Continent during Boreal Winter. Monthly Weat Rev 133: 489-503.

14. Prakosa SH (2013) Kajian Dampak Borneo Vortex Terhadap Curah Hujan Di Indonesia Selama Musim Dingin Belahan Bumi Utara, Tesis Magister Institut Teknologi Bandung, Bandung.

15. Ooi SH, Samah AA, dan Braesicke P (2011) A Case Study of the Borneo Vortex Genesis and Its Interaction with the Global Circulation. J Geophys Res 116: D21116.

16. Syahidah M, Dupe ZL, dan Aldrian E (2015) Keterkaitan Borneo Vortex dengan Curah Hujan di Benua Maritim. Indonesian Undergrad Res J for Geosci 2: 1-9.

17. Wibianto A (2015) Kajian Pembentukan Borneo Vortex Berdasarkan Analisis Cold Surge, Skripsi Sekolah Tinggi Meteorologi Klimatologi dan Geofisika, Jakarta.

10

Mathematical Explanation of Earlier Dissipation of the Energy of Tilted Cyclone

Sukumar Lala*, Nabojit Chakraborty and Milan Kanti Das

Regional Meterological Centre, Kolkata, India

Abstract

The paper presents the mathematical explanation of the fact that the cyclone after entering the landmass, that is, when it's energy is not being constantly fed by the favorable parameters like sea surface temperature, etc., dies out earlier when it's axis which is normal to the ground surface gets inclined to the normal of the plane of the ground surface then when it's axis is normal to the ground surface. Other related derivation has been made and probably it gives a sound logic for studying some insights related to the cyclone and also may help in deriving the time difference of complete energy reduction of the cyclone after entering the landmass between the normal cyclone structure and the tilted cyclone structure. It may help in predicting the extent of the devastation that it may cause after following the track within the landmass if the landform is studied.

Keywords: Tangential velocity; Average pressure gradient; Stokes's law; Centrifugal force; Buoyant force

Introduction

We know that a cyclone when entering a land mass slowly reduces its energy as favorable parameters for its development are being stopped and ultimately it dies out. But few have tried to study the effect of its reduction of energy by the varied landforms. In this paper a humble attempt has been made to study the effect of an inclined plane on the reduction of energy of the cyclone and it has been shown mathematically that when a cyclone traverses an inclined plane then it dissipates it's energy more faster compared to the cyclone traversing the plane land assuming their energy content being same. This assumption has a sound footing as we know that all the favorable parameters required for development of the cyclone is cut off.

Data and Methodology

We suppose that an spherical parcel of air of mass "m_{rz}" is rotating having a tangential velocity "V_1" due to the average pressure gradient " P " w.r.t the centre of the cyclone in the plane in which the spherical parcel of air mass lies (Figure 1). Then according to equation 1 [1].

$$-P\alpha - fV_1 + F_r = \frac{V_1^2}{r} \tag{1}$$

If we consider the left hand side of the circulation assuming the air parcel to be of unit mass. (Figure 1)

{Coriolis force f = 2Ω Sin Ø where Ø is the latitude of the place and Ω is the angular velocity of the earth at that latitude }

(Also α is the cross section area of spherical parcel of air of unit mass m_{rz} on which P acts normally, F_r is the force of friction, "z" is the height of this parcel from the level of the ground and r is the radial vector at which the air parcel m_{rz} lies).

Now from equation 1 it can be further derived that when the axis of rotation of the cyclone inclines by an angle θ for cyclone climbing an inclined plane at an angle θ. Then we can write

$$(-P\alpha - fV_1 + F_r) = \frac{V_2^2}{r} + g\sin\theta \tag{2}$$

or

$$\frac{V_1^2}{r} = \frac{V_2^2}{r} + g\sin\theta \tag{3}$$

$$\frac{V_1^2}{r} - \frac{V_2^2}{r} = g\sin\theta \ (\text{ for } 0 < \theta < 90)$$

Therefore $V_1 > V_2$ as 0<θ<90, except $V_1 = V_2$ when θ=0

This shows that as the radial force towards the centre is constant so to maintain dynamic equilibrium the tangential velocity changes accordingly.

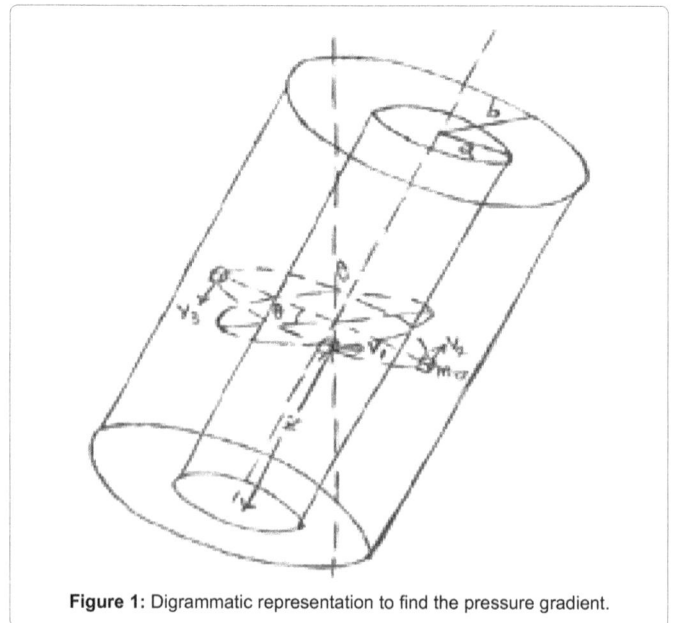

Figure 1: Digrammatic representation to find the pressure gradient.

---

*Corresponding author: Sukumar Lala, Regional Meterological Centre, Kolkata, India, E-mail: sukumarlala@gmail.com

Similarly we can prove that $V_3 > V_1$ as follows

$$\left(P\alpha + fV_1 - F_r \right) + g\sin\theta = \frac{V_3^2}{r} \qquad (3)$$ as radial force towards the centre is constant

as $\left(P_\alpha + fV_1 - F_r \right) = \frac{V_1^2}{r}$ if we consider the left hand side of the circulation in the Figure 1.

$$\frac{V_1^2}{r} + g\sin\theta = \frac{V_3^2}{r} \qquad (4)$$

or $\frac{V_3^2}{r} - \frac{V_1^2}{r} = g\sin\theta$

i.e $V_3 > V_1$ as $0 < \theta < 90$, except $V_1 = V_3$ when $\theta = 0$

Also Energy ' E_{v1} ' possessed by the infinitesimal spherical air mass 'm_{rz}' positioned at a perpendicular distance "r" from the axis of rotation at a height "z" from the level of the ground and rotating at angular velocity ω_1 (where $\omega_1 = V_1$) is

$$\frac{m_{rz}}{r} \cdot \frac{r^2\omega_1^2}{2}$$

As $\omega_1^2 r = \omega_2^2 r + g\sin\theta$ ----- from Eq.3, since $v = \omega r$

or $\omega_1^2 = \frac{\omega_2^2 r + g\sin\theta}{r}$

So the $E_{v1} = m_{rz}\dfrac{r\left(\omega_2^2 r + g\sin\theta\right)}{2}$

or $E_{v1} = \dfrac{m_{rz}r^2\omega_2^2}{2} + \dfrac{m_{rz}rg\sin\theta}{2}$

or $E_{v1} = E_{v2} + \dfrac{m_{rz}rg\sin\theta}{2}$

i.e $E_{v1} - E_{v2} = \dfrac{m_{rz}rg\sin\theta}{2} \qquad (4)$

i.e. Energy will be reduced when the m_{rz} rotating in the horizontal plane with tangential velocity V_1 drops down to the inclined plane and starts rotating in the horizontal plane with tangential velocity V_2. Similarly energy will be gained when the m_{rz} rotating in the horizontal plane with tangential velocity V_1 rises up the inclined plane and starts rotating in the horizontal plane with tangential velocity V_3.

The total energy reduced by the vertical slice portion (lower the horizontal plane) or gained (higher the horizontal plane) by half the cyclone volume will be

$$\sum_{z=0}^{h}\sum_{r=0}^{b} m_{rz}\frac{rg\sin\theta}{2}$$

or $\displaystyle\int_{0}^{h}\int_{a}^{b} \frac{m_{rz}rg\sin\theta}{2}$

The 'h' being the height of the cyclone and 'a' being the inner radius and 'b' being the outer radius of the cyclone (Figure 1).

Also as in a cyclone air mass is in dynamic motion against a essentially a two form of fluid system , fluid mechanics and particle transport equations can be used to describe the behaviour of a cyclone [2-5]. We have already assumed the tangential velocity of the spherical air mass in the horizontal plane is V_1 and assume further that the radial velocity be Vr at the radius r, then assuming Stoke's law, the drag force on any particle moving tangentially in a cyclic path is given by the following equation.

$$F_d = 6\pi r \mu V_t \qquad (5)$$

F_d is the frictional force acting on the interface between the fluid and the particle (in N),

μ is the dynamic viscosity (N s/m^2),

r is the radius of the spherical object (in m), and

V_t is the particle's tangential velocity (in m/s).

If one considers an isolated particle circling in the cylindrical component of the cyclone at the rotational radius of r from the cyclone's central axis , the particle will experience the centrifugal force F_c which counteracts the force $m(-P\alpha - fV_1 + F_r)$ and is equal to $m_{rz} V_1^2$

i.e. $$F_c = m_{rz}\frac{V_1^2}{r} \qquad (6)$$

$$= \frac{4\pi\rho_L r_P^3 V_1^2}{3_r}$$

where ρ_L is the density of lighter spherical air mass and r_P is the radius of the lighter spherical parcel of air mass.

The buoyant force component is obtained by the difference between the lighter air mass and denser air mass which are denoted by ρ_L and ρ_D respectively :

$$F_b = -\frac{4\pi r_P^3 \rho_D g}{3_r}$$

The force balance can be created by summing the forces together.

i.e $\dfrac{d_r}{d_t} = F_d + F_c + F_b$

This rate is controlled by the radius of the revolving air mass around the central axis of the cyclone. A spherical air mass in the cyclone flow will move towards either towards the wall of the cyclone or the central axis of cyclone until the drag, the uplift and the centrifugal forces are balanced. Assuming the system has reached a steady state, the particles will assume a characteristic radius dependent upon the force balance [6-9]. Heavier, denser particles will assume a solid flow at some larger radius than the light particles. The steady state balance assumes that for all particles, the forces are equated, hence

$$F_d + F_c + F_b = 0$$

Which expands to

$$6\pi r_p \mu V_1 + \frac{4\pi\rho_L r_P^3 V_1^2}{3r} - 4\pi r_P^3 \rho_D g = 0$$

$$18\pi r_p \mu V_1 r + 4\pi\rho_L r_P^3 V_1^2 - 4\pi r_P^3 \rho_D g = 0$$

$$2\pi r_p \left[9r\mu V_1 + 2\rho_L r_P^2 V_1^2 - 2r_P^2 \rho_D gr \right] = 0$$

i.e. it is a quadratic equation of V_1 and gives

$$V_1 = \frac{-9r\mu \pm \sqrt{\left(81\mu^2 r^2 + 16_p^4 \rho_L \rho_D gr\right)}}{4r_p^2 \rho_L}$$

Thus we see that V1 α r, as other are fixed quantities.

Hence when tangential velocity decreases (i.e.V_2) the radius of the cyclone decreases which subsequently will increase the centrifugal

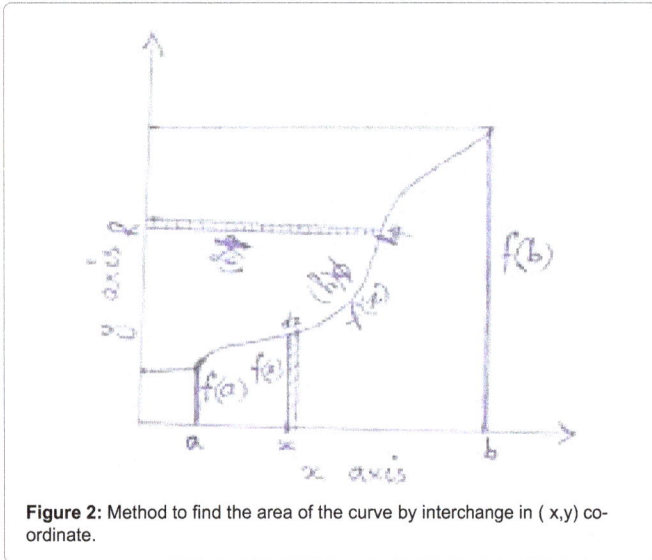

Figure 2: Method to find the area of the curve by interchange in (x,y) co-ordinate.

force but as force balance has to be maintained in the dynamic equilibrium so the drag force and buoyant force will decrease. Similarly when tangential velocity increases (i.e. V_3) the radius of the cyclone increases which subsequently will decrease the centrifugal force but as force balance has to be maintained in the dynamic equilibrium so the drag force and buoyant force will increase. Thus we see that the three forces i.e Fd, Fc & Fb, for the plane of rotating air mass whose inclination lies below the horizontal plane behaves in opposing manner for the plane of rotating air mass whose inclination lies above the horizontal plane. This will trigger wobbling of the system and ultimately the system will die out early compared to smooth horizontal rotation. This is supported by the fact that the Tropical Cyclone (TC) whose intensity during landfall which is proportional to Mean Surface Wind (MSW) speed, is same (i.e. MSW is 65 knot marked in blue color in Figure 3) dies out after moving through less distance (13.875 km) when traversing through relatively steep landforms areas which tilts the axis of the TC than gentle slopes where MSW during landfall is 65 knot but moves through after traveling 101 km, 111 km and 41.625 km respectively (Figures 4-6). The color coded landform scale given below the map specifies the varying altitudes of the land and we can get a idea of the slopes. It is also seen that the TC with more intensity as in Figure 7 where MSW is 135 knot during landfall, marked in red color, also dies out after moving through 20.183 km compared to lower intensities when traversing through steep slopes. This proves the assertion that cyclones whose axis gets tilted more dissipates its energy faster than less tilted cyclones.

To find the length of the arc of the circle if vertical angle is known or vice versa.

Abstract : In this method a formula has been derived for finding the

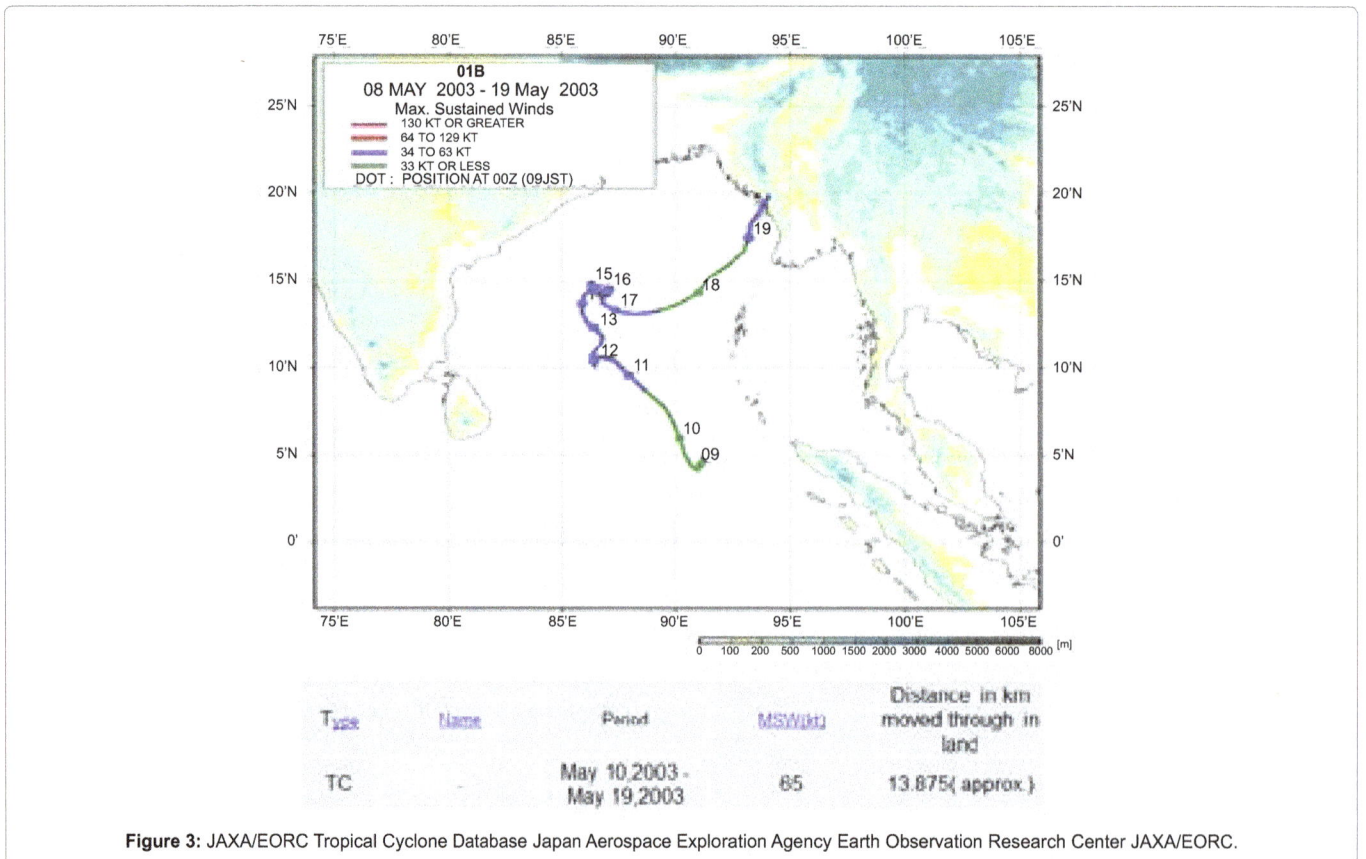

Figure 3: JAXA/EORC Tropical Cyclone Database Japan Aerospace Exploration Agency Earth Observation Research Center JAXA/EORC.

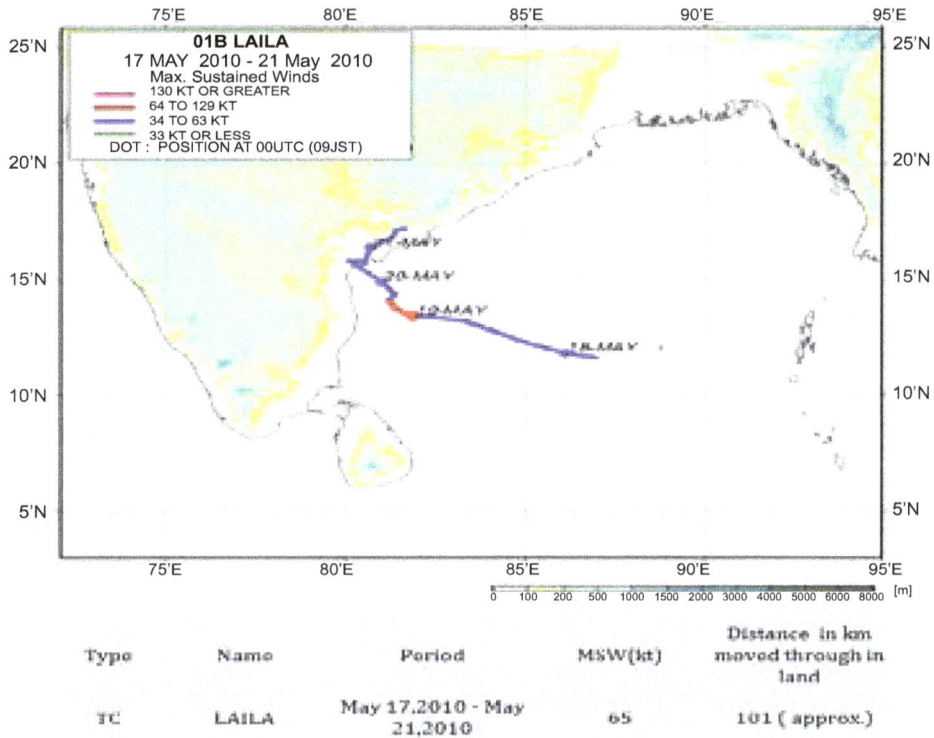

Type	Name	Period	MSW(kt)	Distance in km moved through in land
TC	LAILA	May 17,2010 - May 21,2010	65	101 (approx.)

Figure 4: JAXA/EORC Tropical Cyclone Database Japan Aerospace Exploration Agency Earth Observation Research Center JAXA/EORC.

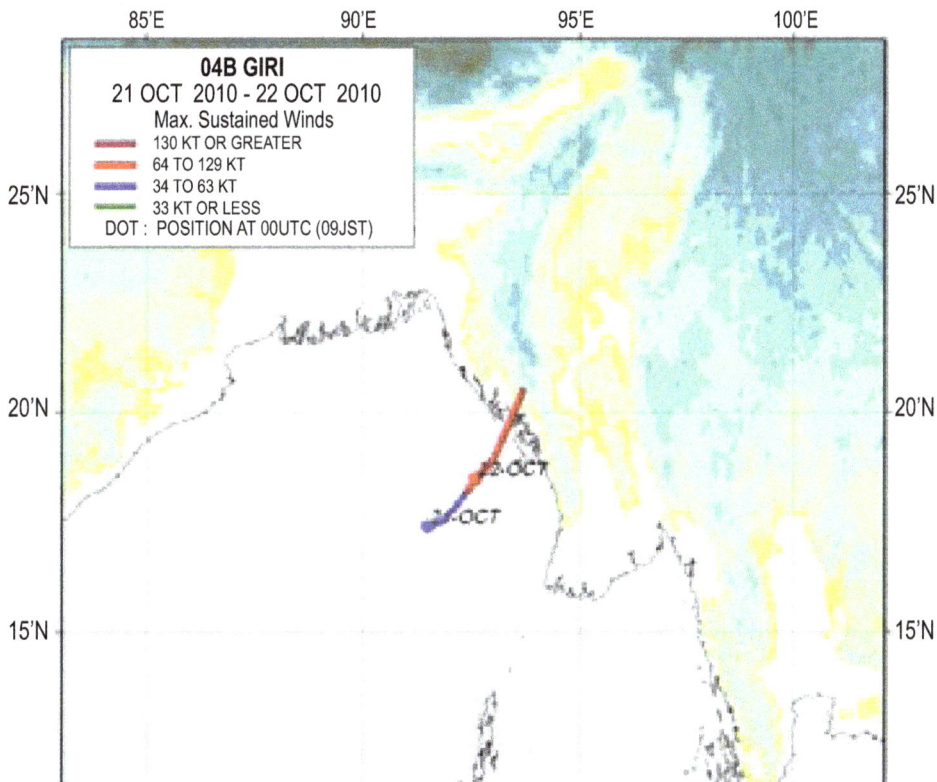

Figure 5: JAXA/EORC Tropical Cyclone Database Japan Aerospace Exploration Agency Earth   Observation Research Center JAXA/EORC

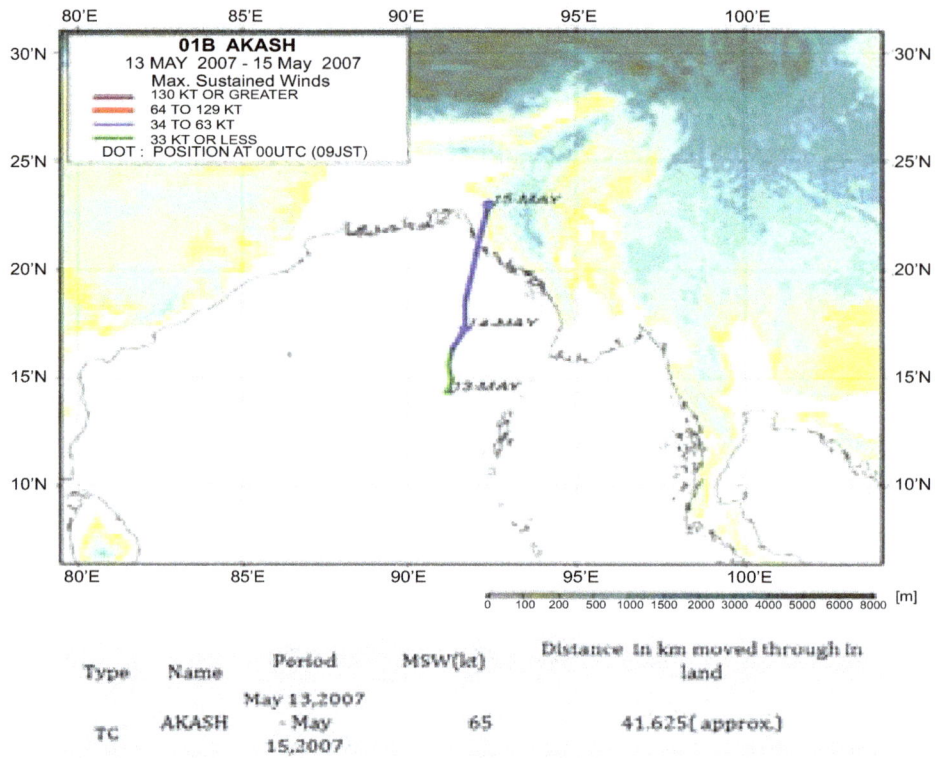

Type	Name	Period	MSW(kt)	Distance in km moved through in land
TC	AKASH	May 13,2007 - May 15,2007	65	41.625(approx.)

Figure 6: JAXA/EORC Tropical Cyclone Database Japan Aerospace Exploration Agency Earth Observation Research Center JAXA/EORC JAXA/EORC.

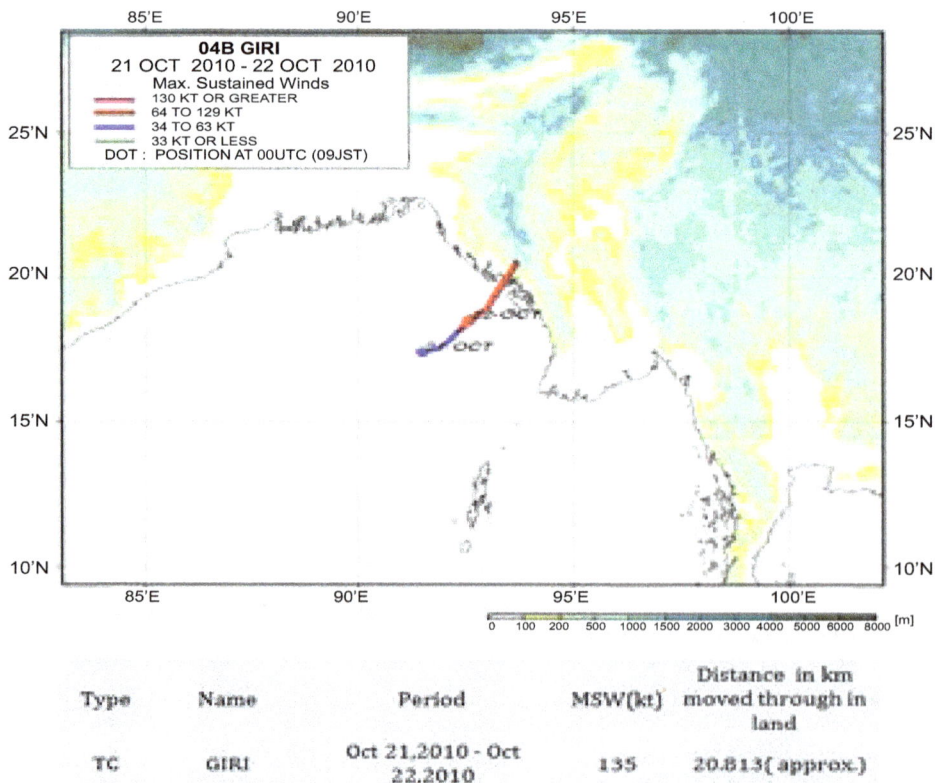

Type	Name	Period	MSW(kt)	Distance in km moved through in land
TC	GIRI	Oct 21,2010 - Oct 22,2010	135	20.813(approx.)

Figure 7: JAXA/EORC Tropical Cyclone database Japan Aerospace Exploration Agency Earth Observation Research Centre JAXA/EORC.

length of the arc of the circle by knowing the vertical angle or to find the vertical angle if the length of the arc is known. This will help to find the length of the arc or vertical angle for any point on the circle and can form a general formula.

Theory: Consider a circle of any radius and two perpendicular diameters CD and AB. By vertical angle α we mean the angle that line AP makes with straight line AOB. The arc length, by taking COD as the reference line , can be found out from the following method.

As we know the circumference which encircles 360^0 is of length $2\pi r$ where r is radius. Therefore if we take COD as reference line then arc CAD and CBD encircles 180^0 each i.e. the length of arc CAD or CBD is $2\pi r/2 = \pi r$ so each degree from centre subtends an arc length $\pi r/180$. Also as AOB is perpendicular bisector of COD therefore triangle CAD is isosceles triangle and as $\angle$CAD is an angle in a semicircle hence 90^0 (from cyclic property) therefore $\angle$ CAO = 45^0 and $\angle$ DAO = 45^0.

Therefore to find the arc length from the reference line COD for any vertical angle say α in this case. The angle subtended by the arc at the circumference becomes (α – 45). So the angle subtended by the same arc at the centre becomes 2(α – 45) (from cyclic property). Hence the arc length above below the reference line COD for α>45 and α<45 respectively is given by $\underline{\pi r}$ X 2 (α – 45)=$\underline{\pi r}$ X (α – 45).

It may be noted for α>45 then arc length will be +ve and for α < 45 it will be -ve. The sign factor has been craftily introduced to find the vertical angle if the arc length with relevant sign is being put. We can also find the arc length of the other side if the vertical angle on the other side is known.

To find the interchanging area of the curve in (x,y) co-ordinate system with the axis by integration method.

Abstract: A method has been devised to find the area of the curve by interchange in (x,y) co-ordinate system with the axis by integration method (Figure 2).

Let a function y = f(x) such that x = φ(y) and for x = a , y = f(a). and for x=b , y=f(b).

$$= \int_{f(a)}^{f(b)} x dy \int_{a}^{b} x f'(x) dx$$

$$\left[x f(x) \right]_{a}^{b} - \int_{a}^{b} 1 . f(x) dx$$

$$\int_{f(a)}^{f(b)} x dy = b f(b) - a f(a) - \int_{a}^{b} f(x) dx$$

Conclusion

Thus we see that when the axis of cyclone tilts then for the same energy possessed by the cyclone when it is not tilted which is the case especially when the cyclone enters the land and traverses the slopes, then it dies out earlier in comparison to the non-tilted cyclone. It also proves the fact as to why the low pressure area having a vertical column tilted does not get stronger as compared to straight vertical columns which gets converted to depression. This will help in analyzing the tracks of cyclone when it traverses the landforms especially in the light of inclination of land.

References

1. An introduction to Dynamic Meteorology by James R. Holton

2. Eliassen, A. (1952) Slow thermally or frictionally controlled meridional circulation in a circular vortex. Astrophysica Norvegica 2, 19-60.

3. Haque, S. M. A. (1952). The initiation of cyclonic circulation in a vertically unstable stagnant air mass. Quart. J . R. meteor. SOC7.8 .

4. Kasahara, A.(1961). A numerical experiment on the development of a tropical cyclone. J. Meteor. 18, 259- 282.

5. Kuo, H. L. (1961). Convection in conditionally unstable atmosphere. Tellus 13, 441-459.

6. Lilly, D. K. (1960). On the theory of disturbances in a conditionally unstable atmosphere. Mon. Wea. Rev. 88, 1-17.

7. Ogura, Y. (1964). Frictionally controlled, thermally driven circulation in a circular vortex with application to tropical cyclones. J. Atm. Sci. 21.

8. Ooyama, K. (1964). A dynamical model for the study of tropical cyclone development. Geofisica International 4, 187-198.

9. Palm Bn E. & Newton C. W. (1969). Atmospheric circulation systems. International Geophysics series Vol. 13. Academic Press 471-560.

11

The Use of Seismic Attributes to Enhance Structural Interpretation of Z-Field, Onshore Niger Delta

Adigun AO* and Ayolabi E A

Department of Geosciences, Faculty of Science, University of Lagos, Lagos State, Nigeria

Abstract

An integrated approach to the study of the structural patterns and seismic attributes was carried out on the Z-field using a 3D seismic data covering approximately 56 Km^2 of western belt of the Niger Delta, check shot data for well to seismic tie and log information for six wells from the field. This study aimed at investigating the available prospects, the responses of the basic seismic attributes to structural and stratigraphic elements within the study area. In all, six hydrocarbon bearing levels were delineated from well logs and correlated across the field. These horizons were analysed and petro physical parameters estimated. An integration of well and seismic data was done by applying a time-depth relationship to identify where the hydrocarbon bearing sands from well posts on the seismic lines. These studies revealed two major regional growth faults (F1 and F5) dipping in the NE-SW directions and crestal faults (F4) dipping in the northern direction. The north dipping crestal fault F4 is responsible for holding the hydrocarbon in the proven closure A. The northernmost regional growth fault F1 is responsible for trapping the hydrocarbon in the prospect closure B in the north eastern part of the field. The closure C prospect is a four way dip closure to the spill point but becomes fault assisted on regional growth faults F1 and F5 at deeper levels. The trapping mechanism identified from our interpretation indicates that the field is characterized mostly by fault assisted closures and a few four way closures. The fault zones are identified by distinct displacement of walls and amplitude distortions towards the fault zones identified on the seismic sections and the extracted amplitude maps. At the south-western part of the field, using only seismic Stratigraphic approach, a part of fault F5 was too subtle to be identified, and this may lead to missing out on the potentials of the identified closure C prospect. Thus, where conventional seismic interpretation has failed, seismic attribute analysis complements.

Keywords: Integrated; Attributes; Faults; Closure; Amplitude.

Introduction

Conventional seismic stratigraphic interpretation of data has been the popular way of interpreting seismic data for the purpose of mapping geological structures, subsurface stratigraphy and reservoir architecture. The geometrical expression of seismic reflectors is qualitatively mapped in time with little or no emphasis on the inherent seismic amplitude variations [1]. However, the introduction of the 3D seismic revolution has made the use of amplitudes an integral part of seismic interpretation and has allowed more valuable geological information to be discerned as seismic attributes (i.e. phase, amplitude, instantaneous frequency, etc.).

Seismic attributes form an integral part of qualitative interpretative tool that facilitates structural and stratigraphic (channels, pinch out, meanders, etc) interpretation as well as offer clues to lithology type and fluid content estimation with a potential benefit of detailed reservoir characterization [2]. For instance, fault structures which have been classified into seismically resolvable and sub-seismic scale (subtle) faults [3] can be interpreted more effectively with the aid of seismic attribute. Though the seismically resolvable faults may be interpreted using traditional diagnostic criteria (e.g. abrupt reflector cut off, kinks etc.), but the subtle faults which are often of geological or exploration significance are usually not visibly imaged by the conventional seismic sections and time slices displays. This is because they have smaller throws relative to the resolution limit of the seismic survey, which is a factor dependent on the frequency content, signal to noise ratio (SNR) of the dataset and also the depth to the reflecting horizon. The subtle fault identified in the southwestern part of the field along fault F5 motivated this study.

In this paper, we have subjected interpreted horizons (using conventional structural seismic interpretation technique) to seismic attribute generation and analysis. Seismic attribute analysis helps to identify structural features missed using the conventional method of interpretation. It is important to note that not only does seismic attribute analysis help to identify structural features; it also helps to increase the chances of success and development of new prospect areas in the study area.

Theoretical Background

Seismic attributes are defined as all the information obtained from seismic data, either by direct measurements or by logical or experience based reasoning [4]. Seismic attributes are essentially derivatives of the basic seismic measurements i.e. time, amplitude, frequency and attenuation which also form the basis of their classification [5]. It was also defined by [6]; as a measurement based on seismic data such as envelope, instantaneous phase, instantaneous frequency, polarity, dip and dip azimuth, etc. It is important to note that attribute interpretation supplements conventional structural interpretation and the discriminating properties of the attributes set may be critically checked for its relevance for a particular problem of a prospect.

The seismic data is treated as an analytic trace which contains real components (original input trace) and the complex (imaginary)

---

*Corresponding author: Adigun AO, Department of Geosciences, Faculty of Science, University of Lagos, Lagos State, Nigeria, E-mail: akeemadio@yahoo.com

component, usually generated from the Hilbert transforms from which various amplitudes, phase and frequency attributes can be deduced [7]. This complex trace allows the amplitude, phase, frequency and reflector polarity attributes of a seismic data to be calculated in a rigorous mathematical sense. Assuming a seismic trace of the form:

$$g(t) = A(t)\cos 2\pi vt, \qquad (1)$$

A(t) is the envelope of g(t) and varies slowly with respect to $\cos 2\pi vt$

For constant A(t), the Hilbert transform of g(t) is given by:

$$g(t) \Leftrightarrow g \perp (t) = -A(t)\sin 2\pi vt, \qquad (2)$$

Thus we can form a complex signal h(t) where:

$$h(t) = g(t) + jg \perp (t) = A(t)e^{-j2\pi vt}, \qquad (3)$$

h(t) is known as the analytical or complex trace [8].

$g\perp(t)$ is the quadrature trace of g(t) (Figure 1). If v is not constant but varies slowly, we define the instantaneous

frequency v(t) as the time derivative of the instantaneous phase γ(t) thus:

$$2\pi vi(t) = \frac{d\gamma(t)}{dt} = \frac{d(2\pi vt)}{dt}, \qquad (4)$$

The quantities A(t), γ(t) and v(t) and other measurements derivable from the seismic data are referred to as attributes [9].

A simpler mathematical description of the complex trace g(t) of a seismic trace is stated thus:

$$g(t) = X(t) + iY(t), \qquad (5)$$

Hence the amplitude a(t) is given by:

$$a(t) = \left[X^2(t) + Y^2(t) \right]^{1/2}, \qquad (6)$$

The phase Φ(t) is given by:

$$\Phi(t) = \tan^{-1}\left[Y(t) / X(t) \right], \qquad (7)$$

While the instantaneous frequency (f) is the time derivative of the instantaneous phase thus:

$$f = \frac{d(\Phi(t))}{dt}, \qquad (8)$$

Location and geology of study area

The study area is within the Niger Delta Basin in southern Nigeria. The Niger Delta is one of the most prolific hydrocarbon provinces in the world. It has a regressive clastic succession of about 10-12 km thick, comprising a shelf, broad slope area and basin floor [10]. The Delta lies on a thick prism of clastic sediments which forms the prominent seaward bulge in the continental margin off southern Nigeria [11]. The Niger Delta is an active Paleocene to Recent, wave dominated delta situated on the Atlantic coast of West Africa. Sedimentation in the Niger Delta began following the Albian rift fill deposits [12], between latitudes 3° and 6° N and longitudes 5° and 8° E and extends throughout the Niger Delta Province as defined by Klett et al. [13].

Generally, sediment inflow is North-South as the clastic sediments are sourced from the high mountains from the north into rivers Niger and Benue. These two rivers merge at a confluence and continue the sediment transportation to form the Delta as it enters into the Atlantic Ocean. The Niger Delta Province contains only one identified petroleum system [14,15]. This system is referred to here as the Tertiary

Niger Delta (Akata –Agbada) Petroleum System. The base of the sequence, consist of massive and monotonous marine shales (Akata Shales) except at the deepwater of the north western area where it has been found to overlie and onlap an older progradational package [10]. This grades upward into interbedded shallow marine and fluvial sand, silts, and clays, which form the typical paralic facies portion of the delta (Agbada Formation). The uppermost part of the sequence is a massive non-marine sand section (Benin Formation). The total thickness of this composite sequence is not precisely known, but may reach 12 km in the basin center with an estimated area of about 75000 km² as defined officially by the Nigerian government. Gravity and magnetic data suggest that the maximum thickness lies in the area between Warri and Port Harcourt. The oldest continental sand in the Benin Formation is Oligocene. The Agbada Formation ranges in age from Eocene to Pleistocene and forms the hydrocarbon prospective sequence in the Niger Delta. The Akata Formation ranges in age from Palaeocene to Holocene [16]. However, studies from a producing deepwater field have ascribed upper Oligocene age to the lowermost part of the Agbada unit and inferred the Akata interval to be pre-Miocene in the lower delta slope [17].

The study area is located in the onshore region of the Niger Delta (Figure 2). It covers about 56 km² of the coastal swamp belt onshore eastern Niger Delta. The oil mining license was obtained in June 1967, development and production of Oil commenced immediately. The real name of the field is not given for proprietary sake but named Z-field for the purpose of this study.

Materials and Methodology

The materials provided for the study include; Well data (LAS files)

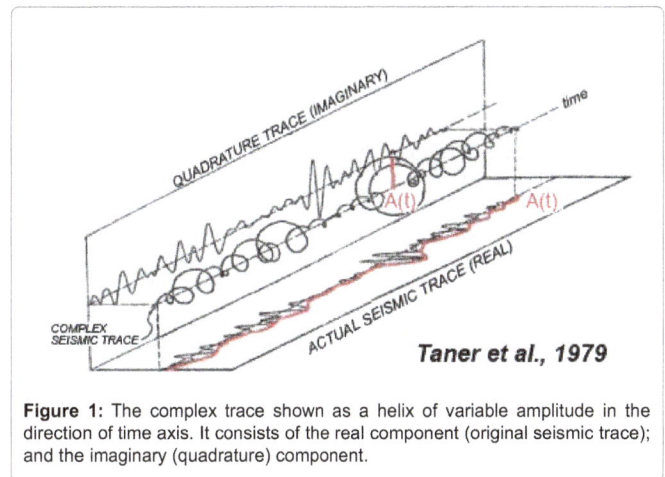

Figure 1: The complex trace shown as a helix of variable amplitude in the direction of time axis. It consists of the real component (original seismic trace); and the imaginary (quadrature) component.

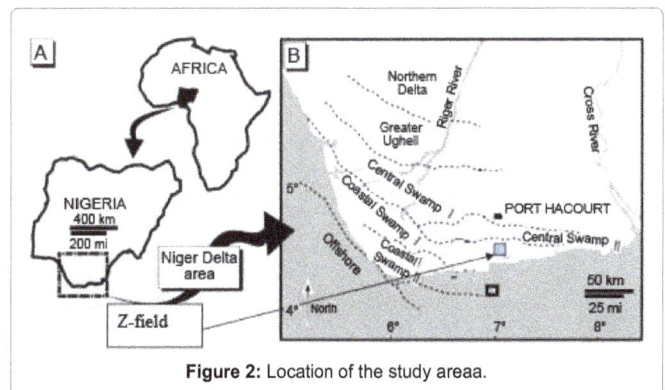

Figure 2: Location of the study areaa.

for six wells named as Z1, Z2, Z3, Z4, Z5 and Z6 in this project, check shot data for Well Z1, deviation data for Wells Z4 and Z6, 56km² 3D seismic data with good resolution processed as 32-bit integer and base map of the study area.

The datasets were loaded into the interpretative tools, in this case, Landmarks Geographix and Schlumberger Petrel. Interpretation of the well data was carried out from correlation to evaluation of petro physical parameters such as water saturation, net pay, net to gross, volume of shale and effective porosity. These results were later integrated into seismic using the available check shot data. Six hydrocarbon bearing reservoirs were identified and correlated across the field. These reservoir levels were tied to seismic and mapped across the field. Time and depth structure maps were generated from the interpretation. Seismic attributes such as amplitude, envelope, dip, azimuth and variance were extracted based on horizon interpretation.

The extracted attributes were analyzed and it was observed that variance showed better response to faults than any other attribute used for the study.

Results and Discussion

Petro physical evaluation

Wire line logs for six wells were used to delineate six hydrocarbon bearing reservoirs in the field (Figure 3). Table 1 below shows the summary of the petro physical evaluation carried out on three of the reservoirs penetrated by the six wells Table 2.

The identification of reservoirs shows that only Well Z1 encountered hydrocarbon in the six reservoir levels. Only three reservoirs were penetrated by more than one well as summarized in Table 3.

The provided check shot for well Z1 was used to tie well picks to seismic. The plot of time against depth from the check shot is shown in Figure 4 & 5.

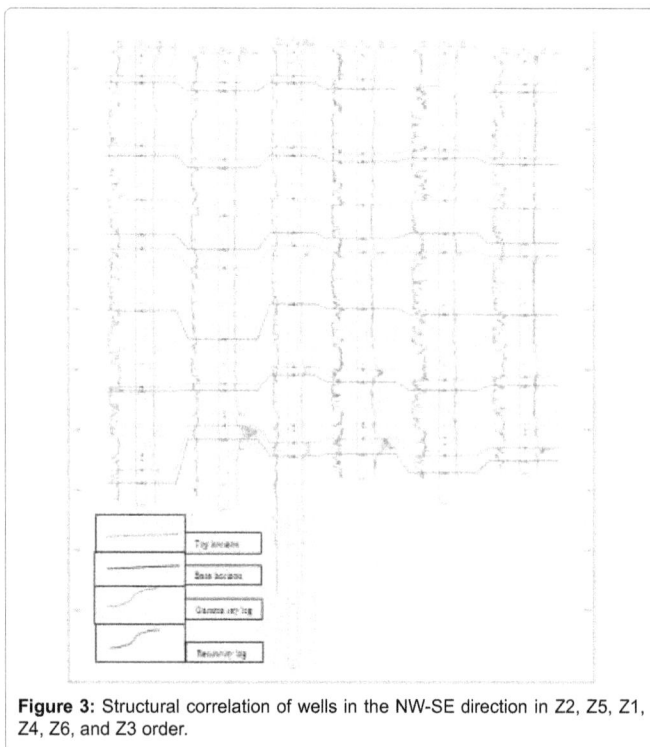

Figure 3: Structural correlation of wells in the NW-SE direction in Z2, Z5, Z1, Z4, Z6, and Z3 order.

Wells	Depth registration	Available Logs
Z1	308ft-13019ft	CALI, GR, LITH, LLD, PHIE, PHI_SNS, SP, SW, SWARCH, VSH, VSHSTB, checkshot data
Z2	46ft- 12996ft	DT, GR, LITH, LLS, RES, SP, VSH.
Z3	51ft-13000ft	CALI, DRHO, DTU, GR, LITH, LLS, MSFL, NPHI, PHIE, PHI_SNS, RES, RHOB, SP, SW, SWARCH, VSH, VSHSTB, and deviation survey.
Z4	10ft- 11541.50ft	CALI, CILD, DT, DTL, GR, LITH, LLD, LLS, NPHI, PHIE, RHOB, SGR, SP, SW, VSH
Z5	3849ft-11675ft	CALI, DT, DTL, GR, LITH, LLD, LLS, MSFL, NPHI, PHIE, RHOB, RT, SGR, SP, SW, VSH
Z6	2800ft- 13089ft	CALI, GR, HDRS, HMRS, LITH, NPHI, PHIE, RES, RHOB, SP, SW, VSH and deviation survey.

Table 1: Summary of the available well data and corresponding logs.

Reservoirs	Net pay	N/G pay	N/G res	Phi pay	Sw	Vshl
D	16	0.03	0.90	0.230	0.44	0.27
E	34	0.32	0.80	0.20	0.37	0.16
F	75	0.60	0.81	0.22	0.24	0.20

Table 2: Average values for parameters in reservoirs D, E and F.

Horizons	Z1	Z2	Z3	Z4	Z5	Z6
A	HC	wet	wet	wet	wet	missing
B	HC	wet	wet	wet	wet	wet
C	HC	wet	wet	wet	wet	wet
D	HC	wet	HC	HC	wet	HC
E	HC	wet	wet	HC	missing	wet
F	HC	HC	HC	HC	HC	wet

Table 3: Showing horizons in different wells with their corresponding fluid content

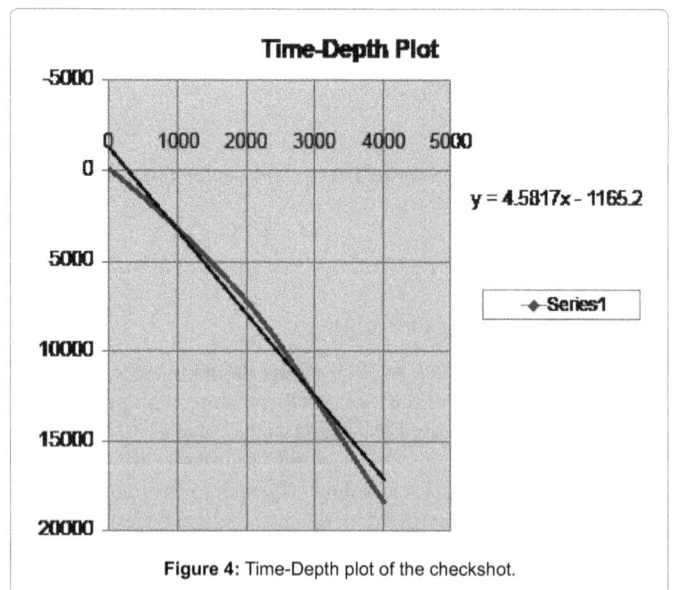

Figure 4: Time-Depth plot of the checkshot.

Linear equation relating depth to time; y= mx+c, Y=4.5817x-1165.2

Where Y= depth information from well data, X= time information from check shot, M=slope of plot, C= intercept.

Seismic interpretation

Seismic to well tie was done by using the check shot provided for Well Z1. Table 4, Figures 6b and 7 shows the horizons and corresponding time events on seismic.

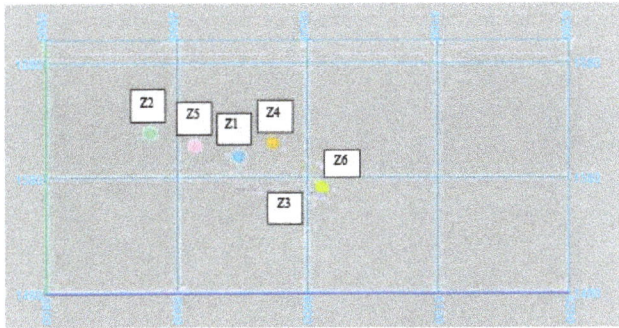

Figure 5: Base map of the study area showing well locations of Z2, Z5, Z1, Z4, Z6 and Z3 respectively

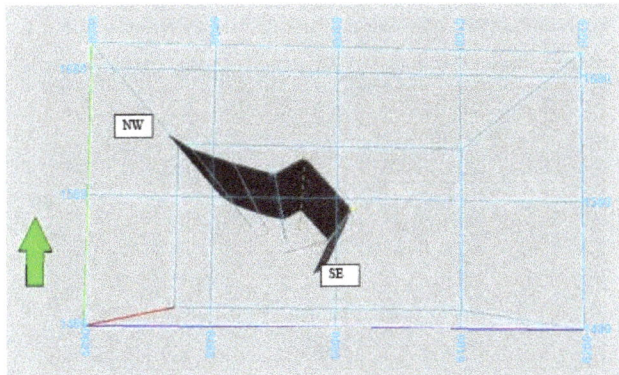

Figure 6a: An arbitrary line along Wells direction and positions.

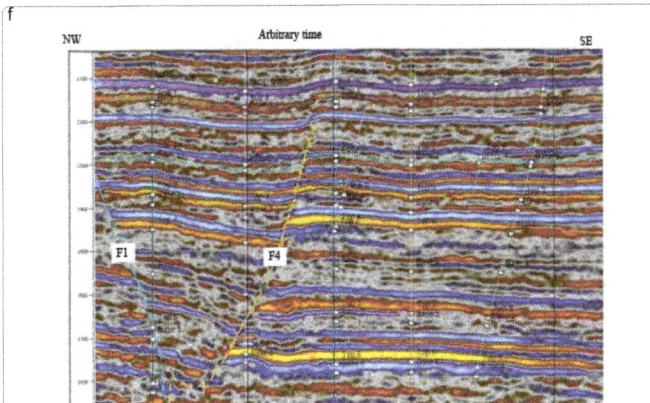

Figure 6b: An arbitrary section showing interpreted horizons with hydrocarbon trapping fault 4

Fault interpretation

The structural framework was done by picking assigned fault segments on inline sections of seismic with the trace appearing on the corresponding cross lines. These faults are represented on the seismic sections as a discontinuous reflection along a preferred orientation of reflectors or as distortion of amplitude around the fault zones. A total of ten faults coded as F1, F2, F3, F4, F5, F6, F8, F9, F10 and F11 were identified (Figures 8a-8c), some extending through the extent of the field known as major regional growth faults, few flank faults appearing on few of the lines and listric crestal faults appearing within the seismic extent (Table 5). Two major

regional growth faults F1 and F5 (Figure 7) were identified and correlated across the field forming the boundaries to the north and south of the field. The two major regional growth faults F1 and F5 and some other faults are dipping to the south away from direction of sediment supply, thus they are both regional faults while some of the other faults are dipping north, northwest, northeast, etc (Table 5).

North dipping crestal fault F4 is an important trapping fault responsible for holding the hydrocarbon in Wells Z1, Z3, Z4 and Z6 named closure A. The upthrown block to the north of the field forms a prospect named closure B, trapped by the northern most regional growth fault F1 towards north eastern part of the field. The closure C prospect is a four-way closure to the spill point at shallow depth but becomes fault assisted on fault F1 and possibly fault F5 at deeper levels.

The hydrocarbon traps are basically fault assisted. Two prospects named closures B and C were identified based on closures formed in the northeastern part of the field and within the two regional growth faults F1 and F5.

Horizon interpretation

A total of six horizons were interpreted across the field (figure 7) with both time and depth maps generated for each of the horizon. Figures 9a and 9b shows the top surface maps of horizon 1 both in time and depth.

Attribute analysis

Five attributes (i.e. dip, azimuth, amplitude, envelope, frequency, and variance) were extracted and displayed as flattened maps (slice at 2155, 2277, 2368, 2446, 2644 and 2764 ms) for each of the interpreted horizons. Structures respond to acoustic wave in different ways thus, the five attributes extracted are best used to study different subtle and

Horizon	Corresponding event	Time (ms)
1	Peak	2155
2	Trough	2277
3	Peak	2368
4	Peak	2446
5	Trough	2644
6	Trough	2764

Table 4: Horizons with their corresponding event on seismic.

FAULTS	DIP DIRECTION	INLINE COVERED	FAULT TYPE
F1	SOUTH	5800-6200	MAJOR REGIONAL GROWTH FAULT
F2	SOUTH	5800-5880	CRESTAL FAULT
F3	NORTH	5800-5860	FLANK FAULT
F4	NORTHWEST	5800-5950	CRESTAL FAULT
F5	SOUTH	5800-6200	MAJOR REGIONAL GRWOTH FAULT
F6	NORTH	5835-5990	COUNTER REGIONAL FAULT
F8	SOUTH	6080-6200	BACK TO BACK FAULT WITH F9
F9	NORTH	6140-6200	BACK TO BACK FAULT WITH F8
F10	SOUTH	5800-5820	FLANK FAULT
F11	SOUTH	5800-5820	FLANK FAULT

Table 5: Showing interpreted faults with corresponding seismic coverage.

sub-seismic structures missed by conventional seismic interpretation. Structural features such as closures and faults were studied in the course of this investigation. Tables 6 and 7 show the summary of the attributes and corresponding structural features they enhance. Figures 10 and 11 below shows that amplitude as an attribute can be used to identify prospect on a green field where no prior exploration work has taken place and also in areas where further exploration work is required.

During the course of the research, it was observed that, using conventional seismic interpretation method, fault 5 diminished to the southeast of the field. This fault was interpreted to have terminated as shown in Figure 12 but upon extracted variance attribute, it was clearly visible that F5 did not terminate as previously interpreted.

Attributes	Time slice(s)(ms) associated with the attribute
Amplitude	2155, 2277ms
Azimuth	2155, 2277, 2368, 2446ms
Dip	2277ms
Variance	2155, 2277, 2368, 2446, 2644, 2764ms
Envelope	2155, 2277, 2446ms

Table 6: Attributes and corresponding response to structural closures.

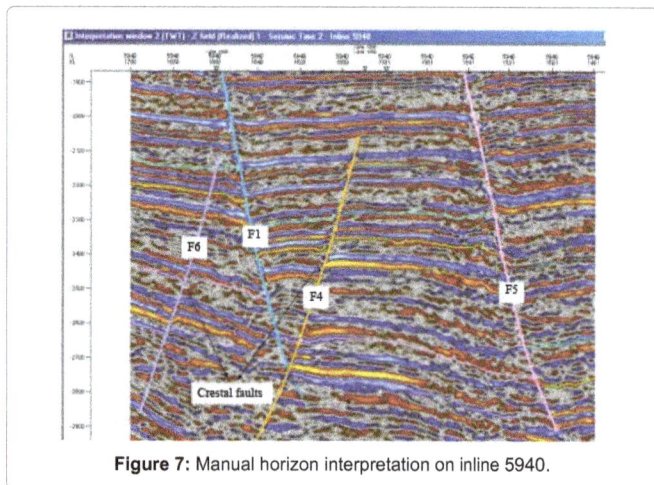

Figure 7: Manual horizon interpretation on inline 5940.

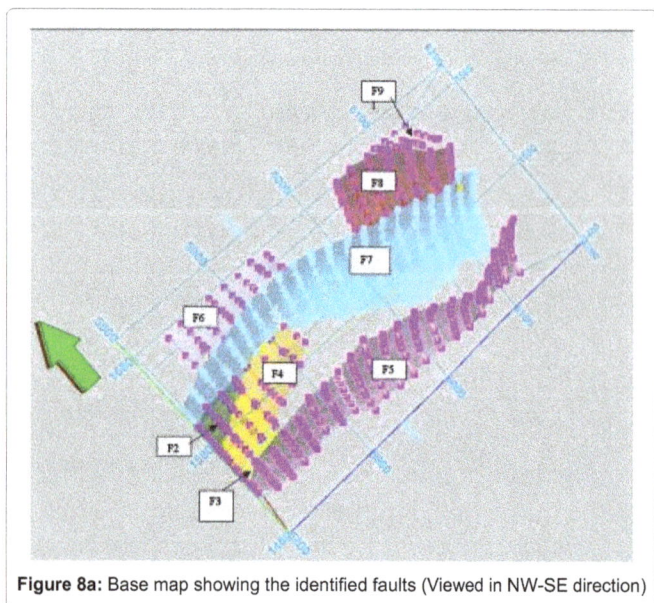

Figure 8a: Base map showing the identified faults (Viewed in NW-SE direction)

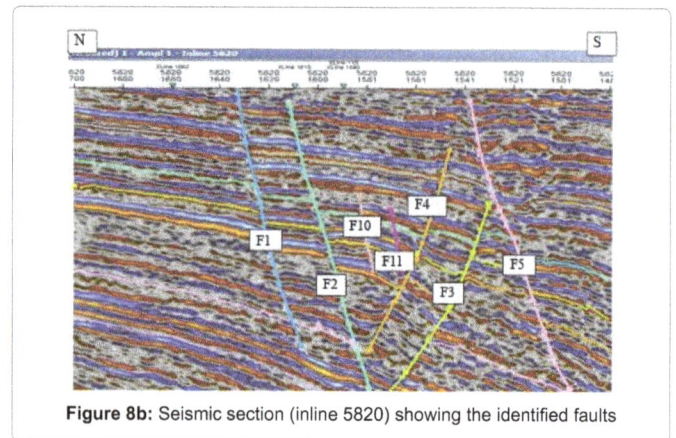

Figure 8b: Seismic section (inline 5820) showing the identified faults

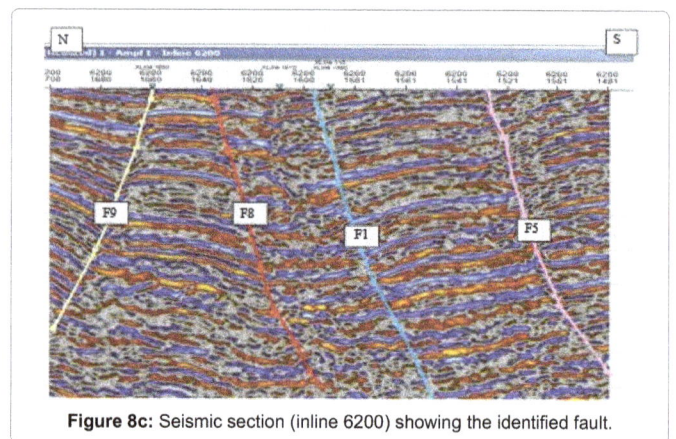

Figure 8c: Seismic section (inline 6200) showing the identified fault.

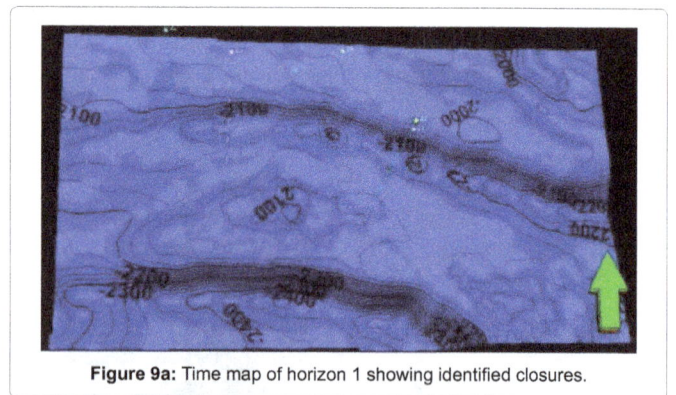

Figure 9a: Time map of horizon 1 showing identified closures.

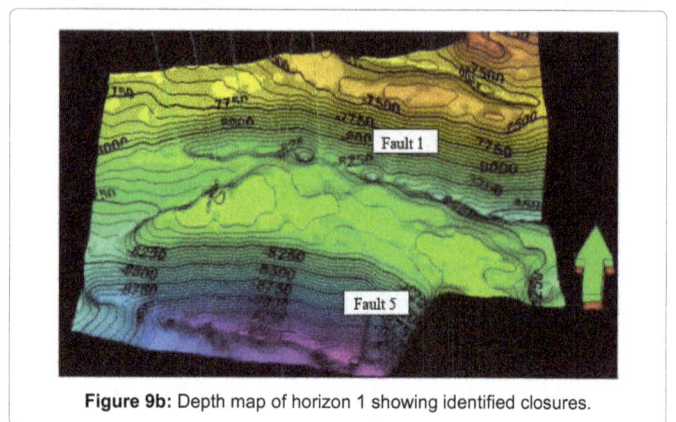

Figure 9b: Depth map of horizon 1 showing identified closures.

Conclusion

Interpretation of the 3D seismic data for locating both seismic scale and sub-seismic scale structural and stratigraphic elements has been demonstrated to be more efficient by the use of seismic attribute mapping and analysis. Variance attribute map has proved to be an appropriate tool to study fault architecture than dip attribute or any other attribute map in the study area. Therefore, to reduce the risk of drilling dry hole, resulting from missed fault by conventional seismic interpretation, seismic attribute analysis can be integrated into the standard practice of hydrocarbon Exploration and Production Company.

Acknowledgement

Thanks are due to the Department of Geosciences, University of Lagos for the opportunity to use their workstation facilities.

Attributes	Time slice(s)(ms) associated with the attribute
Amplitude	2155, 2368, 2446, 2764ms
Azimuth	No distinct pattern shown in all the slice
Dip	2155, 2277, 2368, 2446, 2644, 2764ms
Variance	2277, 2368, 2446ms
Envelope	2277, 2368, 2446, 2764ms

Table 7: Attributes and corresponding response to structural closures.

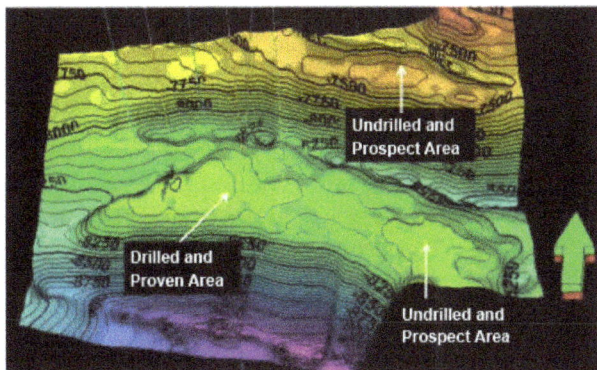

Figure 10: Depth map showing drillable prospects based on structural geometry

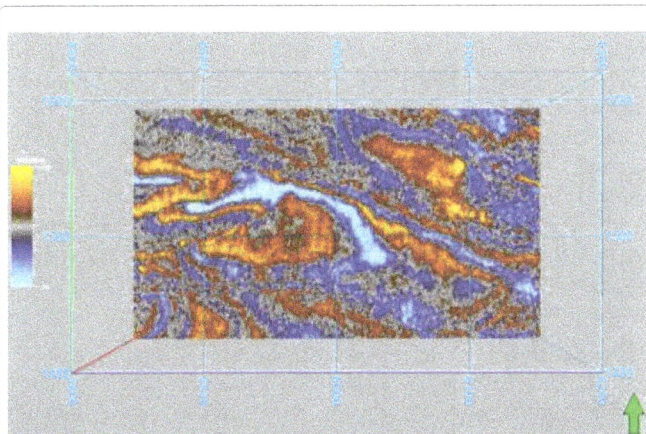

Figure 11: Amplitude map showing drillable prospects based on attribute extraction

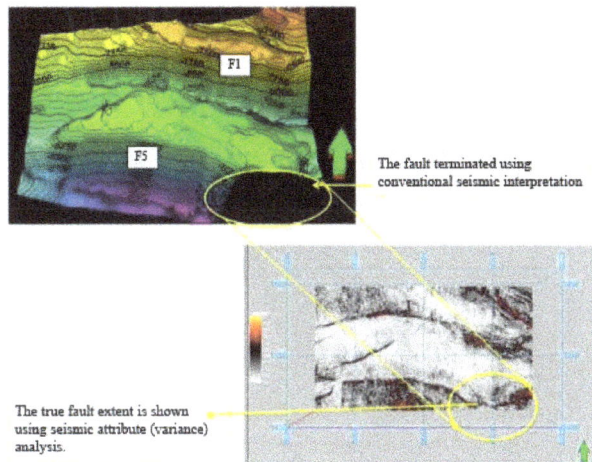

Figure 12: Variance and top structure map showing true fault extent of F5

References

1. Avseth P, Mukerji T, Mavko G (2005) Quantitative seismic interpretation. Cambridge University Press.

2. Strecker MR, Carrapa B, Hulley GE, Scoenbohm L, Sobel ER (2004) Erosional control of Plateau evolution in the Central Andes. Geological Society of America 36: 48.

3. Gauthier DM, Lake LW (1993) Prediction of sub-seismic faults and fractures.

4. Taner MT (2001) Seismic attributes: Canadian Society of Exploration Geophysicists Recorder 26: 48-56.

5. Brown AR (2004) Interpretation of three-dimensional seismic data. (6th edn), American Association of Petroleum Geologists Memoir.

6. Sheriff RE (1992) Reservoir geophysics, Society of Exploration Geophysicist.

7. Nissen SE (2002) Seismic Attributes for the Explorationist. Kansas Geological Survey,USA.

8. Bracewell RN (1965) The Fourier Transform and Its Applications. McGraw-Hill, New York, USA.

9. Sheriff RE, Geldart LP (1995) Exploration Seismology. (2nd edn), Cambridge University Press, New York, USA.

10. Morgan R (2004) In at the Deep End: Geology and Exploration of the Niger Delta Ultra Deep-water Slope. PESGB London Evening Meeting.

11. Damuth JE (1994) Neocene Gravity Tectonics and Depositional Processes on the Deep Niger Delta Continental Margin. Marine and Petroleum Geology 11: 321-346.

12. Doust H (1989) The Niger delta: hydrocarbon potential of a major tertiary delta province. Proceedings KNGMG Symposium Coastal Lowlands, Geology and Geotechnology, Dordrecht, Kluwer 203-212.

13. Klett TR, Ahlbrandt TS, Schmoker JW, Dolton JL (1997) Ranking of worlds's Oil and Gas Provinces by known Petroleum volumes. U.S. Geological Survey Open-file report 97-463.

14. Kulke H (1995) Regional geology of the world, part 2: Africa, America, Australia and Antarctical: Berlin, Gebruder borntraeger.

15. Ekweozor CM, Daukoru EM (1994) Niger Delta Depobelt Portion of the Akata-Agbada Petroleum System, Niger delta, Nigeria. AAPG Memoir 60: American Association of Petroleum Geologist, Tulsa, USA.

16. Doust H, Omatsola E (1990) Niger Delta In: Divergent and Passive Margin Basins. American Association of Petroleum Geologists, Memoir 48: 201-238.

17. Chapin M, Swinburn P, Weiden VR, Dieter Skaloud,Sola Adesanya et al. (2002) Integrated seismic and subsurface characterization of Bonga Field, offshore Nigeria. The Leading Edge 21: 1125-1131.

12

Prediction Experiment of Regional Drought over Korea Using the Similarity of Spatiotemporal Patterns of Past Droughts

Jae-Won Choi*, Yumi Cha and Jeoung-Yun Kim

National Institute of Meteorological Sciences, Seohobuk-ro, Jeju, 63568, Korea

*Corresponding author: Jae-Won Choi, National Institute of Meteorological Sciences, 33, Seohobuk-ro, Jeju, 63568, Korea
E-mail: choikiseon@daum.net

Abstract

This study investigated a drought prediction method on the basis of similarity of spatiotemporal patterns of past droughts in Korea. The method was implemented in the following steps: First, drought areas in Korea were divided into four drought subregions by means of hierarchical clustering analysis. Second, spatiotemporal drought statistics for each subregion for the period from 1926 to 2008 were established. Temporal statistics involve the drought onset, end dates, duration, and regional drought intensities, measured by the Effective Drought Index (EDI). These statistics were collected over the four subregions, and spatial patterns of drought were examined. Third, the analogous drought events that had spatiotemporal patterns similar to those of the current or subject drought were selected. Fourth, the progress of the subject drought and the selected drought were compared. Finally, the progress pattern of the subject drought was predicted on the basis of the hypothesis that it will progress in a way similar to the selected analogous case. We applied this predicted method to several previous drought cases and evaluated the prediction efficiency. The results showed that this method was efficient in predicting droughts for about 1 year.

Keywords Drought; Korea; Hierarchical clustering analysis; Effective drought index

Introduction

Droughts have occurred since many centuries, and the damage caused by or influence of droughts has increased in recent years. However, because droughts progress slowly unlike other natural disasters and the damage caused by them is gradual, they give us enough time to determine ways in which they can be prevented. Thus, the damages can be reduced if a prediction and warning system is established. However, thus far, no theories have been fully recognized as useful for predicting droughts. Previous studies on droughts were limited to recognizing the signs of droughts before their occurrence; this was achieved by determining their causes through case analysis. As such studies were limited in their scopes; it was not easy to determine the causes of droughts. Hence, in some cases, antithetical theories as well as different causes or results have been suggested for the same droughts. For example, three case studies on droughts in Korea in 1994 [1-3] presented different conclusions on the causes of these droughts. Furthermore, a number of case studies on the drought that occurred in the USA in 1988 [4-7] pointed out different causes.

With such limited resources, researchers tried to predict droughts because of the urgency of doing so. However, because of insufficient studies having been conducted on the causes of droughts, statistical approaches were mainly applied for prediction. In such efforts, statistical approaches such as the probability distribution [8], Markov chain model [9,10], neural network [11,12], low-order discrete autoregressive moving average models [13], seasonal autoregressive integrated moving average model [14], and rotated empirical orthogonal function analysis [15] were used. Nevertheless, such approaches were merely restricted to reporting the results rather than being actually applied to predict droughts.

This study also attempted to predict droughts using an analog method, in which a statistical approach was employed. The analog method is based on the assumption that the progress of a drought will be similar to that of the previous one if the climatic conditions at present are similar to those at the time at which the previous drought occurred. The analog method has been adopted mainly when it is required to predict droughts in spite of poor understanding of their causes. The analog method has been particularly applied to predict the path of typhoons and atmospheric circulation, precipitation, and temperature in the long term [16-19]. Furthermore, previous studies pointed out that the analog method was more useful for long-term predictions than for short-term predictions [20,21]. Hence, the analog method is expected to be a breakthrough to predict droughts in terms of the features of droughts that progress gradually in the long term.

Thus far, no studies were conducted using the analog approach to predict droughts, since it is very difficult to objectively quantify the previous data because of the obscure definition of the period or spatiality of droughts. This study partially attempts to resolve this difficulty by using the Effective Drought Index (EDI) [22]. Thus, this study substantially depends on the fact that the EDI is more precise and accurate than other drought indices [23,24]. The EDI is the useful for worldwide application, because it is independent of climatic characteristics of the locations [22]. The previous studies concluded that the EDI is reasonable enough to assess the intensity and duration of a drought based on the comparison between the existing drought index and the EDI. The EDI calculation and the corresponding strengths are further described in section 2.2.

The spatiality of drought was obtained by the clustering analysis. The clustering analysis is a method to divide subregions by similar characteristics. Most of the previous studies relied on precipitation to define climatological clusters [25,26]. However, these are not suitable for climatological drought clustering, because the precipitation data comes from a short period or a single season. Therefore, this study performed a clustering analysis relying on drought intensity using the EDI collected over the past 35 years. The spatiality of drought was used as fundamental data for the analog method.

However, as it is difficult to predict if cases identical to previous cases would occur again, the prediction of droughts using the analog method cannot be expected to be highly reliable in scientific aspects. This is the common problem of all analog methods that have been studied thus far. Nevertheless, the practical value of the analog method is considered to be significant, which is why it has been applied thus far.

Spatiotemporal Pattern of Drought Study area

We used the precipitation data from 1925 to 2008. The operation of each observation station started in different years. There were only 6 observation stations in 1925 and the number of observation stations added up ever since. In 1973, a total of 61 observation stations were in operation. As shown in Figure 1, the observation stations are distributed over nationwide.

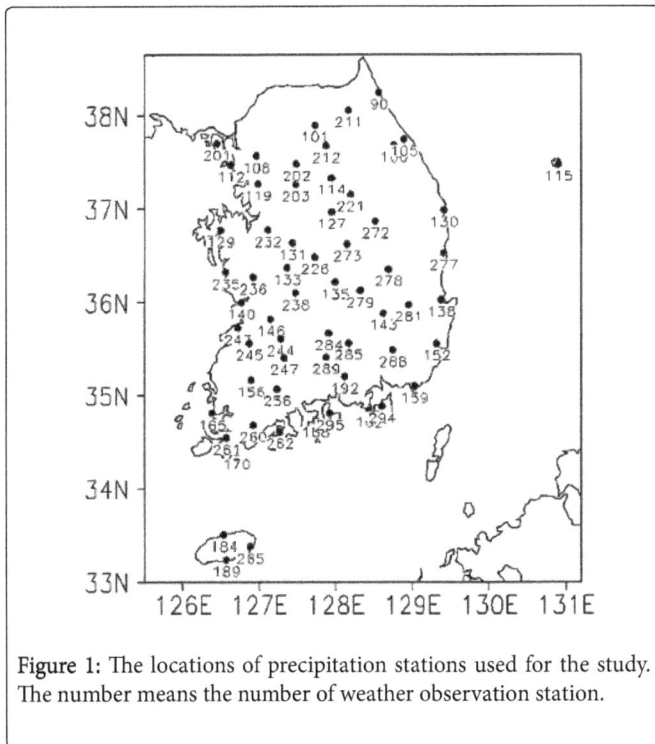

Figure 1: The locations of precipitation stations used for the study. The number means the number of weather observation station.

Effective Drought Index (EDI)

The EDI is used to measure an intensity of drought. The calculation process of the EDI is as follows:

$$EP_i = \sum_{n=1}^{i}\left[\left(\sum_{m=1}^{n} P_m\right)/n\right] \qquad (1)$$

$$DEP = EP\text{-}MEP \qquad (2)$$

$$EDI = DEP / ST (DEP) \qquad (3)$$

Equation 1 expresses Effective Precipitation (EP) expresses currently available water resources generated by past precipitation and reflects the depletion by runoff and evaporation with time. Here, Pm denotes the precipitation in m days before a particular day and it denotes the number of the days whose precipitation is summed for calculation of drought severity. Equation 2 is used to calculate the deviation of EP (DEP) from the climatological mean of EP (MEP) for each calendar day. Finally, Equation 3 is used to calculate the standardized value of the DEP (EDI). Here, ST(DEP) denotes the standard deviation of the DEP of each day. For further details, refer to Byun and Wilhite [22].

The main advantages of the EDI are as follows: (1) It gives a reasonable measure of the current level of water resources by considering daily precipitation accumulation with a weighting function with passing time; (2) It is effective to define the starting day, ending day, and duration of the drought because the EDI is expressed in the unit of days; (3) The EDI has universal applicability because it is independent of the climatic characteristics of a particular region. The feasibility of the EDI has been proved in previous studies [24,27-30].

This study calculated the EDI using the daily precipitation at 61 stations in Korea from 1925 to 2008 (84 years). The EDI values are listed in Table 1.

Effective Drought Index	Classification	Drought class (n)
EDI>2.5	Extreme wet	0
1.5<EDI ≤ 2.5	Severe wet	0
0.7<EDI ≤ 1.5	Moderate wet	0
0<EDI ≤ 0.7	Weak wet (Normal)	0
-0.7<EDI ≤ 0	Weak drought (Normal)	1
-1.5<EDI ≤ -0.7	Moderate drought	2
-2.5<EDI ≤ -1.5	Severe drought	3
EDI ≤ -2.5	Extreme drought	4

Table 1: The classification of the Effective Drought Index (EDI).

Drought Subregions

Drought subregions were fixed in order to investigate the spatial pattern of droughts. By using the Statistical Package for Social Science (SPSS), hierarchical clustering analysis was carried out; the between-group linkage method with the Pearson correlation measure was used in this clustering procedure. For the clustering, time series from each station were calculated on the basis of a 35 years (1974-2008) monthly minimum EDI. As shown in Figure 2a, we used a clustering procedure for groups 61 to 1 and calculated the between-group correlations for each linkage. If the point where the coefficient between merged clusters increases markedly can be discerned, the clustering process can be stopped at this point. Thus, this study selected five as the appropriate number of clusters. Figure 2b shows the spatial distribution of five clusters: central district (cluster 1; A), southern district (cluster 2; B), eastern district (cluster 3; C), Jeju, Island (cluster 4; D), and Ulreung Island (cluster 5; E).

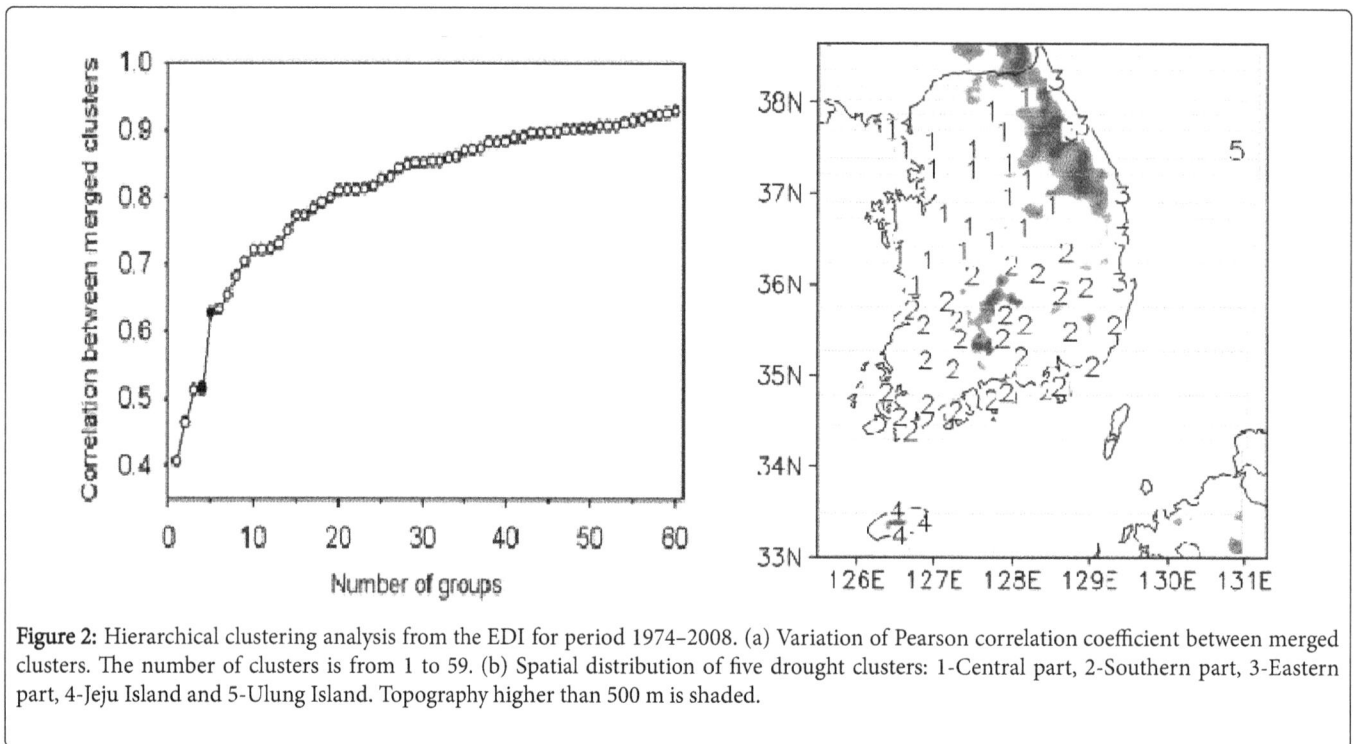

Figure 2: Hierarchical clustering analysis from the EDI for period 1974–2008. (a) Variation of Pearson correlation coefficient between merged clusters. The number of clusters is from 1 to 59. (b) Spatial distribution of five drought clusters: 1-Central part, 2-Southern part, 3-Eastern part, 4-Jeju Island and 5-Ulung Island. Topography higher than 500 m is shaded.

Experments on the Drought Predictability Using an Analog Method

Construction of drought code

The daily drought codes were generated for the intensity, duration, and spatial distribution of droughts acquired using the EDI. The temporal patterns of drought codes were obtained from the days of duration of the drought and the mean EDI at all stations. The spatial patterns were collected using the drought class on a cluster-by-cluster basis. The code type is ±Em Co AnBnCnDn, and the meaning of each type is described in Table 2. Cluster E was excluded from the analysis because the analysis of this cluster was difficult owing to presence of just one station in this cluster.

For example, the drought code on December 31, 2008, was -1.60 131 A2B3C2D1, which indicates that the mean EDI was -1.60, the duration of the droughts was 131 days, clusters A and C experienced moderate drought, cluster B experienced severe drought, and cluster D experienced weak drought. As precipitation data on all stations in Korea exist from 1973, the drought indices for each cluster from January 1, 1986, to December 31, 2008, were acquired using the mean EDI of existing stations.

Code	Meaning
± Em	Mean EDI of total station
Co	Drought duration (± Em ≤ 0)
AnBnCnDn	Drought class of cluster A, B, C, D

Table 2: The explanation of each part of the daily drought code.

Selection of subject cases

Figure 3 shows the time series of the annual minimum EDI in four subregions (A, B, C, and D) and the deviation of annual precipitation averaged over 60 stations for the past 30 years (1979–2008). The years 1988, 1994, and 2001 were selected as the subject years because they showed an anomalously low EDI and precipitation at the same time. For 1988, cluster B showed the minimum EDI, -2.10, on December 30. In 1994, cluster C showed the minimum EDI, -2.35, on October 10; in 2001, cluster A, showed the minimum EDI, -2.14, on June 11. The days on which the minimum EDI below -1.0 just before the annual minimum value was observed were selected as the subjected days. The dates August 19, 1988, August 12, 1994, and May 10, 2001, were selected as the subject dates since they met the abovementioned criterion.

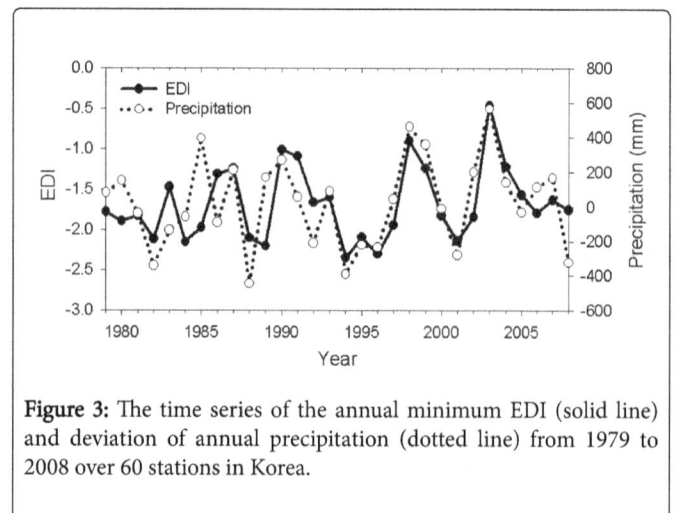

Figure 3: The time series of the annual minimum EDI (solid line) and deviation of annual precipitation (dotted line) from 1979 to 2008 over 60 stations in Korea.

Selection of Analogous Cases

We selected only one analogous case, one that is most similar to the subject case among several drought cases, by considering the daily drought codes using the following steps:

Step 1: Select cases similar to the subject case by considering the spatial distribution and seasonality of droughts.

Sort out the cases that have the same spatial drought code (AnBnCnDn).

Choose the cases that have similar seasonality (± 30 days from the subject case).

Step 2: If several similar cases are found in the same year, select only one case with the most similar intensity and duration.

2-1. Select the cases with the most similar mean EDI (Em).

2-2. Select the cases with the most similar duration of drought (Co).

Step 3: If similar cases are observed in several different years, select the year with the highest correlation coefficient by comparing the time series.

In accordance with the application of the analog method to the subject drought cases selected as described, August 19, 1988 (-0.93 023 A2B2C1D1) was similar to July 23, 1992 (-1.14 049 A2B2C1D1); August 12, 1994 (-0.96 076 A2B2C2D0) was similar to September 11, 1937 (-1.30 082 A2B2C2D0); and May 10, 2001 (-0.94 036 A2B1C2D1) was similar to June 7, 1996 (-0.95 034 A2B1C2D1).

Results of Experiment

The subject cases and their analogous cases were examined to determine the similarity in their progresses. Figure 4 shows the comparison between the times series of the subject cases (black line) and analogous cases (gray line) in the subregion where the minimum EDI value was obtained for the subject drought case.

For the case on August 19, 1988, both the subject case and the analogous case showed drought conditions for about 4 months (September–December 1988), although the intensity was stronger in the subject case. Two cases showed similar behavior for about 1 year (January 1989–January 1990). For the cases on August 12, 1994, and May 10, 2001, it was confirmed that the subject and analogous cases made similar progress for about 2 years (August 1994–May 1996) and 3 years (June 2001–June 2004), respectively.

After the periods in which the analogous cases showed similar behavior, no more significant similarity was detected in all three cases because they showed different progresses for more than 6 months.

The root mean square error (RMSE) was calculated to examine the change in prediction accuracy with time.

The RMSE is widely used to measure the difference between the actual case and the predicted case. It is calculated using the following formula:

$$RMSE = \sqrt{\frac{1}{n}\sum_{k=1}^{n}\left(p_k - o_k\right)^2} \qquad (4)$$

Where o denotes the actually observed value, p denotes the predicted value, and n denotes the period of prediction test. The larger the value of the RMSE, the larger is the difference between a subject case and an analogous case.

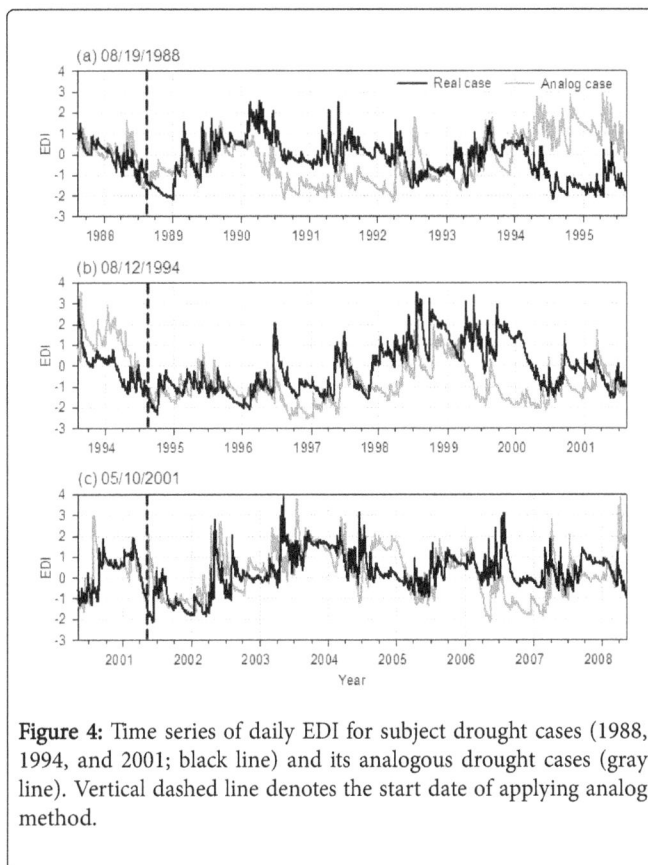

Figure 4: Time series of daily EDI for subject drought cases (1988, 1994, and 2001; black line) and its analogous drought cases (gray line). Vertical dashed line denotes the start date of applying analog method.

This study investigated the change in the RMSE on the basis of the monthly minimum EDI with increasing prediction period in annual units (Figure 5).

For the cases on August 19, 1988, and August 12, 1994, the prediction errors increased rapidly after 1 year and 2 years, respectively. In contrast, the case on May 10, 2001, showed low errors compared with other cases, and the errors increased as the prediction period increased.

We extended the application period to 76 years (1926–2008). A drought event was defined as the period of consecutive negative EDIs with the minimum EDI of below -1.

As a result, a total of 75 drought cases were selected, and the analog method was applied to them. In accordance with the mean RMSE of the 75 drought cases, it was found that the prediction errors increased rapidly when the prediction period increased to beyond 2 years (Figure 6).

In other words, the prediction using the analog method cannot be extended to beyond 2 years.

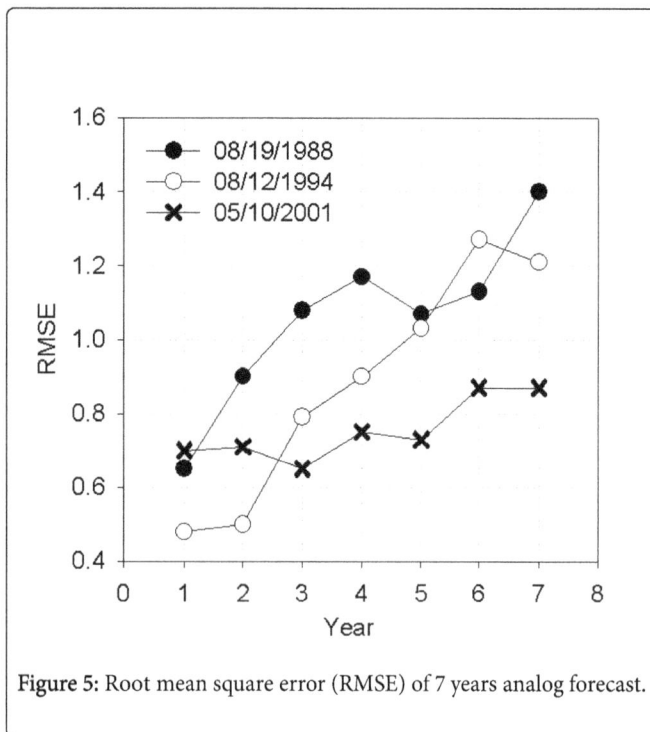

Figure 5: Root mean square error (RMSE) of 7 years analog forecast.

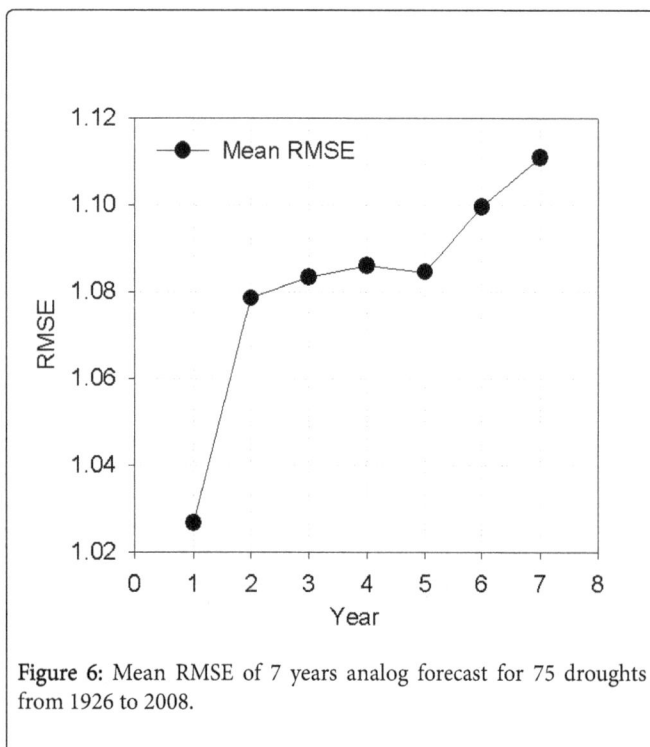

Figure 6: Mean RMSE of 7 years analog forecast for 75 droughts from 1926 to 2008.

Summary and Discussion

This study investigated the prediction of droughts using the analog method. Although the analog method has some limitations as a statistical approach, it is useful to effectively predict droughts for about 1 year. We selected analogous cases on the basis of the similarity of variation in the EDI time series. However, the prediction results may be improved when the analog method is applied in consideration of several factors such synoptic-scale atmospheric circulation and climatic elements highly correlated with droughts. Further, the analog method could be successfully used to predict days on which the analogous and subject cases were not similar. The drought subregions and the daily drought codes presented in this study could probably be used as the fundamental data for drought pattern classification or case selection in future studies. This study is expected to prompt research on the general causes of droughts. If we classify drought cases in accordance with the spatiotemporal patterns of droughts based on daily drought codes and identify the general causes and features from all cases, we might obtain more valuable predictions of droughts.

Conflict of Interest

The author confirms that this article has no conflict of interest.

Acknowledgement

This work was supported by the R&D Project of the Korea Meteorological Administration "Development and application of technology for weather forecast".

References

1.   Park CK, Schubert SD (1997) On the nature of the 1994 East Asian summer drought. J Clim 10: 1056-1070.

2.   Guan Z, Yamagata T (2003) The unusual summer of 1994 in East Asia: IOD teleconnections. Geophys Res Lett 30.

3.   Yoo SH, Ho CH, Yang S, Choi HJ, Jhun JG (2004) Influences of tropical western and extratropical Pacific SST on East and Southeast Asian climate in the summers of 1993-94. J Clim 17: 2673-2687.

4.   Trenberth KE, Branstator GW, Arkin PA (1988) Origins of the 1988 North American drought. Science 242: 1640-1645.

5.   Palmer TN, Brankovic C (1989) The 1988 United States drought linked to anomalous sea surface temperature. Nature 338: 54-57.

6.   Qu J, Sestak ML, Riebau AR, Smith LR, Ouren D (1994) A study of El Nino and Southern Oscillation (ENSO) impact on drought and wetness in the Western United States. 6th Con Clim Var pp: 101-104.

7.   Namias J (1991) Spring and summer 1988 drought over the contiguous United States-Causes and prediction. J Clim 4: 54-65.

8.   Yevjevich VM (1967) An objective approach to definitions and investigations of continental hydrologic droughts. Hydrology Papers p: 23.

9.   Lohani VK, Loganathan GV (1997) An early warning system for drought management using the Palmer drought index. Journal of American Water Resources Association 33: 1375-1386.

10. Cancelliere A, Mauro GD, Bonaccorso B, Rossi G (2007) Drought forecasting using the Standardized Precipitation Index. Water Resources Management 21: 801-819.

11. Kim TW, Valdes JB (2003) Nonlinear model for drought forecasting based on a conjunction of wavelet transforms and neural networks. Journal of Hydrological Engineering ASCE 8: 319-328.

12. Smakhtin SMV, Bagherzadeh CK (2006) Drought forecasting using artificial neural networks and time series of drought indices. Int J Climatol 27: 2103-2111.

13. Chung CH, Salas JD (2000) Drought occurrence probabilities and risks of dependent hydrological processes. Journal of Hydrological Engineering ASCE 5: 259-268.

14. Mishra AK, Desai VR (2005) Drought forecasting using stochastic models. Stochastic Environmental Research Risk Assessment 19: 326-339.

15. Kim S, Park CK, Kim MK (2005) The Regime Shift of the Northern Hemispheric Circulation Responsible for the Spring Drought in Korea. Asia Pac J Atmos Sci 41: 571-585.

16. Radinovic D (1975) An analogue method for weather forecasting using the 500/1000 mb relative topography. Monthly Weather Review 103: 639-649.

17. Christensen RA, Eilbert RF, Lindgren OH, Rans LL (1981) Successful hydrologic forecasting for California using an information theoretic model. J Appl Meteorol 20: 706-713.

18. Bergen RE, Harnack RP (1982) Long-range temperature prediction using a simple analog approach. Monthly Weather Review 110: 1083-1099.

19. Gutzler DS, Shukla J (1984) Analogs in the wintertime 500mb height field. Journal of the Atmospheric Sciences 41: 177-189.

20. Park WH, Lee JG (2003) Long-range Forecast using Analog/Anti-analog Method with GDAPS Forecasts. Asia-Pac J Atmos Sci 39: 491-501.

21. Xavier PK, Goswami BN (2007) Analog method for real time forecasting of summer monsoon subseasonal variability. Monthly Weather Review 135: 4149-4160.

22. Byun HR, Wilhite DH (1999) Objective quantification of drought severity and duration. J Clim 12: 2747-2756.

23. Kim DW, Byun HR, Choi KS (2009) Evaluation, Modification, and Application of the Effective Drought Index to 200-Year Drought Climatology of Seoul, Korea. J Hydrol 378: 1-12.

24. Morid S, Smakhtin V, Moghaddasi M (2006) Comparison of seven meteorological indices for drought monitoring in Iran. Int J Climatol 26: 971-985.

25. Moon YS (1990) Division of precipitation regions in Korea through the cluster analysis. J Korean Meteor Soc 26: 203-215.

26. Qian WH, Qin A (2008) Precipitation division and climate shift in China from 1960 to 2000. Theoretical and Applied Climatology 93: 1-17.

27. Byun HR, Lee DK (2002) Defining three rainy seasons and the hydrological summer monsoon in Korea using available water resources index. J Meteorol Soc Jpn 80: 33-44.

28. Yamaguchi Y, Shinoda M (2002) Soil moisture modeling based on multiyear observations in the Sahel. J Appl Meteorol 41: 1140-1146.

29. Smakhtin VU, Hughes DA (2007) Automated estimation and analyses of meteorological drought characteristics from monthly rainfall data. Environ Model Software 22: 880-890.

30. Akhtari R, Morid S, Mahdian MH, Smakhin V (2009) Assessment of areal interpolation methods for spatial analysis of SPI and EDI drought indices. Int J Climatol 29: 135-145.

13

Dynamical downscaling of temperature variability over Tunisia: evaluation a 21-year-long simulation performed with the WRF model

Bilel Fathalli[1*], Benjamin Pohl[2], Thierry Castel[2] and Mohamed Jomâa Safi[1]

[1]Université de Tunis El Manar, Ecole Nationale d'Ingénieurs de Tunis, Tunis, Tunisie

[2]Centre de Recherches de Climatologie, Biogéosciences, CNRS / université de Bourgogne Franche-Comté, Dijon, France

*Corresponding author: BilelFathalli, Université de Tunis El Manar, Ecole Nationale d'Ingénieurs de Tunis, Tunisie, E-mail: bilelfathalli@yahoo.fr

Abstract

This study evaluates the capabilities of the Weather Research and Forecasting (WRF) model to reproduce the space-time variability of near-surface air temperature over Tunisia. Downscaling is based on two nested domains with a first domain covering the Mediterranean Basin and forced by 21 years of ERA-Interim reanalysis (1991-2011), and a second domain (12 km spatial resolution) centered on Tunisia. Analyses and comparisons are focused on daily average (T_{avg}), minimum (T_{min}) and maximum (T_{max}) near-surface air temperatures and are carried out at the annual and seasonal timescales. WRF results are assessed against various climatological products (ERA-Interim, E-OBS and a local network of 18 surface weather stations).

The model correctly reproduces the spatial patterns of temperature being significantly superimposed with local topographic features. However, it broadly tends to underestimate temperatures especially in winter. Temporal variability of temperature is also properly reproduced by the model although systematic cold biases mostly concerning T_{max}, reproduced throughout the whole simulation period, and prevailing during the winter months. Comparisons also suggest that the WRF errors are not rooted in the driving model but could be probably linked to deficiencies in the model parameterizations of diurnal/nocturnal physical processes that largely impact T_{max} / T_{min}.

Keywords: WRF; Near-surface air temperature; Tunisia; Downscaling

Introduction

Downscaling is a regionalization approach for obtaining high-resolution climate or climate change informations from relatively coarse resolution global models [1]. This strategy allows to account for more realistic surface features (such as topography, land-use complexity and heterogeneity) and small scale atmospheric processes (e.g., convective systems) which are not properly represented or resolved by General Circulation Models (GCMs) owing to their coarse resolution (typically run at 150-300 km [2]).

This is of primary importance for decision makers demanding climate informations at fine spatial scales (about 10-50 km) in order to address climate change risks and potential impacts, and subsequently to implement adaptive measures to reduce or to avoid these risks [3,4]. Two widely used downscaling techniques can derive climate informations initially provided by GCMs at the needed regional or local scales [5,6]. While statistical downscaling techniques establish statistical relationships between variables at different spatial scales [7], the dynamical downscaling approach avoids relying on empirical observed relationships [8] and is based on coupling large scale climate dynamics and local climate and hydrological futures by using regional climate models (RCMs) [2].

As they use finer surface parameters and more elaborated physical parametrization schemes, RCMs can potentially improve the simulation ability for regional climates over various regions of the world and thus their use has been steadily increased over the last decades. In this paper, we aim at evaluating the capability of a non-hydrostatic limited area model, namely the Weather Research and Forecasting (WRF) model, to dynamically downscale near-surface air temperature (temperature at 2m) over Tunisia. WRF has been largely used by the scientific community and showed good skills in reproducing climate variability [9-12], justifying its use as an RCM in this study.

The need to examine the Tunisian present space-time variability of surface temperature at high-resolution (about 12 km) is motivated by the complex geomorphology of Tunisia with high mountains in the north, land depressions in the west central and desert in the south implying large climate gradients, and also by the extreme vulnerability of Tunisia to climate change given its location within the Mediterranean Basin (considered as one the "Hotspots" projected to encounter major climatic changes in the twenty-first century as a result of the global warming).

Lying in a contact zone marking the transition between the temperate humid Mediterranean climate and the dry Saharan climate, the Tunisian climate varies from extremely arid in the south with extreme warm temperatures, significant interannual variability in rainfall and severe drought episodes to a Mediterranean climate in the north. Meteorological records, derived climate indices and satellite products show that the Tunisian climate is getting hotter, drier and more variable. For instance, average temperature rose by about 1.4°C in the twentieth with the most rapid warming rate since the 1970's, particularly in summer. The northern and southern regions are experiencing the greatest warming rates [13].

High resolution data provided by RCMs is then strongly needed to help understand the current Tunisian climate variability and its impacts on specific sectors (e.g., agriculture, water resources, disease incidence, etc.).

In this study, we focus on the near-surface air temperature regarding its essential role in controlling large variety of environmental processes involved in the water, matter and energy cycles [14]. The main futures of the near-surface temperature space-time variability are described as well as the model skills and errors.

Data and methods

Data

To assess the performance of the regional model to downscale near-surface air temperature variability over Tunisia, we first compare WRF to its driving model. The European Centre for Medium-Range Weather Forecasts (ECMWF) ERA-Interim [15] (ERA-I, hereafter) reanalysis is used to force the WRF model and also to evaluate the WRF-simulated temperature. ERA-I covers the period from 1979 onwards. Its data assimilation and modeling system is based on the Integrated Forecast System (IFS Cy31r2) model. This system includes a 4-dimensional variational analysis (4D-Var) with a 12-hour analysis window. ERA-I spatial resolution is approximately 80 km (T255 spectral truncation) on 60 vertical levels from the surface up to 0.1 hPa.

The other used gridded product is the European (ECA&D) land-only E-OBS daily gridded dataset [16], provided by the European Climate Assessment and Dataset project (ECA&D) and covers the period 1950 to present. E-OBS has been developed as part of the ENSEMBLES European project and is based on daily observations from a gradually expanding network of over 3500 stations [17] interpolated onto a regular grid using thin-plate splines and kriging [16]. ERA-I and E-OBS are linearly interpolated onto WRF grid for direct comparison purposes.

Simulated surface temperature is also compared to local surface observations of daily average temperature belonging to the Tunisian National Institute of Meteorology and obtained from NOAA's National Climatic Data Center (NCDC). Available observations were checked for continuity, retaining only 18 stations for model validation by using the nearest grid point of the model to the observation. The low density of surface stations does not permit in fact interpolation of the data to a regular grid.

Methods

The model used in this work is the Weather Research and Forecasting/Advanced Research WRF (ARW) model [18] in its version 3.3.1 (WRF hereafter). WRF is a next-generation limited area, fully compressible, non hydrostatic with terrain following eta coordinate mesoscale modeling system. It has a rapidly growing user community and has been designed to serve both operational forecasting and climate research purposes. Our WRF simulation was setup with two domains, one at 60 km and a second at 12 km horizontal grid spacing, using the two-way nesting technique. The coarse (or parent) domain (120 x 60 grid points) extends over the Mediterranean Basin (Southern Europe and North Africa) while the high-resolution nest (46 x 71 grid points) covers the most part of Tunisia (Figure 1).

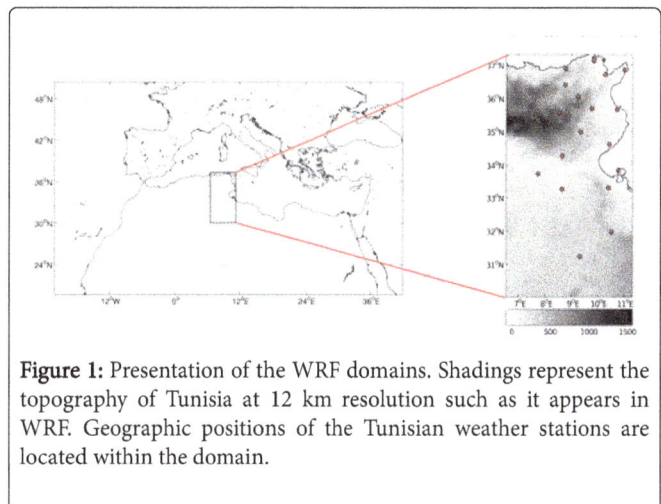

Figure 1: Presentation of the WRF domains. Shadings represent the topography of Tunisia at 12 km resolution such as it appears in WRF. Geographic positions of the Tunisian weather stations are located within the domain.

Both domains have 28 vertical levels between the surface and 50 hPa. The initial and lateral boundary conditions for the parent domain are provided by ERA-I. Lateral forcings are provided every 6 h at a 0.75° horizontal resolution and 19 pressure levels. A buffer zone composed of five grid points (1-point specified zone and 4-point relaxation zone) on the periphery of the domain is chosen and allows for a smooth transition between the model's prognostic variables and the driving reanalysis. Sea Surface Temperature (SST) fields are prescribed every 24 h by linear interpolation of monthly ERA-I SST.

The physical parameterizations chosen for the two domains include the WRF Single Moment 6-class (WSM6 [19]) cloud microphysics, the Yonsei University parameterization of the Planetary Boundary Layer (PBL, [20]), the MM5 similarity surface layer scheme [21], the Rapid Radiative Transfer Model (RRTP) scheme [22] for longwave and the Dudhia scheme for shortwave radiation [23]. The Kain-Fritschscheme [24] is used to parameterize atmospheric convection. Over the continent, WRF is coupled with Noah LSM 4-layer soil temperature, soil and canopy moistures model [25]. Surface data is derived from the 20-category Moderate Resolution Imaging Spectroradiometer (MODIS) land use data with inland water bodies [26].

The WRF run started at 0000 UTC 1 January 1991 and ended at 2400 UTC 31 December 2011, with the first year being discarded for spin-up. Temperature is archived each hour.

Analyses and comparisons are focused on daily average, minimum and maximum near-surface air temperatures (respectively T_{avg}, T_{min} and T_{max}) and are carried out at the annual and seasonal. Different verification metrics (Mean Error (ME), Root Mean Square (RMSE) and Pearson correlation coefficient (R)) are used to evaluate the WRF simulation. They are defined as:

$$ME = \frac{1}{N}\sum_{i=1}^{N}(S_i - O_i) \qquad (1)$$

$$RMSE = \sqrt{\frac{1}{N}\sum_{i=1}^{N}(S_i - O_i)^2} \qquad (2)$$

$$R = \frac{\sum(S_i - \bar{S}).(O_i - \bar{O})}{\sqrt{\sum(S_i - \bar{S})^2.\sum(O_i - \bar{O})^2}} \qquad (3)$$

(S: simulated Temperature, O: observed temperature and N: number of grid-points)

ERA-I is however compared to the same references (E-OBS and TN_OBS) in order to know if WRF outperforms or not its driving model (i.e. whether the model improves or not the ERA-I driving temperature). Error measures are computed for common WRF, ERA-I and E-OBS landmasses (i.e after applying a mask to exclude WRF and ERA-I grid points within the Mediterranean Sea), or for the nearest grid points of gridded datasets to the Tunisian weather stations.

To better understand temperature variability and also the WRF biases, we separately examine the daily maximum and minimum near-surface temperatures (T_{max} and T_{min}). The simulated T_{min} and T_{max} obtained from the WRF hourly temperature output are then compared to the same gridded datasets.

Results and discussion

Spatial variability of temperature

Daily average temperature (Tavg): Long-term (1992-2011) climatologies of annual mean T_{avg} simulated by WRF and derived from ERA-I and E-OBS are computed on the basis of the annual means and are shown in Figure 2a. Overall, the spatial distribution of T_{avg} is heterogeneous, significantly superimposed with local orographic features and showing a clear altitudinal gradient.

Indeed, the minimum simulated (~ 12°C) and observed (~ 14°C according to E-OBS) T_{avg} are observed along the main Tunisian mountain ranges, particularly along the Tunisian Dorsale (a southwest-northeast trending mountain range that mostly constitutes the eastern end of the Atlas Mountain and runs across Tunisia from the Algerian border in the west to the Cap Bon Peninsula in the east) and secondarily along the low sandstone Dahar mountain chain bisecting the south of Tunisia. Maximum simulated T_{avg} (~ 23°C) occurs in the west central of Tunisia particularly in the Tunisian salt depressions (the Chotts region), and south of the country on the margins of the Sahara

(Grand Erg Oriental) and is in good agreement with the comparing datasets.

Figure 2b shows the spatial distribution of the 20-year average annual T_{avg} biases. Near-surface temperature is slightly overestimated by WRF over the north of the country. Nonetheless, the model underestimates temperature mainly along the aforementioned Tunisian mountain ranges where the cold bias can reach ~ -4°C eventually denoting strong orographic forcing, and less intensively elsewhere. Similar findings are reported by [11] for the North Western Mediterranean Basin where WRF (driven by ERA-40) particularly underestimates surface temperature along the Pyrenees. When compared to E-OBS, ERA-I slightly overestimates/underestimates air temperature in the center of Tunisia/south of the Tunisian Dorsale. Mean annual errors are summarized in Table 1.

Pearson's correlation coefficients quantifying the spatial matching between WRF and the comparing datasets are also listed in same table. Results show that ERA-I roughly performs better than WRF as indicated by the lower values of ME and RMSE. This result is somewhat expected since various surface observations are assimilated within the ERA-I reanalysis system. The high correlation coefficients between WRF and the other datasets (e.g., R = 0.97 with regard to E-OBS) denote the good ability of the model in correctly reproducing temperature spatial variability.

Spatial distributions of seasonal mean T_{avg} climatologies from WRF, ERA-I and E-OBS are displayed in Figure 3. Seasons are defined as winter [December–January–February (DJF)], spring (MAM), summer (JJA), and autumn (SON). The atitudinal temperature gradient prevails throughout the year. Minimum simulated T_{avg} (~ 3°C) is in fact recorded in DJF over the Tunisian Dorsale, while the maximum (~ 34°C) is registered in JJA over the land depression of "Chott el-Gharsa" which constitutes the lowest altitudes of Tunisia (~ -17 m).

Model	Ref	Annual			DJF			MAM			JJA			SON		
		ME (°C)	RMSE (°C)	R	ME (°C)	RMSE (°C)	R	ME (°C)	RMSE (°C)	R	ME (°C)	RMSE (°C)	R	ME (°C)	RMSE (°C)	R
WRF	ERA-I	-1.02	1.28	0.95	-1.46	1.62	0.91	-0.97	1.2	0.97	-0.83	1.37	0.94	-1.08	1.32	0.93
	E-OBS	-1.12	1.28	0.97	-2.15	2.36	0.84	-0.93	1.08	0.98	-0.29	0.9	0.96	-1.35	1.53	0.95
	TN_OBS	-1.32	1.68	0.89	-1.8	1.95	0.81	-1.48	1.1	0.93	-1.28	1.84	0.9	-1.2	1.65	0.87
ERA-I	E-OBS	-0.1	0.57	0.97	-0.69	0.97	0.94	0.04	0.6	0.98	0.58	1.1	0.95	-0.4	0.63	0.97
	TN_OBS	-0.82	1.43	0.89	-0.6	1.22	0.79	-0.83	1.5	0.88	-1.3	1.86	0.82	-0.8	1.38	0.85

Table 1: Annual and seasonal T_{avg} errors.

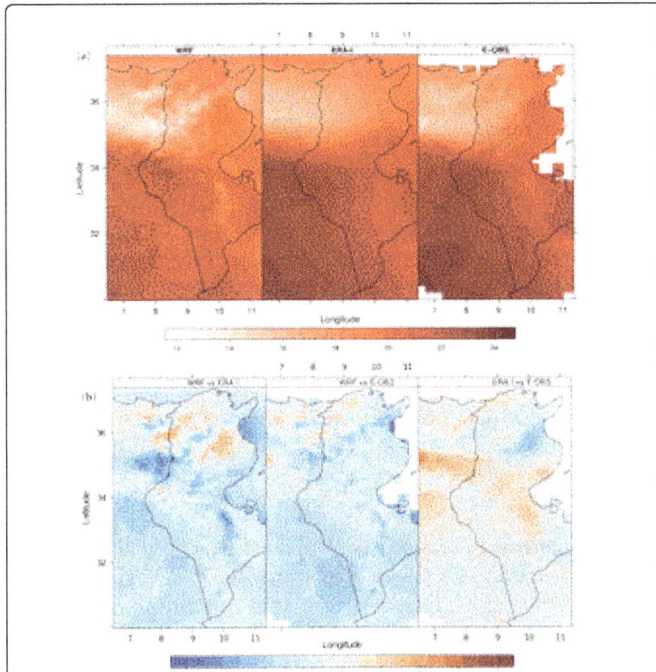

Figure 2: (a) Annual mean T_{avg} (°C) according to WRF, ERA-I and E-OBS and over the period 1992-2011. (b) Annual mean WRF T_{avg} biases (°C) against ERA-I (left-hand column) and E-OBS (middle column), annual mean ERA-I biases (°C) against E-OBS (right-hand column).

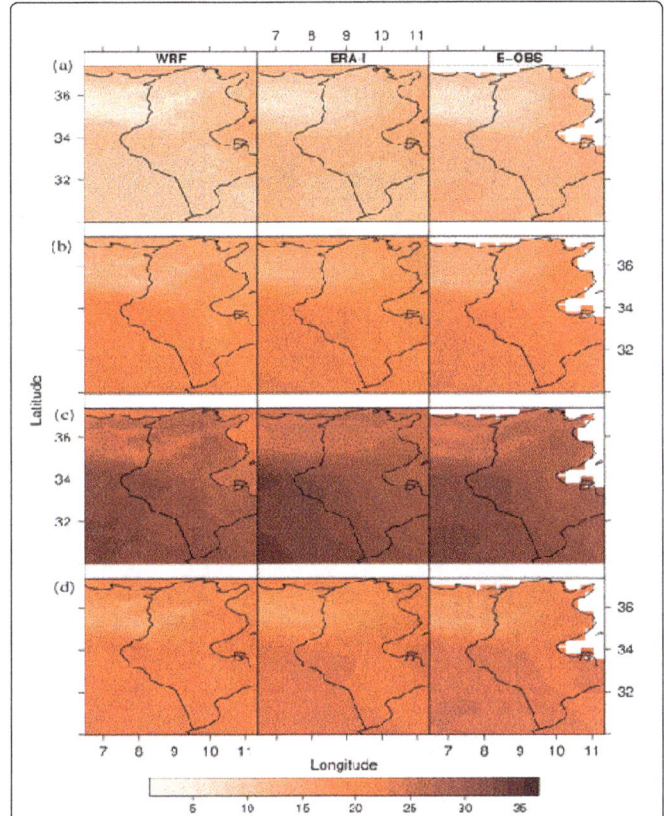

Figure 3: Seasonal mean T_{avg}(°C) according to WRF, ERA-I and E-OBS, for DJF (a), MAM (b), JJA (c) and SON (d).

Seasonal WRF biases (Figure 4) reveal a moderate seasonal dependency. From one season to another, the spatial distribution of the biases remains roughly similar, the main differences concerning the magnitude of the errors. This magnitude itself depends on seasons, particularly for JJA when the hot biases north of Tunisia are strongest. ERA-I, compared to E-OBS, largely overestimates JJA temperature (hot bias reaching 4°C) in the center and south of Tunisia. Seasonal errors are listed in Table 1, showing that the largest WRF errors occur in DJF. The model is however closer to ERA-I then to the observational datasets in DJF (the lowest/highest spatial correlations/RMSE against observation are indeed obtained in winter).

Contrary results are obtained in JJA when WRF is, on the one hand, much closer to observation (R = 0.96 and RMSE = 0.9°C with regard to E-OBS) and slightly outperforms its driving model, on the other hand. This WRF seasonal behavior can be explained by the synoptic patterns. In DJF large scale forcing is indeed prevailing (intense prevailing westerlies), temperature is then constrained by large scale forcing (ERA-I). The connection between wintertime Tunisian climate and large-scale patterns is particularly verified for observed precipitation [27]. Contrariwise, the synoptic forcing is weak in summer, anti-cyclonic conditions are in fact predominant over the Mediterranean Basin [28] and the model then could be more responding to strong local (small-scale) forcings and thus better capturing details of the regional climate.

Model	Ref	Annual			DJF			MAM			JJA			SON		
		ME (°C)	RMSE (°C)	R	ME (°C)	RMSE (°C)	R	ME (°C)	RMSE (°C)	R	ME (°C)	RMSE (°C)	R	ME (°C)	RMSE (°C)	R
WRF	ERA-I	-0.97	1.53	0.88	-0.88	1.39	0.8	-0.82	1.17	0.95	-1.08	1.9	0.9	-1.24	1.93	0.79
	E-OBS	-0.54	1.01	0.91	-0.92	1.45	0.77	-0.46	0.82	0.97	-0.18	1.17	0.9	-0.73	1.45	0.83
ERA-I	E-OBS	0.43	0.83	0.95	-0.04	0.78	0.89	0.36	0.74	0.97	0.9	1.41	0.95	0.51	0.89	0.95

Table 2: Annual and seasonal T_{min} errors.

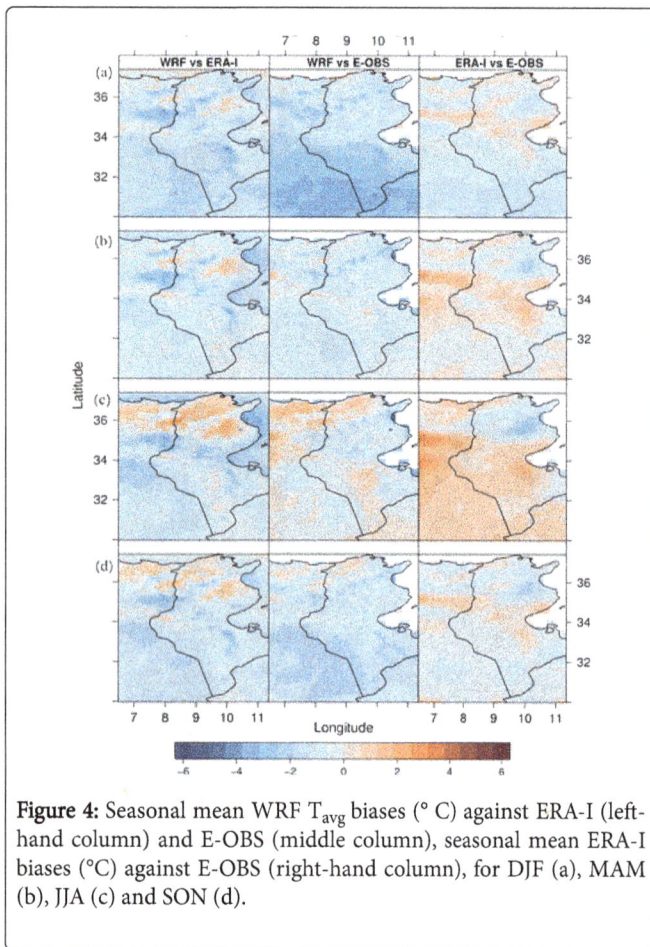

Figure 4: Seasonal mean WRF T_{avg} biases (° C) against ERA-I (left-hand column) and E-OBS (middle column), seasonal mean ERA-I biases (°C) against E-OBS (right-hand column), for DJF (a), MAM (b), JJA (c) and SON (d).

Daily maximum and minimum temperatures

Annual and seasonal spatial patterns of simulated, analyzed and observed 20-years average T_{min} and T_{max} (not shown) over landmasses are in agreement with the broad altitudinal pattern of the Tunisian mean temperatures.

A seasonal sea-land thermal contrast can however be noticed for both T_{min} and T_{max}. Indeed, T_{min} (ranging in WRF from ~1°C to ~26°C) is higher over the sea (particularly over the Gulf of Gabeseast of Tunisia) then over landmasses in all seasons, excepting in JJA when the maximum occurs in the Chotts depression.

The opposite pattern is observed for T_{max} (ranging in WRF from ~7°C to ~41°C) being lower over the sea in all seasons, excepting in DJF when the minimum occurs along the Tunisian Dorsale. Annual correlation coefficients between WRF and the comparing datasets (listed in Tables 2 and 3) are higher for T_{max} than for T_{min}. Similar findings are reported by other authors who used WRF to downscale temperature over Portugal [9] and over the US North pacific [29].

Spatial patterns of seasonal WRF and ERA-I T_{min} and T_{max} mean biases with regard to E-OBS are respectively shown in Figures 5 and 6. These patterns are broadly consistent with those of T_{avg} previously described.

The largest ERA-I errors concern T_{min} and are occurring in summer. WRF particularly underestimates T_{max} in DJF when the cold biases are prevalent and can locally exceed -10°C. Seasonal errors (listed in Tables 2 and 3) show indeed that the largest WRF errors are obtained in DJF and mostly concern T_{max} (RMSE ~ 2.6°C). WRF T_{min} shows a slight improvement on ERA-I only in summer. However, ERA-I always performs better than WRF (particularly in DJF and less in JJA) when reproducing the T_{max} spatial variability.

Model	Ref	Annual			DJF			MAM			JJA			SON		
		ME (°C)	RMSE (°C)	R	ME (°C)	RMSE (°C)	R	ME (°C)	RMSE (°C)	R	ME (°C)	RMSE (°C)	R	ME (°C)	RMSE (°C)	R
WRF	ERA-I	-1.5	1.86	0.95	-1.92	2.15	0.89	-1.86	1.94	0.96	-1.2	1.87	0.92	-1.25	1.63	0.94
	E-OBS	-1.54	1.77	0.95	-2.47	2.57	0.94	-1.56	1.84	0.96	-0.81	1.59	0.92	-1.56	1.73	0.96
ERA-I	E-OBS	-0.04	0.96	0.95	-0.56	0.79	0.97	0.3	1.12	0.96	0.32	1.54	0.92	-0.31	0.91	0.96

Table 3: Annual and seasonal T_{max} errors.

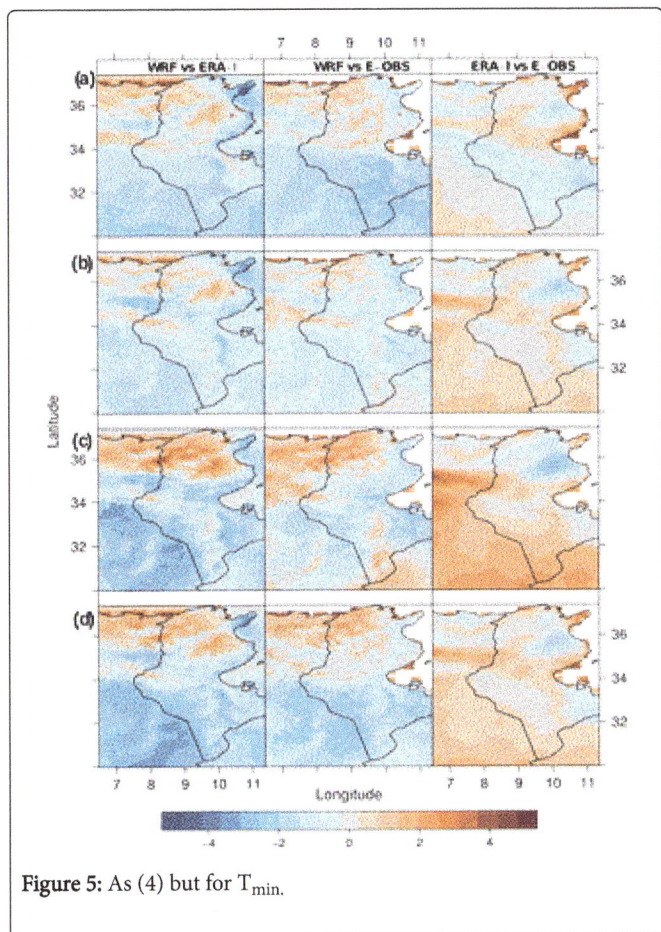

Figure 5: As (4) but for T_{min}.

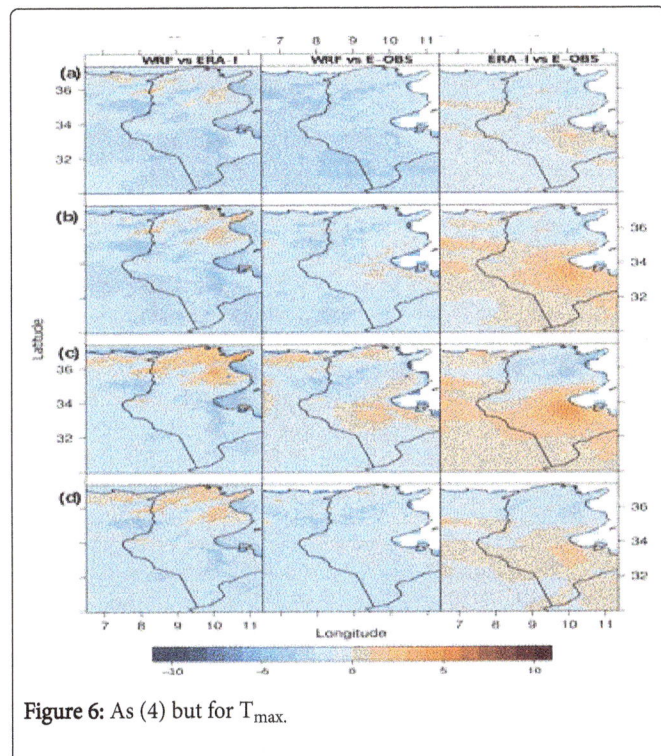

Figure 6: As (4) but for T_{max}.

Temporal variability

Figure 7a shows the interannual variability of WRF, ERA-I and E-OBS T_{avg} spatially averaged over the landmass grid-points of the domain. A good agreement between ERA-I and E-OBS is noteworthy supporting the relevance of this reanalysis product. Concerning WRF errors, in spite of a systematic cold bias prevailing throughout the whole simulation period (reaching a maximum of -1.36°C against E-OBS), the year-to-year variations of T_{avg} is well reproduced by the model. Interannual correlation coefficients are indeed very high against ERA-I (R = 0.95) and E-OBS (R = 0.93).

Compared to the Tunisian stations (Figure 7b), all gridded datasets underestimate temperature. The maximum bias is recorded in WRF and ranges from -1 to -1.7°C.

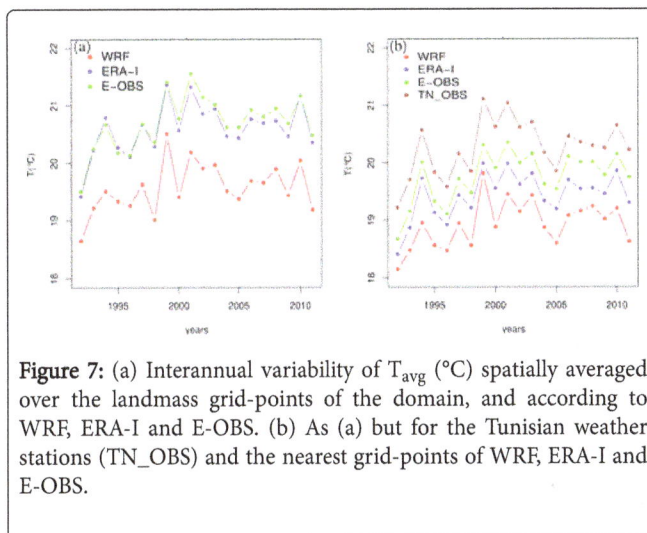

Figure 7: (a) Interannual variability of T_{avg} (°C) spatially averaged over the landmass grid-points of the domain, and according to WRF, ERA-I and E-OBS. (b) As (a) but for the Tunisian weather stations (TN_OBS) and the nearest grid-points of WRF, ERA-I and E-OBS.

Long-term mean annual cycles of T_{avg} in WRF and the comparing datasets are plotted in Figure 8a for the landmasses of the whole domain and in Figure 8b for TN_OBS and the nearest grid points of gridded datasets to the local weather stations.

The strong seasonality of the Tunisian temperature and its unimodal distribution are well captured by the model. The minimum and maximum simulated T_{avg} are respectively recorded in January (~ 9°C) and August (~ 30°C) and are in quite agreement with the other datasets. However, WRF underestimates spatially averaged T_{avg} throughout the whole annual cycle but mainly in winter.

In fact, WRF shows a quite strong cold bias with respect to E-OBS in DJF (especially in January when the cold bias exceeds -2°C) unlike summer when the model is notably close to E-OBS so the cold bias barely exceeds -0.2°C. Compared to TN_OBS, WRF also shows a systematic cold bias ranging from -1.8°C in February to -0.8°C in September.

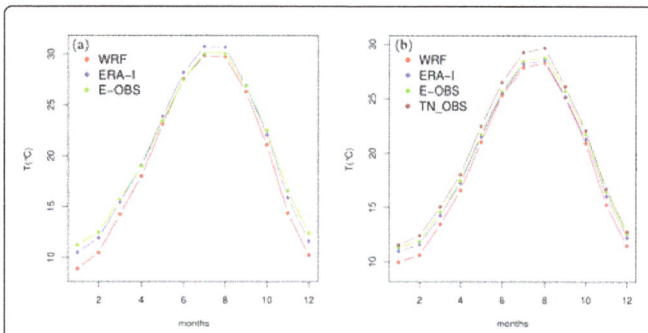

Figure 8: (a) T_{avg} (°C) mean annual cycle spatially averaged over the landmass grid-points of the domain, and according to WRF, ERA-I and E-OBS. (b) As (a) but for the Tunisian weather stations (TN_OBS) and the nearest grid-points of WRF, ERA-I and E-OBS.

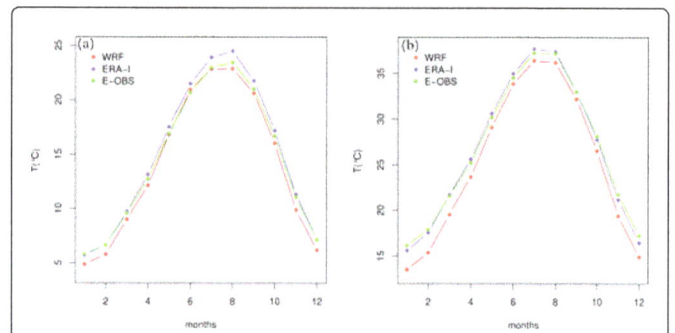

Figure 10: (a) As (8a) but for T_{min}. (b) As (8a) but for T_{max}.

Interannual variabilities of the spatially averaged T_{min} and T_{max} are shown in Figures 9a and 9b. WRF systematically underestimates both T_{min} and T_{max} throughout the whole simulation period. Nevertheless, the largest errors concern T_{max}, thereby ranging from -1.9°C to -1°C with respect to E-OBS. ERA-I systematically overestimates T_{min} unlike T_{max} being in perfect agreement with the observation (R = 0.98).

Mean annual cycles of T_{min} and T_{max} are plotted in Figures 10a and 10b. Consistently with our previous findings, the main WRF errors concern T_{max} and occur in wintertime, particularly in January when the systematic cold bias against E-OBS reaches its maximum (~ -3°C).

The model T_{min} is slightly underestimated against E-OBS except in June when WRF marginally overestimates temperature (bias ~ 0.2°C). Furthermore, WRF T_{min} is rather in better agreement with E-OBS in JJA pointing out again the important role of regional small-scale processes being resolved by the regional model in summer. Contrariwise, the main ERA-I errors concern T_{min} being overestimated in summer (a hot bias reaching ~ 1°C in August).

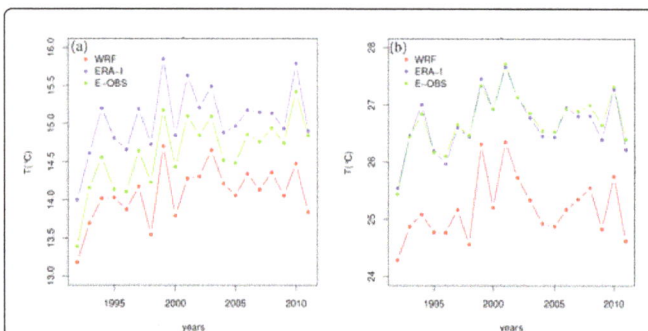

Figure 9: (a) As (7a) but for T_{min}. (b) As (7a) but for T_{max}.

Our results may suggest that the WRF errors are not inherited from its driving model but are rather internal to the model itself. Indeed, underestimation of the daily average temperature seems to be a consequence of systematic underestimation of T_{max} (mainly in winter) and less of T_{min}, which is not verified in ERA-I. Underestimation of temperature then could be linked to deficiencies in the model parameterizations of diurnal/nocturnal physics that particularly control T_{max} / T_{min} [29]. The surface radiation budget (latent and heat fluxes at the surface), the PBL physics as well as the downward/outgoing shortwave / longwave radiations at the top of the atmosphere are physical processes that strongly impact T_{max} / T_{min}. Evaluating the capability of WRF in simulating these processes is beyond the scope of this study.

Conclusion

The present study contributes for the first time to the evaluation of a near-surface air temperature simulation over Tunisia at 12 km spatial resolution. A current state-of-the-art non-hydrostatic regional (WRF) model driven by the most recent ECMWF climate reanalysis (ERA-Interim) has been used to downscale temperature spatial and temporal variabilities over a 21-year long period (1991-2011). Analyses of the simulation are carried out at the annual and seasonal timescales and the WRF skills in simulating daily average (T_{avg}), maximum (T_{max}) and minimum (T_{min}) temperatures are assessed against gridded climatological datasets (ERA-I and E-OBS) and against high-quality in-situ data from 18 Tunisian weather stations. Overall, the model proves good performances in reproducing the spatial variability of mean annual and seasonal temperatures showing significant topographic signatures. Indeed, WRF correctly captures the minimum temperatures always recorded along the main Tunisian mountain ranges (particularly the Tunisian Dorsale), as well as the maximum mainly occurring in the salt chott depressions. Temporal variability of temperature is satisfactorily reproduced by WRF, despite cold biases obtained for T_{avg}, T_{min} and T_{max}, and systematically reproduced throughout the whole simulation period or the mean annual cycle. WRF tends indeed to underestimate temperature especially in DJF when the strongest cold biases are obtained and are mostly linked to an underestimation of T_{max}. Results also show that the model skills depend on seasons. In fact, WRF is closer to its driving model in winter then to observation, unlike summer when WRF is notably in good agreement with E-OBS and it exclusively (but slightly) outperforms ERA-I when simulating T_{avg} and T_{min}. This model seasonal behaviour could be linked to the Mediterranean (generally the mid-latitudes) large-scale atmospheric patterns. In winter, the synoptic forcing is active (intense westerlies) and the model then could be constrained by the ERA-I forcing. Contrariwise in summer, the low synoptic forcing could enhance the model freedom in developing its own climate thus allowing it to better capture details of the regional climate. ERA-I errors are quite different from WRF since they mainly concern T_{min} being overestimated in JJA, then suggesting that the WRF errors are not inherited from its driving model but could rather be linked to deficiencies in the regional model, especially in the model parameterizations of physical processes (surface radiation budget, PBL physics and downward/outgoing shortwave / longwave radiations at the top of the atmosphere) that have a large impact on daytime / nighttime temperatures. A further study focusing on evaluating the capabilities of WRF in simulating these physical processes and also investigating other physical options seems to be worthy in order to better understand the origin of the WRF errors. Finally, this study presents a high-quality climatological dataset which can be used as reference for climate impact studies.

Acknowledgments

We wish to thank anonymous reviewers for their constructive comments that helped improve the paper. WRF was provided by the University Corporation for Atmospheric Research website (for more information see http://www.mmm.ucar.edu/wrf/users/download/). ERA-Interim data were provided by the ECMWF. Calculations were performed using HPC resources from DSI-CCUB, université de Bourgogne. We acknowledge the E-OBS dataset from the EU-FP6 project ENSEMBLES (http://ensembles-eu.metoffice.com), and the data providers in the ECA&D project (http://www.ecad.eu) and the NCDC (http://www.ncdc.noaa.gov/) for making available the Tunisian weather station data.

References

1. Giorgi F, Jones C, Asrar GR (2009) Addressing climate information needs at the regional level: the CORDEX framework. WMO Bull 5: 175-183.

2. IPCC (2007) Climate Change 2007, the Fourth Assessment Report AR4 of the United Nations Intergovernmental Panel on Climate Change IPCC, Working Group I, The Physical Science Basis of Climate Change.

3. Chen J, Brissette FP, Leconte R (2012) Coupling statistical and dynamical methods for spatial downscaling of precipitation. Clim Change 114: 509-526.

4. Brown C, Greene AM, Block PJ, Giannini A (2008) Review of downscaling methodologies for Africa climate applications. IRI Technical Report 08-05.

5. Hay LE, Clark MP (2003) Use of statistically and dynamically downscaled atmospheric model output for hydrologic simulations in three mountainous basins in the western United States. J Hydrol 282: 56-75.

6. Fowler HJ, Blenkinsop S, Tebaldi C (2007) Linking climate change modelling to impacts studies: recent advances in downscaling techniques for hydrological modeling. Int J Climatol 27: 1547-1578.

7. Teutschbein C, Wetterhall F, Seibert J (2011) Evaluation of different downscaling techniques for hydrological climate-change impact studies at the catchment scale. ClimDyn 37: 2087-2105.

8. Caldwell P, Chin HNS, Bader DC, Bala G (2009) Evaluation of a WRF dynamical downscaling simulation over California. Clim. Change 95: 499-521.

9. Soares PMM, Cardoso RM, Miranda PMA, de Medeiros J, Belo- Pereira M, et al. (2012) WRF high resolution dynamical downscaling of ERA-Interim for Portugal. ClimDyn 39: 2497-2522.

10. Heikkilä U, Sandvick A, Sorteberg A (2010) Dynamical downscaling of ERA-40 in complex terrain using the WRF regional climate model. ClimDyn 37: 1551-1564.

11. Gonçalves M, Barrera-Escoda A, Guerreiro D, Baldasano JM, Cunillera J (2014) Seasonal to yearly assessment of temperature and precipitation trends in the North Western Mediterranean Basin by dynamical downscaling of climate scenarios at high resolution (1971–2050) Climatic Change 122: 243-256.

12. Flaounas E, Drobinski F, Vrac M, Bastin S, Lebeaupin-Brossier C et al (2013) Precipitation and temperature space–time variability and extremes in the Mediterranean region: evaluation of dynamical and statistical downscaling methods. ClimDyn 40: 2687–2705.

13. Verner D (2013) Tunisia in a Changing Climate : Assessment and Actions for Increased Resilience and Development.

14. Gao L, Bernhardt M, Schulz K (2012) Elevation correction of ERA-Interim temperature data in complex terrain. Hydrol Earth Syst Sci 16: 4661-4673.

15. Dee DP, Uppala SM, Simmons AJ, Berrisford P, Poli P, et al. (2011) The ERA-Interim reanalysis: configuration and performance of the data assimilation system. Q J R Meteorol Soc. 137: 553-597.

16. Haylock MR, Hofstra N, Klein Tank AMG, Klok EJ, Jones PD, et al. (2008) A European daily high resolution gridded data set of surface temperature and precipitation for 1950-2006. J Geophys Res 113: 119.

17. van den Besselaar EJM, Haylock MR, van der Schrier G, Klein Tank AMG (2011) A European daily high-resolution observational gridded data set of sea level pressure. J Geophys Res 116: 1-11.

18. Skamarock WC, Klemp JB, Dudhia J, Gill DO, Barker DM, et al. (2008) A Description of the Advanced Research WRF Version 3.125.

19. Hong SY, Lim JOJ (2006) The WRF single-moment 6-class microphysics scheme (WSM6). J Kor MeteorolSoc 42: 129-151.

20. Hong SY, Noh Y, Dudhia J (2006) A new vertical diffusion package with an explicit treatment of entrainment processes. Mon Weather Rev 134: 2318-2341.

21. Paulson CA (1970) The mathematical representation of wind speed and temperature profiles in the unstable atmospheric surface layer. J Appl Meteor 9: 857-861.

22. Mlawer E, Taubman S, Brown P, Iacono M, Clough S (1997) Radiative transfer for inhomogeneous atmosphere: RRTM, a validated correlated-k model for the long-wave. J Geophys Res 102: 16663-16682.

23. Dudhia J (1989) Numerical study of convection observed during the winter monsoon experiment using a mesoscale two-dimensional model. J AtmosSci 46: 3077-3107.

24. Kain JS (2004) The Kain-Fritsch convective parameterization: an update. J ApplMeteorol 43: 170-181.

25. Chen F, Dudhia J (2001) Coupling an advanced land-surface/hydrology model with the Penn State/NCAR MM5 modeling system. Part II: Preliminary model validation. Mon Weather Rev 129: 587-604.

26. Friedl MA, McIver DK, Hodges JCF, Zhang XY, Muchoney D, et al. (2002) Global land cover mapping from MODIS: algorithms and early results. Remote Sens Environ 83: 287-302.

27. Ouachani R, Bargaoui Z, Ouarda T (2013) Power of teleconnection patterns on precipitation and streamflow variability of upper Medjerda Basin. Int J Climtol 33: 58-76.

28. Elizalde A (2011) The water cycle in the Mediterranean region and the impacts of climate change. Max Planck Institute for Meteorology Rep on Earth System Science 103: p 128.

29. Zhang Y, Dulière V, Mote P, Salathé EP (2009) Evaluation of WRFand HadRM mesoscale climate simulations over the UnitedStates Pacific Northwest. J Clim 22: 5511-5526.

14

Bias Correction for RCM Predictions of Precipitation and Temperature in the Chaliyar River Basin

Raneesh KY[1]* and Thampi SG[2]

[1]Ex-research Scholar, Department of Civil Engineering, National Institute of Technology, Calicut - 673601, Kerala, India
[2]Associate Professor, Department of Civil Engineering, National Institute of Technology, Calicut - 673601, Kerala, India

Abstract

Global climate models (GCMs) relate greenhouse gas (GHG) forcing to future potential climate states and enable development of climate projections for the future. GCMs exhibit some limitations when focusing on smaller scales (regional to local) or to resolve the processes caused by topography and land use. To overcome this problem, regional climate models (RCMs) and other downscaling methods have been developed. The Indian Institute of Tropical Meteorology (IITM), Pune, in collaboration with the Hadley Centre, UK has developed future climate scenarios for India. Climate data (precipitation and temperature) for the Chaliyar river basin in Kerala for both A2 and B2 scenarios were utilized in this study. Bias correction was performed to ensure that important statistics (coefficient of variation, mean and standard deviation) of the downscaled output matched the corresponding statistics of the observed data. This method of bias correction does not correct for the fraction of wet and dry days and lag inverse autocorrelation. But, it was observed that a relatively simple non-linear correction, adjusting both the biases in the mean and variability, leads to better reproduction of observed extreme daily and multi-daily precipitation amounts. A marked improvement was achieved with a nonlinear transformation, adjusting the mean as well as the coefficient of variation of daily precipitation. Predictions show that annual rainfall in the Chaliyar river basin may decrease by about 20-25% from the present day annual average value in the A2 scenario. In the B2 scenario, the decrease was in the range of 10-15%. Annual average temperature may increase by 3.3°C and 1.8°C from the present day average values in the A2 and the B2 scenarios respectively.

Keywords: Indian future climate; PRECIS; Precipitation; Regional climate models; Temperature

Introduction

Climate change is recognized as one of the most serious challenges facing mankind today [1]. Driven by anthropogenic activities, it is projected to be a direct threat to our food and water supplies and an indirect threat to world security. It is believed that increase in the concentration of carbon dioxide and other greenhouse gases in the atmosphere and consequent global warming will influence hydrological regimes and will have serious implications on water resources management [1]. Global climate models (GCMs) are fundamental tools for predicting future climate to enable developing a better understanding of climate change. At the regional scale, there is an urgent need for relevant, targeted projections of regional climate change [2]. There has been a rapid change in climate in response to human influences caused by local, national and global social, economic, industrial, and land use developments [1]. These changes continue to have impacts on different aspects of society, including health, agriculture, water resources, and energy demand. Therefore, it is important to investigate observed changes in the present climate so that future climate predictions can be validated and put into context. Most studies of national climate variabilities have been made using long term series of data from a small number of stations [3,4]. Also these studies concentrate on analysing data for one or two climate variables, usually temperature or precipitation. Rainfall, the principal input to freshwater systems, simulated in the available GCMs has certain uncertainties which have to be taken into account while using for impact assessments [2].

GCMs reconstruct important details of the climate at smaller scales (regional to local), but further downscaling is much needed. Hence techniques such as regional climate models (RCMs) and other downscaling methods have been developed [5]. RCMs are an effective method of adding fine-scale detail to simulated patterns of climate variability and change as they resolve better the local land-surface properties such as orography, coasts and vegetation and the internal regional climate variability better through improved resolution of atmospheric dynamics and processes [6]. A typical RCM grid is of the order of 0.44° to 0.44°, although some climate simulations employ smaller grids, usually only for a shorter temporal horizon of simulations [5]. Downscaled data from the RCM, PRECIS (Providing Regional Climates for Impacts Studies), has been validated extensively at both regional and catchment scales in different parts of the world, and represents the mean climate reasonably well [7,8]. In India, dynamical downscaling has been implemented by the Indian Institute of Tropical Meteorology, Pune using the RCM, PRECIS. This model has depicted the surface climate over the Indian region, particularly the summer monsoon precipitation, both in terms of mean and extremes reasonably well [9].

RCMs are generally run for a baseline period and the results are compared with observed values for the same period. Normally, PRECIS runs with 50km horizontal resolution for the present climate (1961-1990) using different baseline lateral boundary conditions (LBCs) and for future scenarios (2070-2100) using the Special Report on Emission Scenarios of the Intergovernmental Panel for Climate Change. This research work focuses on the application of an RCM, PRECIS (Providing

*Corresponding author: Raneesh. K. Y, Ex - research Scholar, Department of Civil Engineering, National Institute of Technology, Calicut 673601, Kerala, India
E-mail: kyraneesh@yahoo.co.in

Regional Climates for Impacts Studies) for making projections of future climate (2071-2100) in the Chaliyar river basin in terms of rainfall and temperature in order to assess its likely impacts. Also, meteorological data, rainfall and temperature for the present period (1981-2010), were analysed to identify and study variabilities, if any, present in the data.

Literature Review

Emission scenarios

The Intergovernmental Panel on Climate Change (IPCC) published a new set of emission scenarios in the Special Report on Emissions Scenarios (SRES) [10] to serve as a basis for the assessment of future climate. Among all the SRES scenarios, four marker scenarios (A1, A2, B1 and B2) are often used [11]. The A1 and B1 scenarios emphasize ongoing globalization and project a homogeneous world, while the A2 and B2 scenarios lay emphasis on social, economic, and environmental development on regional and local basis and project a heterogeneous world. The A1 scenario family has been developed into three groups describing alternative directions of technological change in the energy system viz., AIFI (fossil intensive), A1T (non-fossil energy sources) and A1B (balance across all sources). In the case of A1B scenario, balanced is defined as not relying too heavily on one particular energy source, on the assumption that similar improvement rates apply to all energy supply and end-use technologies.

In this study, the projected changes in rainfall and temperature in the A2 and B2 are analysed. The A2 scenario projects high population growth and slow economic and technological development, whereas the B2 scenario projects slower population growth, rapid economic development, and lays more emphasis on environmental protection. Emission of green house gases (GHGs) and other gases and the driving forces were quantified in the Third Assessment Report of the IPCC [12] for use in climate simulations using General Circulation Models (GCMs).

PRECIS

The Hadley Centre for Climate Prediction and Research, with the sponsorship of the United Kingdom Department for Environment, Food and Rural Affairs (DEFRA), the United Kingdom Department for International Development (DFID) and the United Nations Development Programme (UNDP) developed a flexible RCM, PRECIS (Providing Regional Climates for Impacts Studies), to serve as a practical tool for making projections of national patterns of climate change and hence to estimate its possible impacts and to assess vulnerability. RCMs do not replace GCMs, but are powerful tools to be used together with GCMs in order to add fine scale detail to their broad-scale projections [6]. PRECIS is freely available for use by scientists from developing countries, priority being given to those involved in vulnerability and adaptation studies conducted by their governments, to be reported in National Communications to the United Nations Framework Committee on Climate Change. Adaptation decisions made for running PRECIS shall be based on a range of climate scenarios accounting for the large number of uncertainties associated with projecting future climate. National climate change scenarios can be created locally for use in impact and vulnerability studies using local knowledge and expertise. Carrying out this type of work at a regional level will lead to much more effective dissemination of scientific expertise and awareness on assessing climate change impacts [13].

PRECIS is a hydrostatic, primitive equation grid point model with 19 levels described by a hybrid vertical coordinate system [14]. The version of PRECIS used in this study has a horizontal resolution of 0.44°×0.44°

[15]. The model domain is so selected that a sufficiently large area is covered so that synoptic and mesoscale circulations generated within the RCM are not undesirably damped. Simultaneously, the chosen domain shall be sufficiently small so that the deviation of the large scale seasonally averaged circulations predicted by the RCM from the driving Atmosphere Ocean General Circulation Model is not overwhelmingly large to imply a significant perturbation to planetary scale divergent circulation. These are necessary to ensure consistency between the RCM solution and the pre-determined GCM solution external to the RCM domain [6].

The anticipated adverse impacts of projected climate change are of paramount importance in planning future development and utilisation strategies in the water resources and agricultural sectors. Simulations employing climate models under scenarios of increasing greenhouse gas concentrations and sulphate aerosols indicate marked increase in both rainfall and temperature over India in the 21st century. According to the downscaled climate projections by PRECIS for the Indian peninsula, an average temperature increase of about 1°C in winter and 2°C in summer and an average decrease in precipitation by about 10-20% are predicted. The decrease in precipitation is not uniform throughout the year; a decrease of up to 10% in spring, and between 10 to 15% in autumn, is likely to occur, whereas an increase of about 10-20% in winter is also predicted [9]. The predicted change in rainfall under the B2 scenario is relatively less than that under the A2 scenario. There are substantial spatial differences in the projected changes in rainfall. The maximum expected increase in rainfall (10 to 30%) occurs over central India. There is no clear evidence of any substantial change in the year-to-year variability of rainfall over the next century [9]. Surface air temperature shows comparable increasing variabilities in the A2 as well as B2 scenarios. The temperatures are projected to increase by as much as 3°C to 4°C towards the end of the 21st century. The warming is widespread over the country and relatively more pronounced over the northern parts of India [9]. Overall, these variabilities point towards increased concentrated precipitation events and longer drought spells. The changes are not uniform across the sub-continent too; the sharpest reduction in precipitation is predicted in the southern region, while small areas in the north might have a small increase in precipitation [9].

Methodology

Study area-Chaliyar river basin

The Chaliyar is the fourth longest river in Kerala with a length of about 170km. The geographical area of the part of the river basin in Kerala is 2530km^2 it lies between latitudes 11° 06'N and 11° 36'N and longitudes 75° 48'E and 76° 33'E. Figure 1 presents the Chaliyar river basin in Kerala along with its location in the PRECIS domain. The main river starts from the Elambalari hills at an altitude of 2,067m above mean sea level (MSL). The climate in the basin is typically that of humid tropics [16]. Meteorological data including daily rainfall, temperature, relative humidity and solar radiation at the Kottaparamba observatory of the Centre for Water Resources Development and Management (CWRDM), Kunnamangalam, Calicut were collected for the period 1981 to 2010. From the analysis of the rainfall data, it is observed that the southwest monsoon (June to August) contributes about 60% and the northeast monsoon (September to November) contributes about 25% of the annual rainfall respectively. The remaining 15% is received as pre-monsoon showers during the months April to May. December to March is the driest period. The average annual precipitation in the basin is 3012.61mm and the maximum and minimum temperatures are 34°C and 24°C respectively. The annual average relative humidity

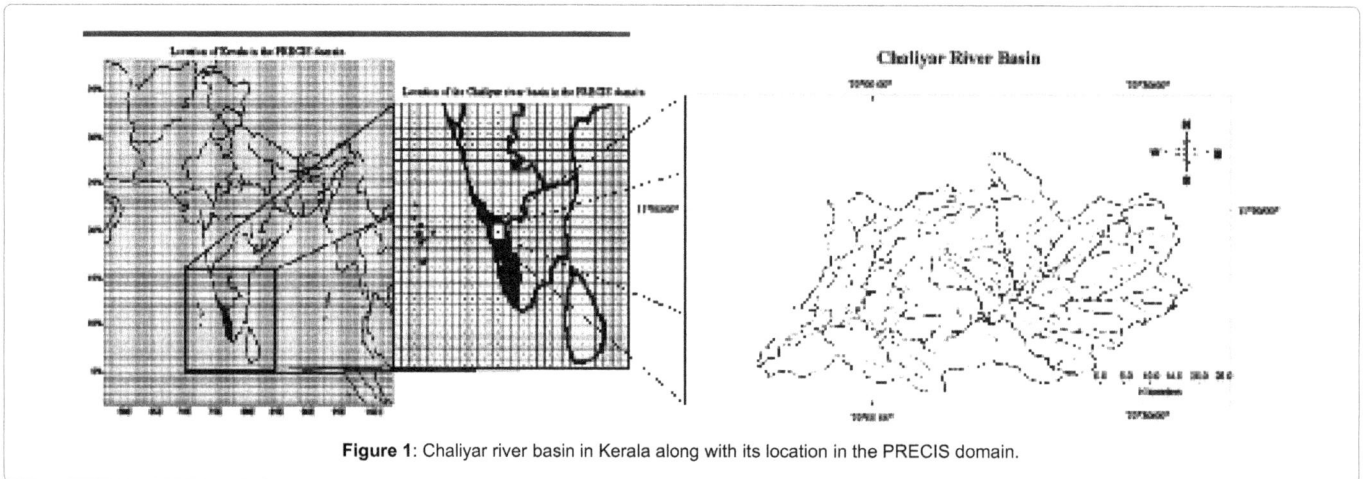

Figure 1: Chaliyar river basin in Kerala along with its location in the PRECIS domain.

ranges from 60% (minimum) to 90% (maximum) in summer; the corresponding values for winter are 65% and 85%.

Downscaling climate data using PRECIS

PRECIS was run with a horizontal resolution of 0.44° x 0.44° for the baseline period (1981-2010) using different baseline lateral boundary conditions (LBCs) and for a future period, 2071-2100, for the scenarios A2 and B2 outlined in the SRES of the IPCC [10]. The model was forced at its lateral boundaries by the results of simulations of a high-resolution global model (HadAM3H) with a horizontal resolution of 3.75° x 2.5°. HadAM3H is an atmosphere-only GCM derived from the atmospheric component of HadCM3, the Hadley Centre's state-of-the-art coupled model which has a horizontal resolution of 3.75° longitude by 2.5° latitude. The Indian Institute of Tropical Meteorology (IITM), Pune, in collaboration with the United Kingdom Hadley Centre for Climate Prediction and Research has developed future scenarios for India [9]. The data generated in transient experiments by the Hadley Centre for Climate Prediction, United Kingdom, at a regional climate model resolution of 0.44° x 0.44° latitude by longitude grid points was obtained from IITM, Pune. Climate data for both A2 and B2 scenarios were collected and utilized in this study. The Chaliyar river basin having a drainage area of 2530km^2 falls in one grid of 0.44° x 0.44° in the PRECIS domain. The downscaled climate data, collected from IITM, used in this study are precipitation and temperature (maximum and minimum) for the period 2071 to 2100.

Uncertainties associated with the projected climate data

There exist uncertainties in the projected climate change data collected from IITM, Pune. These uncertainties must be taken into account when assessing the impacts, vulnerability and adaptation options. However, all aspects of these uncertainties cannot be quantified yet [1]. There are inherent uncertainties in the key assumptions about and the relationships between future population, socio-economic development and technical changes that form the basis of the IPCC SRES. The imperfect understanding of some of the processes and physics in the carbon cycle and chemical reactions in the atmosphere generates uncertainties in the conversion of emissions to concentrations. Many things pertaining to the working of the climate system are not clearly understood yet, and hence uncertainties arise because of our incorrect or incomplete description of key processes and feedbacks in the model. This is clearly illustrated by the fact that current global climate models, which are based on different representations of the climate system, project different patterns and magnitudes of climate change

for the same period in the future when using the same concentration scenarios. Climate varies on time scales of years and decades due to natural interactions between the atmosphere, ocean and land, and this natural variability is expected to continue into the future.

Bias correction of PRECIS predictions

A problem with the use of regional climate model output directly for hydrological purposes is that the computed precipitation and temperature differs systematically from the observed precipitation and temperature [17]. Bias is defined as the time independent component of the error [17]. Bias arises because of several reasons. It has a high spatial component as well. Also, the biases in the output subsequently influence other hydrologic processes like evapotranspiration, runoff, snow accumulation and melt [18-21]. Some form of pre-processing is necessary to remove biases present in the computed climate output fields before they can be used for impact assessment studies [22-24].

Several studies have been carried out on correction of bias in RCM output. In the simplest formulations of bias correction, only the changes in a specific statistical aspect of the computed field are used. Often the change in mean value or the variance is employed. This is equivalent to correcting the observations with an additive or multiplicative constant. Hay et al. [25] made use of the gamma distribution to match the distribution of modelled daily precipitation with that of observed daily precipitation. After bias correction, the corrected data can be used for stream flow simulations using a semi-distributed hydrologic model. It was observed that the corrected precipitation data did not have the day-to-day variability which was present in the observed data set. Apart from the bias in the computed precipitation, the estimation of flow suffers from the limited length of the RCM simulations (usually no longer than 30 years). Extrapolation of the distribution of computed discharges is then needed to estimate the extreme flood quantiles if the hydrological model is run directly with the RCM output. Leander and Buishand [20] applied a power law transformation to correct for the coefficient of variation (CV) and the mean of the precipitation values. They found that a relatively simple non-linear correction, adjusting both the biases in the mean and its variability, leads to better reproduction of observed extreme daily and multi-daily precipitation amounts than the commonly used linear scaling correction. This method of bias correction does not correct for the fraction of wet and dry days and lag inverse autocorrelation. A marked improvement was achieved with nonlinear transformation, adjusting the mean as well as the coefficient of variation of daily precipitation.

For this reason the method developed by Leander and Buishand [20] for bias correction was used in this study to correct for bias in temperature and precipitation data. The temporal and spatial resolution at which bias correction is applied is extremely important. In this study, this was performed at a temporal resolution of one day over a grid size 50km x 50km. Bias correction was performed on the downscaled PRECIS data for the period 1981-2010 (30 years). The most important statistics (coefficient of variation, mean and standard deviation on a scale of 5 days of the PRECIS data) were matched with corresponding quantities computed from the observed values. The daily precipitation P is transformed to a corrected value P^* using:

$$P^* = aP^b \qquad (1)$$

where a and b are constants.

The effect of sampling variability was reduced by determining the parameters a and b for every five-day period of the year [20]. Determination of the parameter b was done iteratively, so that the coefficient of variation of the daily precipitation values predicted by PRECIS matches the coefficient of variation of the observed daily precipitation. The coefficient of variation thus becomes a function of parameter b according to:

$$CV(P) = function(b) \qquad (2)$$

where, P is the precipitation in a block of 5 days (total of 73 blocks x 30 years = 2190 blocks). After evaluating the parameter b, the transformed daily precipitation values are calculated as:

$$P^* = P^b \qquad (3)$$

Thereafter, the parameter a was determined such that the mean of the transformed daily values of precipitation matched with the observed mean. The parameter a depends on b. Parameter b depends only on the coefficient of variation and is independent of the value of parameter a. The values for a and b determined by this method for the A2 scenario were 0.1506 and 1.0344 and the corresponding values for the B2 scenario were 0.1679 and 1.1062. For correcting daily temperature, a different technique was used. Correction for temperature involves shifting and scaling to adjust the mean and the variance [20]. The corrected daily temperature T^* is given by:

$$T^* = AVG(T_O) + \frac{SD(T_O)}{SD(T_M)}[(T_U - AVG(T_O)) + (AVG(T_O) - T_M)] \qquad (4)$$

where T_U is the uncorrected daily temperature from PRECIS, T_O is the observed daily average temperature (from the meteorologic data), T_M is the corresponding daily average temperature obtained from PRECIS and SD is the standard deviation. Again both the statistics, the coefficient of variation and the mean, were determined for each 5-day block of the year separately.

Results

Variability in annual and seasonal rainfall

A consistent decrease in annual rainfall was observed from 4159.90 mm in 1981 to 2724.20 mm in 1990. From 1990 to 1998, there was an increase in annual rainfall (up to 3585.80 mm). Again a decreasing inter-annual variability was observed till 2003 as the annual rainfall value reduced to 2536.60 mm. This was followed by a period that showed an increasing change in annual rainfall. The annual rainfall in 2006 was 2998.73 mm. This was followed by a decreasing variability till 2009. The annual rainfall in the year 2010 was 2691.94mm. The pattern which is observed in the variation of rainfall with a period of decreasing

variability followed by one of increasing variability in annual rainfall is shown in Figure 2.

Rainfall during the southwest monsoon (June-September) also showed variability similar to that of annual rainfall. The average rainfall during this period was 2245.70 mm (1981 to 2010). The rainfall during this period was 3360.90 mm in the year 1981 and 2213.60 mm in 2010. The northeast monsoon rainfall (October-December) showed a slightly increasing variability. The average rainfall during this season over the 30 year period was 434.92mm. Northeast monsoon sets in the first week of October and retreats by the start of December. Except during the year 1986, which was a dry year, rainfall during this season was close to the average value in most of the years. The rainfall during this period was 570.40 mm in 1981 and 478.34 mm in 2010. The results of the analysis of rainfall during these seasons are presented in Figure 3.

After studying the rainfall data during the southwest monsoon, it is observed that rainfall during the months of June and July showed a declining variability whereas this variability is not seen in the rainfall during the months of August and September. A seasonal shift in the rainfall pattern is observed with a significant decrease in southwest monsoon rainfall in recent years where as an increase in rainfall is observed during the northeast monsoon. Rainfall decline is more predominant in the months of June and July but not so in August and September. The decreasing variability in southwest monsoon rainfall over Kerala is supported by other researchers [26,16]. The decline in frequency of the weather systems in recent years over the peninsula may be an important reason for the reduction in southwest monsoon rainfall over Kerala. In addition to this, there is a drastic change in the

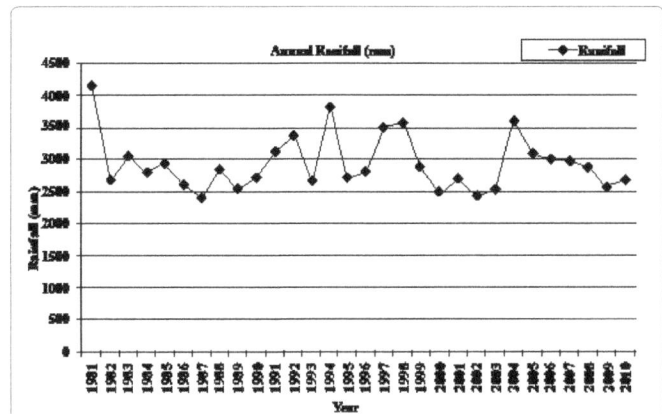

Figure 2: Annual rainfall data at Kottaparamba (Source : CWRDM, Calicut)

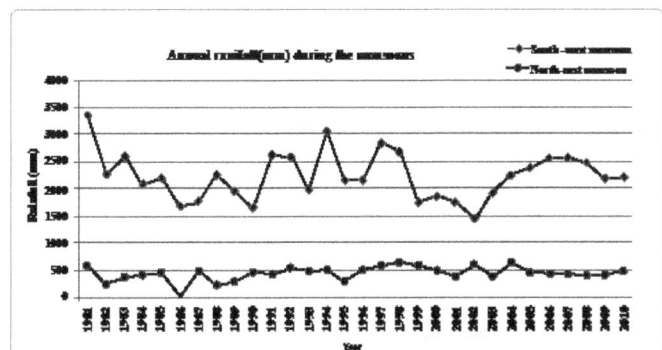

Figure 3: Rainfall during the southwest and northeast monsoons (Source : CWRDM, Calicut)

biophysical resources of Kerala due to human interventions in recent decades [16]. It indirectly affects the physical processes in the earth-atmosphere continuums and influences the distribution of local rainfall during winter and pre-monsoon seasons.

Variability in annual temperature

Analysis of temperature data reveals that, the average maximum temperature in the basin is 31.99°C and the average minimum temperature is 23.70°C. Figure 4 presents the annual average maximum and minimum temperatures in the Chaliyar river basin during the period 1981-2010. The average maximum temperature during the period 1981 to 1985 is 31.28°C, 1986 to 1990 is 31.90°C, 1991 to 1995 is 32.01°C, 1996 to 2000 is 32.11°C, 2001 to 2005 is 32.62°C and from 2006 to 2010 is 32.65°C. The corresponding values of average minimum temperature are 23.52°C, 23.61°C, 23.71°C, 23.78°C, 23.79°C and 23.89°C. From the above analysis it can be observed that there is a steady increase in temperature in the basin; the increase in maximum temperature is higher when compared to the increase in minimum temperature. Figure 5 is a plot of the average monthly temperature, average monthly maximum temperature and average monthly minimum temperature computed from the temperature data for the period 1981 to 2010. The maximum temperature is the highest in the month of April; the average value for the study period is 34.54°C. This is followed by the month of March with an average maximum temperature of 34.35°C. The lowest value of average minimum temperature is observed in the month of January (21.16°C) followed by the month of December (21.25°C). During the southwest monsoon season, the highest average maximum

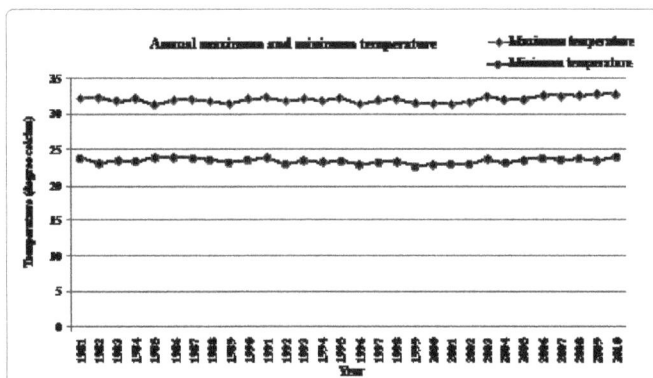

Figure 4: Annual temperature data at Kottaparamba (Source : CWRDM, Calicut)

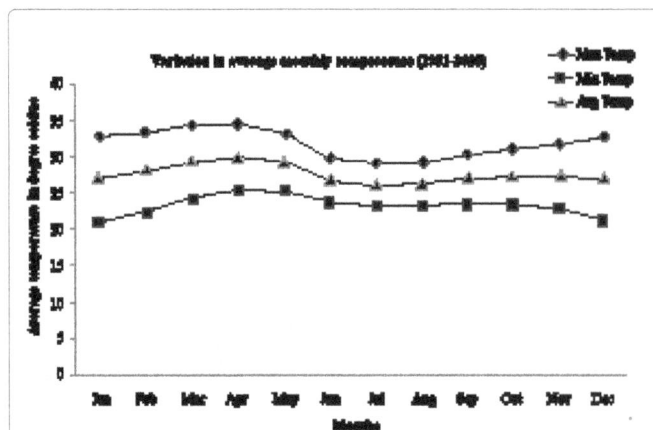

Figure 5: Monthly temperature data at Kottaparamba (Source : CWRDM, Calicut)

temperature is 29.67°C and the lowest average minimum temperature is 23.46°C. The corresponding values for the northeast monsoon season are 31.80°C and 22.60°C.

Analysis of the bias corrected temperature and precipitation data

The meteorologic data (precipitation and temperature) for the post climate change period, obtained from PRECIS was corrected for bias. The bias corrected precipitation and the bias corrected temperature for the future periods (2071-2100) along with the observed values are presented in the Figures 6 and 7 respectively. The projected annual rainfall in the Chaliyar river basin shows a decrease of about 20-25% from the present day annual average value in the A2 scenario. In the B2 scenario, the decrease was of about 10-15%. Annual average temperature shows an increase of about 3.3°C and 1.8°C from the present day values in the A2 and the B2 scenarios respectively. The maximum temperature increased by about 3°C and 2°C from the present average values in the A2 and B2 scenarios respectively. An increase of 3.5°C in the A2 and 2°C in the B2 scenarios was observed in the case of the minimum temperature.

In the A2 scenario, during the southwest monsoon period, an increase in temperature of 2°C and reduction in rainfall by 11.50% from the present day average values was observed. An increase in temperature of about 1°C and decrease in rainfall by about 8.79% was observed in the B2 scenario. Similar variabilities were seen in the northeast monsoon period also. In the A2 scenario, a decrease in rainfall by about 8.70% and an increase in temperature by about 2°C from the present day average values were observed. In the B2 scenario, it was observed that there was an increase in temperature of about 1°C and a decrease of about 4.73% in rainfall. In summer (pre-monsoon period), precipitation showed an increase of about 1.60% from the present day average value; the corresponding value in the B2 scenario was about 1.40%. The average temperature showed an increase of about 3°C and 2°C from the present day average values in the A2 and B2 scenarios respectively.

Climate change is expected to modify the frequency, intensity and duration of extreme events in many regions [24]. It is impossible to attribute single extreme events directly to anthropogenic climate change because of the probabilistic nature of these events. Also, there is always a chance that a given event may be the result of natural climate variability, even though an event of such magnitude has never been recorded [27]. It is clear that a dynamic downscaling experiment would have greater value if boundary conditions from more than one global climate model are used.

Conclusions

The potential changes of the surface climate over the Chaliyar river basin based on a single grid value from the 0.44° x 0.44° resolution PRECIS simulation driven by a high-resolution global model (HadAM3H) has been examined in this study. Results of simulations using the regional climate model, PRECIS for the present period and likely future climate were analyzed to develop an understanding as to how climate extremes may change in the Chaliyar river basin. The analysis was performed using the output from PRECIS after correction for bias. Bias correction was performed to ensure that important statistics (coefficient of variation, mean and standard deviation) of the downscaled output matched the corresponding statistics of the observed data. This method of bias correction does not correct for the fraction of wet and dry days and lag inverse autocorrelation. But, it

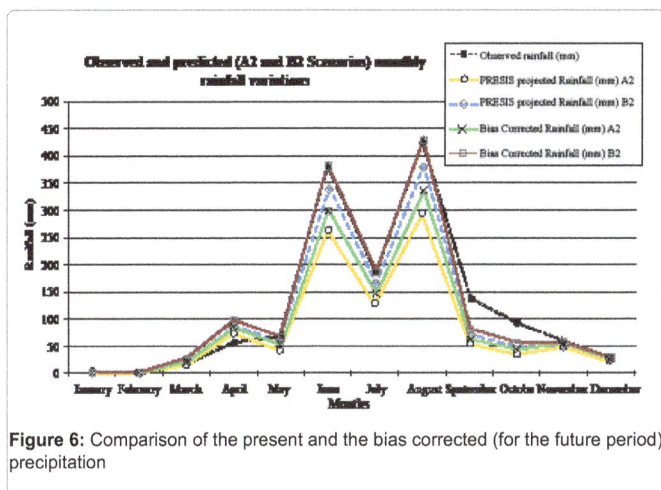

Figure 6: Comparison of the present and the bias corrected (for the future period) precipitation

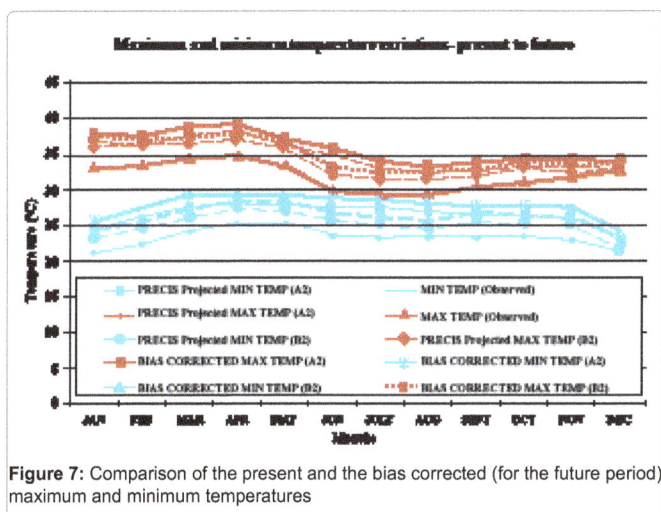

Figure 7: Comparison of the present and the bias corrected (for the future period) maximum and minimum temperatures

was observed that a relatively simple non-linear correction, adjusting both the biases in the mean and variability, leads to better reproduction of observed extreme daily and multi-daily precipitation amounts. A marked improvement was achieved with a nonlinear transformation, adjusting the mean as well as the coefficient of variation of daily precipitation. The general results of this analysis can be incorporated into water resources management plans for the future in order to ensure sustainable water use in the river basin.

References

1. IPCC Intergovernmental Panel on Climate Change (2007) Climate change 2007: The Physical Science Basis. Contribution of WG I to the fourth assessment report of the Intergovernmental Panel on Climate Change, Cambridge University Press, Cambridge.

2. Kundzewicz ZW, Mata LJ, Arnell N, Doll P, Jimenez B, et al., (2008) The implications of projected climate change for freshwater resources and their management. Hydrol Sci J 53: 3-10.

3. Begert M, Schlegel T, Kirchhofer W (2005) Homogeneous temperature and precipitation series of Switzerland from 1864 to 2000. International Journal of Climatology 25: 65-80.

4. Kruger AC, Shongwe S (2004) Temperature variabilities in South Africa: 1960-2003. International Journal of Climatology 24: 1929-1945.

5. Jones RG, Murphy JM, Noguer M (1995) Simulation of climate change over Europe using a nested regional climate model: assessment of control climate,

including sensitivity to location of lateral boundaries. Quarterly Journal of the Royal Meteorological Society 121: 1413-1449.

6. Jones RG, Noguer M, Hassell DC, Hudson D, Wilson SS (2004) Generating high resolution climate change scenarios using PRECIS. Met Office Hadley Centre, Exeter, UK.

7. Gosain AK, Rao S, Debajit B (2006) Climate change impact assessment on hydrology of Indian river basins. Current Science 90: 346-353.

8. Ashfaq MY, Shi WW, Tung RJ Trapp, Gao X, Pal JS, et al. (2009) Suppression of south Asian summer monsoon precipitation in the 21st century. Geophys Res Lett 36: L01704.

9. Rupa Kumar K, Sahai AK, Krishna Kumar K, Patwardhan SK, et al. (2006) High-resolution climate change scenarios for India for the 21st century. Current science, 90: 334-345.

10. Nakicenovic N, Alcamo J, Davis G, DeVries BS, Gaffin K, et al. (2000) IPCC Special Report on Emissions Scenario. Cambridge, UK Cambridge University Press.

11. VanVuuren PD, B C, O'Neill (2006) The consistency of IPCC's SRES scenarios to recent literature and recent projections. Climatic Change 75: 9-46.

12. IPCC Intergovernmental Panel on Climate Change (2001) Climate Change 2001: Impacts, Adaptation and Vulnerability. Contribution of Working Group II to the Third Assessment Report of the Intergovernmental Panel on Climate Change. Cambridge University Press, Cambridge, UK.

13. Omondi et al. (2013) Changes in temperature and precipitation extremes over the Greater Horn of Africa region from 1961 to 2010. International Journal of Climatology

14. Simmons AJ, Burridge DM (1981) An energy and angular-momentum conserving finite-difference scheme and hybrid coordinates. Monthly Weather Review, 109: 758–766.

15. Simon W, Hassell D, Hein D, Jones R, Taylor R (2004) Installing and using the Hadley Centre regional climate modelling system PRECIS. Met Office Hadley Centre, Exeter, UK.

16. Guhathakurta P, Rajeevan M (2007) Variabilities in the rainfall pattern over India, International Journal of Climatology 28: 1453-1469.

17. Frei C, Christensen JH, Deque M, Jacob D, Jones RG, et al. (2003) Daily precipitation statistics in regional climate models: Evaluation and intercomparison for the European Alps. J Geophys Res 108: 4124.

18. Hagemann S, Botzet M, Machenhauer B (2001) The summer drying problem over south-eastern Europe: Sensitivity of the limited area model HIRHAM4 to improvements in physical parameterization and resolution. Phys Chem Earth, Part B 26: 391-396.

19. Wood AW, Leung LR, Sridhar V, Lettenmaier DP (2004) Hydrologic implications of dynamical and statistical approaches to downscaling climate model outputs. Clim Change, 62: 189-216.

20. Leander R, Buishand T (2007) Resampling of regional climate model output for the simulation of extreme river flows. J Hydrol, 332: 487-496.

21. Hurkmans R, Terink W, Uijlenhoet R, Torfs P, Jacob D, et al. (2010) Changes in Streamflow Dynamics in the Rhine Basin under Three High-Resolution Regional Climate Scenarios. J Climate 23: 679-699.

22. Hansen J, Challinor A, Ines A, Wheeler T, Moronet V (2006) Translating Climate forecasts into agricultural terms: advances and challenges. Clim Res 33: 27-41.

23. Sharma D, Das Gupta A, Babel MS (2007) Spatial disaggregation of bias-corrected GCM precipitation for improved hydrologic simulation: Ping River Basin, Thailand. Hydrol Earth Syst Sci 11: 1373-1390.

24. Christensen JH, Boberg F, Christensen OB, Lucas-Picher P (2008) On the need for bias correction of regional climate change projections of temperature and precipitation. Geophys Res Lett 35: L20-709

25. Hay LE, Clark MP, Wilby RL, Gutowski Jr, Leavesley WJ (2002) Use of regional climate model output for hydrologic simulations. Journal of Hydrometeorology 3: 571-590.

26. Rupa Kumar K, Pant GB, Parthasarathy B, Sontakke NA (1992) Spatial and subseasonal patterns of the long term variabilities of Indian summer monsoon rainfall. International Journal of Climatology 12: 257-268.

27. Stott PA, Stone DA, Allen MR (2004) Human contribution to the European heat-wave of 2003. Nature 432: 610-614.

15

Climate-growth Relationship of *Pinus patula* Schldl. et Cham. in Wondo Genet, South Central Ethiopia

Belay TT*

Department of Urban Environment and Climate Change Management, Ethiopian Civil Service University, Ethiopia

***Corresponding author:** Belay TT, Department of Urban Environment and Climate Change Management, School of Graduate Studies, Ethiopian Civil Service University, Ethiopia, E-mail: btiratu@yahoo.com

Abstract

Most tropical regions are facing historical difficulties of generating biologically reconstructed long-term climate records. Dendrochronology (tree-ring studies) is a powerful tool to develop high-resolution and exactly dated proxies for climate reconstruction. In particular, dendroclimatological research is important in understanding both spatial and temporal characteristics of climate influences on tree growth. This project investigated the relationship between climate (precipitation and temperature) and tree rings, and showed its potential for climate reconstruction. Owing to the seasonal variation in rainfall, we expected the formation of annual growth rings in the wood of the forest of Wondo Genet. The project also carried out annual ring assessment and described the basic anatomical features of growth ring boundaries of the species under study. Wood disc samples of 15 trees belonging to the species of *Pinus patula* in the study site served to develop climate sensitive ring-width chronologies using standard dendrochronological techniques. A tree ring chronology was developed for the study species, and this enabled the assessment of the existing relationship between climate and tree growth patterns. The relationship between annual and seasonal precipitation, and mean monthly maximum and minimum temperature and ring-width indices was performed by correlation and simple regression analyses. The study species showed significant relationships with the rainy season precipitation proving the existence of annual tree rings ($r=0.42$). Monthly correlation analysis also showed that rainfall during July had the highest positive and significant correlation with the growth of the study species ($r=0.50$). The results indicated that annual growth of the species is primarily influenced by water availability during the growing (rainy) season. The study generally infers that the presence of anatomically distinct annual growth rings and high-quality cross dating and climate-growth relationships make the study species useful candidate for dendrochronological studies.

Keywords: Climate-growth relationship; Growth rings; *Pinus patula*; Wood anatomy

Background and Justifications

Climate change is of increasing concern to scientists because of its serious ramifications on ecosystems, in particular. Impacts from either global climate change or local climate variation can be imprinted on ecosystems [1]. To understand long-term impacts of climate change both in the past and in the future, it is necessary first to go beyond the time limit of instrumental records of climate. To this end, high resolution climate proxies are used to extend historical climate records back in time, providing a more comprehensive overview of trends in past climate [2,3]. Among these proxies that have the potential to express aspects of climate variability with perfect dating fidelity, at annual resolution, tree rings remain the most important and widely used sources of long-term proxy data [2].

Trees contain some of the nature's most accurate evidence of the past, because their growth largely depends upon local conditions such as temperature, water availability and phenology [4-6]. Tree-ring records are exceptionally valuable proxy for paleo-environmental study because of the following factors: They provide continuous records with annual to seasonal resolution; the existence of large geographic-scale patterns of synchronic inter-annual variability; the increasing availability of extensive networks of tree-ring chronologies; and the

possibility of using simple linear models of climate-growth relationships that can be easily verified and calibrated [7].

Trees forming annual growth rings are found in many regions of the world. In temperate climates, trees add one ring yearly [2,8,9]. With the strong climate seasonality in temperate climates, growth-ring patterns are used as a climate proxy to reconstruct climate for the past [8,10]. In tropical areas, it has been widely assumed that trees do not form annual rings; but many researchers have succeeded in using tree rings in tropical trees to determine tree age, understand growth dynamics and carry out ecological and climate studies [11,12]. Indeed, tropical trees may have more than one growth ring formation per year depending on seasonality of growth [13]. In tropical climates with at least two arid months (bi-modal dry seasons), growth boundaries are usually visible, forming two rings per year [14].

The application of dendrochronology in tropical regions in general and Ethiopia in particular has been limited by the difficulty in finding trees with distinct annual rings that can be cross dated [4,13,15]. In Ethiopia, however, tree ring analysis using tree species, such as *Juniperus procera* from North Gondar [16], *Acacia senegal*, *Acacia seyal*, *Acacia tortilis* and *Balanites aegyptiaca* from the semi-arid savanna woodland [14]; *Podocarpus falcatus*, *Pinus patula*, *Prunus africana* and *Celtis africana* from the tropical mountain forest of Munessa-Shashemene [17]; *Juniperus procera* from Menagesh and Dodola [15,18]; *Boswellia papyrifera* from Lemlem Terara, Metema district, North-Western Ethiopia [19,20]; and *Celtis africana* from

South Central Ethiopia [21] has been successful. Hence, one can suggest that the country has potential species and sites for dendrochronological studies.

Despite this, tree ring analysis using tropical tree species is still challenging owing to the reason that ring boundaries in such species are often less distinct than in species from temperate areas [12,22]. This is because seasonal changes in environmental conditions are often less pronounced in timing and intensity than in temperate areas [22,23]. This is manifested in anomalies like false rings, wedging rings, missing rings, and multiple rings formed per year [12-14,22]. To tackle these problems, knowledge of wood anatomy including investigation of growth rhythms and monitoring of cambial activity; testing the climate signal of additional tree-ring features; and comparison and later combination of chronologies from different species [12,22] are vital techniques.

The species selected for this study, P. patula, is believed to be economically important species which may be affected by the unfolding change in climate. So, understanding the climate variable which is significantly related to the growth dynamics of the study species will enable to generate reliable information for the sustainable management of these species. Besides this, the inevitability of climate change and the insufficiency of long-term instrumental climate data demand dendroclimatological research heading for an assessment of climate and tree growth relationship. In particular, in areas like Wondo Genet where few or no dendrochronological work has been made yet, the role that this study plays in filling the existing gap is unquestionable.

This study looked into wood anatomy and growth responses of P. patula to climate variability at different temporal scales using standard dendrochronological techniques in Wondo Genet, South Central Ethiopia. In this study, assessment of annual ring formation, description of wood anatomy of growth ring boundaries, and statistical analysis of climate-growth relationships were given due consideration.

Objectives of the Study

The specific objectives of this study are to describe wood anatomical features of the growth ring boundaries of the study species; verify whether the growth rings formed by the study species are annual or not; and determine the climate-growth relationship of the study species at different time scales.

Materials and Methods

Description of the study area

Wondo Genet is found in Sidama Zone of Southern Nations Nationality and Peoples Regional States. It is located 263 km southeast of Addis Ababa, and about 13 km from the nearest town Shashemene. It is located within the geographic coordinates of 7° 06′ N to 7° 11′ N latitude, and 38° 05′ E to 38° 07′ E longitude [25]. It occupies the Northeastern portion of Hawassa 'Zuria' district. Shashemene and Kofele districts of Oromia regional state border it in the Northern, North Western and North Eastern, and Eastern directions, respectively [26]. This study was carried out in Wondo Genet forest, found residing WGCF-NR, the major organization of the area where the major forest and the selected study species are found.

The topography of the Wondo Genet has mountains and hills covering 43.5%, flat areas 36.25% and undulating parts 20.25% of the

district [24,25]. The area comprises the hills of Abaro, Bachil Gigissa, Gariramo, Kentere and Cheko, and the depression surrounded by these hills. The height of land varies between 2,580 m a.s.l. at Abaro and 1,600 m a.s.l. around the marshy area. The higher altitudes and steep slopes support natural forest and lower altitudes and the gentle terrain consists mainly of farmlands. The forests found in Wondo Genet occupy most of the Northeastern parts of the area. The hillsides and their foot belong to most of these forests (Figure 1) [26,27].

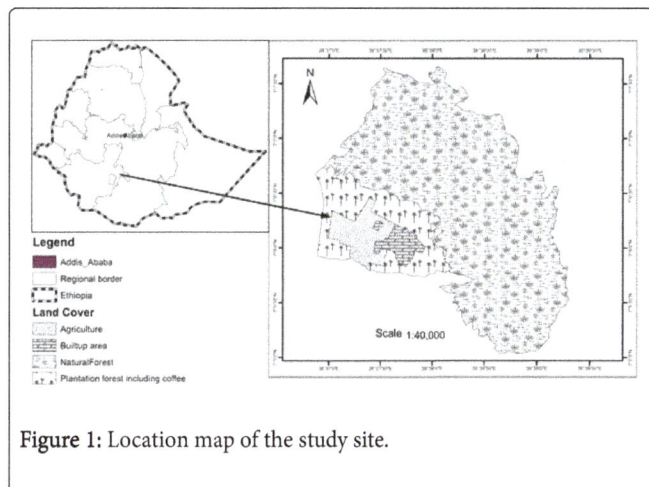

Figure 1: Location map of the study site.

The climate in Wondo Genet is characterized by two rainy seasons, a long rainy season from June to September and a short rainy season from March to May. The total annual rainfall range is between 800 mm to 1600 mm. On the other hand, average annual temperature varies between 18°C and 21°C. In most parts of Wondo Genet area, Woina dega (sub-tropical) agro-climatic type prevails (Figure 2).

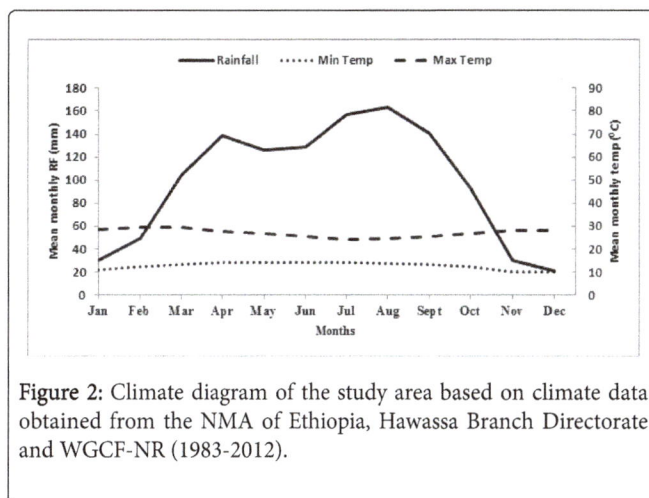

Figure 2: Climate diagram of the study area based on climate data obtained from the NMA of Ethiopia, Hawassa Branch Directorate and WGCF-NR (1983-2012).

Sample collection

A total of 15 stem discs were collected for the study species using a chainsaw. The samples of P. patula (10 stem discs) were collected at breast height (DBH, ~1.3 m). The remaining samples of P. patula (5 stem discs) were collected at a height of 0.5-1 m above the ground [28], so that these samples could be used for verifying annual ring formation. Besides this, some additional tree cross sections were also removed from the branches of trees that had already been felled down for thin section preparation and anatomical studies. Moreover, in cases where the collected samples were checked problematic after the survey in the laboratory, for example, where detecting the growth bands were difficult due to the presence of dark heartwood, and/or where wedging, missing, locally absent and false rings were predominant, the discs were excluded from the sample and instead, additional stem disks were collected.

Preparation of wood anatomy samples

The basic anatomical features that depict the growth ring boundaries of the study species were examined by preparing micro-thin sections from the transverse planes of the samples using a sliding microtome [29,30]. Accordingly, some wood sections that fit the microtome (15 mm length and 2 mm width) were removed with a changing sampling point. Following this, thin sections of 30 μm thickness were prepared with a sliding microtome (type G.S.L.1 light weight microtome). And then, the thin sections were stained with a mixture of Astra-blue and Safranin for 3-5 min to improve visibility/ contrast of the wood anatomical features of the study species. The stained thin sections were rinsed with de-mineralized water and dehydrated with a graded series of Ethanol (50%, 96% and 100%). Then, for permanent fixation, the sections were rinsed with xylol and embedded in Canada-Balsam and dried at 60°C for 12 h in the oven [19]. Subsequently, the thin sections were photographed by a digital camera attached to a light microscope and desk top computer. Different images of a given sample were aligned automatically to a single image and then edited by correcting intensity of colour and focus with the help of Adobe Photoshop CS3 software. The terminology used follows the IAWA List of Microscopic Features for wood identification [31].

Disc processing and ring width measurement

After the discs were collected and transported to the wood-science laboratory of WGCF-NR, Ethiopia, they were air-dried for 2 weeks. Then, their cross-sectional surface was sanded using progressively finer grades of sandpapers (60, 80, 120, 220, 320 and 600 grits) ensuring maximum visibility of anatomical features. Cellular features of each disc were examined up to 40X magnifications using a stereoscopic microscope. If the bark was present, a calendar year was assigned for each ring by counting backwards from the most recent year's growth near the bark. If bark was missing, each ring was assigned a floating number, beginning with year one at the outermost ring.

Wood-anatomical features that depict growth-ring boundaries were studied on micro-thin sections [29,30]. Growth rings were detected by following concentric features around the stem circumference, and then identified when all rings on a disc and ring numbers and characteristics matched along different radii. Then, growth rings were marked and counted under stereo-microscope on four radial directions of the sample discs. Subsequently, the width of each growth ring was measured from the bark to the pith on each radius to compute a chronology. Tree-ring width was measured to the nearest 0.001 mm with a semi-automatic device, a LEICA MS 5 microscope coupled with LINTAB TM 5 digital measuring stage associated to TSAP-dos software, Rinn tech, Heidelberg, Germany [32].

Cross-dating

Cross-dating is a fundamental technique in dendrochronology, whereby distinctive series of narrow and wider tree rings are identified and matched among trees of different ages. In this study, the samples were cross-dated using a combination of visual and statistical techniques following the method of Stokes and Smiley [33]. The visual cross-dating involves documenting characteristic patterns of pointer years (extreme wide and narrow rings) and matching these patterns between samples [34]. Cross-dating allowed to checking and correcting for missing rings or false rings.

Furthermore, the similarity of individual curves were tested statistically with the computer program Time Series Analysis and Presentation (TSAP), which allowed the measurement of "Gleichlaufigkeitskoeffizient" (GLK) and t-values. These were the two statistical indicators used to evaluate the match between two time series. GLK indicates the percentage year-to-year harmony in the fluctuation of two curves within the overlapping period; and the t-values convey the degree of similarity of two curves [8]. T-values and GLK were calculated within and between trees of each species to check whether the ring width measurements cross-date or not and to assess the quality of the final chronologies developed for climate-growth analysis. Consequently, a combination of t-values ≥4.0 and a mean GLK ≥ 85.0 (P<0.001) were considered for further climate-growth analyses [20,35]. Samples that fail to cross-date through these stages were excluded from the subsequent analysis.

The statistical cross-dating was accomplished using COFECHA software [36]. For this, 32-year segments with 16-year overlaps in COFECHA were used to test the correlation between each series and the average of all other series. The critical threshold for such segment and overlap is 0.4093 (P<0.01) and r-values above this were considered best for cross-dating and climate analysis, while segments below this value were flagged as potential cross-dating errors. Each disc, with flagged segments, was visually checked carefully under a microscope and re-measured when necessary.

Standardization and chronology development

Ring widths are not only affected by climate but also change with tree age, height within the stem, site-specific characteristics and site productivity [2,12]. For the study of climate-growth relationships, it is essential to determine the changes in ring widths associated with all other "noises" and remove them from the measurements. This correction is known as standardization, and the transformed values are referred to as ring-width indices [2,37]. The purpose of standardization is to remove an overall growth trend in tree-ring measurement series, and to remove part of the variance at very low frequencies approaching the length of the series [38].

All cross-dated measurement series were standardized to remove growth trends related to all other "noises" and minimize effects of autocorrelation in the time series [2,37]. The program ARSTAN was used to produce the standardized tree-ring index chronology [37]. Each tree-ring series was detrended (double detrending), employed based on the pattern of ring width measurements in TSAP) with a cubic smoothing spline of 32 years, which preserved 124.62% of mean series length to de-trend all series. This detrending curve was more flexible compared with some conservative methods, such as linear regression lines or negative exponential curves, and it removed the long-term non-climatic variations in growth [2]. With this procedure, it was possible to effectively eliminate age-related radial growth patterns so that ring-widths only reflect environmental constraints.

Three types of index chronologies were created using the program ARSTAN: standard, residual, and ARSTAN. To determine which chronology best suited to this study, a preliminary correlation analysis was conducted between climatic variables (temperature and precipitation) and each index chronology. All three index chronologies yield similar results implying that the interpretation of the influences of regional climate on tree growth was the same for all chronologies. The standard chronology, however, showed the highest correlation between tree growth and climatic variables and therefore was used in all further analyses. Then, tree ring indexed chronologies for the study species were constructed from the standard indexed chronology obtained from ARSTAN outputs.

Climate data

The climate-growth relationship for each of the study species was analyzed using monthly climate data (rainfall and maximum and minimum temperature) from 1983 to 2012 obtained from the National Meteorological Agency (NMA), Hawassa Branch Directorate. These data were, then, analyzed to gain reliable climate–growth related information. All climate-related information in this study was calculated from the data obtained from the agency. In some cases, climate data obtained from the NMA of Hawassa Branch Directorate were missing (not available), and hence data obtained from WGCF-NR Meteorological station (class I) daily observations for rainfall and temperature were used.

Correlation and simple linear regression analyses

Correlation and simple linear regression analyses were used to examine how regional climatic variables influence radial growth of the study species. Correlation analysis was used to describe associations between growth variability and climatic variability for the period 1983-2012. Pearson Product-Moment Correlation Coefficient (r) was calculated to describe associations between factor chronologies and monthly climatic variables. And, coefficient of determination from simple linear regression analyses (R^2) was calculated to describe how well a regression line fits the set of data generated. The R^2 attempted to look at the strength of relationships between the indexed ring widths and the climate variable in terms of deviations from the regression line or best-fit line.

Statistically significant coefficients indicated a confident association between a given climate variable in a particular month and tree growth. But more often, the climate effect on tree growth is stronger in the form of a multi-month seasonal signal [39]. In such cases i.e., when two or more consecutive months showed strong positive or negative correlations between growth and climate (temperature or precipitation), seasonal climatic variables were created and included in the correlation analysis. Seasonal variables (dry and rainy seasons) were formed by simple averaging of monthly climatic variables. Unless stated otherwise, all results are statistically significant at $P<0.05$ i.e., at 95% confidence interval.

Results

Description of wood anatomy of growth ring boundaries

P. patula forms distinct growth rings. In some cases, wedging rings and false rings are also prevalent. It undergoes an abrupt transition from early wood to latewood boundaries. The early wood and late wood growth boundary is too visible to distinguish. Thus, tree ring formation is easier to study.

The microscopic cross section showed relatively simple wood anatomical structures. The axial or vertical system is composed mostly of axial tracheids and the radial or horizontal system is composed mostly of ray parenchyma cells (Figure 3). The latewood growth boundaries consist of narrow bands of radially flattened tracheids with slightly thickened cell walls. Intra-annual density variations and resin ducts are also common features of these species.

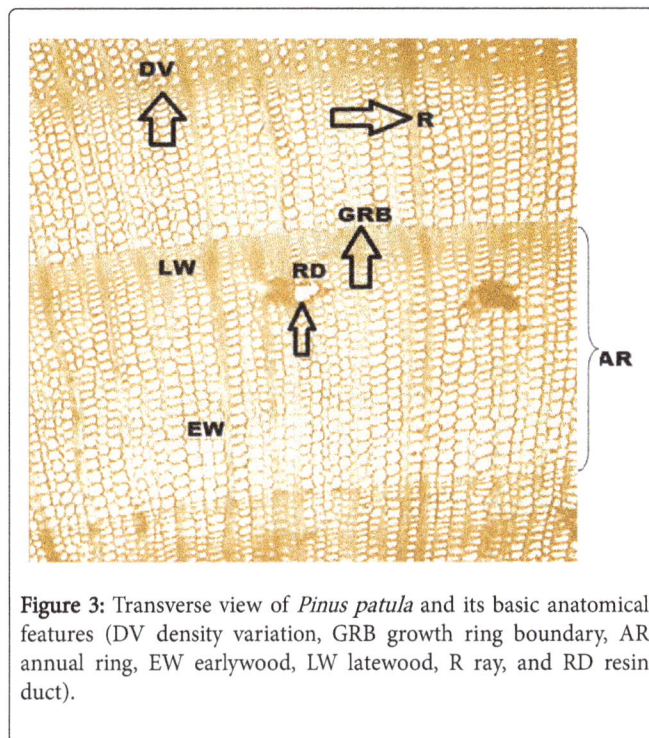

Figure 3: Transverse view of *Pinus patula* and its basic anatomical features (DV density variation, GRB growth ring boundary, AR annual ring, EW earlywood, LW latewood, R ray, and RD resin duct).

Verification of annual ring formation, cross-dating and chronology development

Samples of the studied species demonstrated that the distinct growth rings formed by the species are annual in nature. This was proved by ring counting in trees of known age. To this end, five sample discs that were cut between 0.5 and 1 m above ground level ensuring the visibility of maximum number of growth rings were used. The remaining ten discs were not used for annual ring assessment because they were cut at their DBH (1.3 m) and hence do not ensure the visibility of maximum number of growth rings.

Information on the planting year and hence age of the samples was obtained from the final document of the Management Plan for Forest Plantations of WGCF-NR. Consequently, it was found out that ring counts of the five discs exactly coincided with the known age of the sampled tree species (Table 1). This readily proved the annual nature of tree rings. Here, it was considered that the seedlings had already grown to a height of 0.5-1 m before they were planted in the study site.

The other method used to assess the annual nature of rings was through climatological analysis of tree ring curves. In this regard, the statistical comparison between measured tree ring curves and climate data is helpful. Values of climatic variables of four months (June to September) assumed to be important for tree growth were seasonalized and compared to tree ring widths.

Sampled discs	Comp. No. (new)	Planting year	Age (years)	No. of rings[*]
Pin 1	45	1982/83	30	30
Pin 2	53	1979/80	33	33
Pin 3	53	1979/80	33	33
Pin 4	45	1982/83	30	30
Pin 13	49	1981	31	31

Table 1: Comparison between number of rings and known age of the samples of *Pinus patula*. [*]Rings were counted on the longest radii from the bark to the pith.

Choice of climate variables and periodic limits of the seasons were defined by the experience of the researcher. Accordingly, a more or less concurring shape of seasonal precipitation and tree ring curves that point to the formation of annual rings for the study species was found (Figure 4).

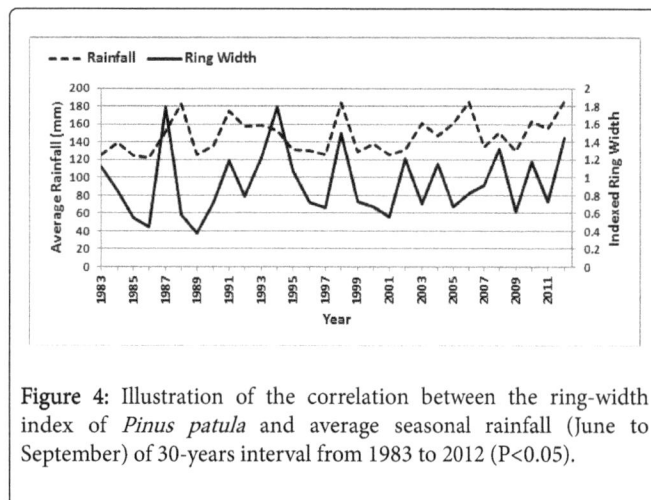

Figure 4: Illustration of the correlation between the ring-width index of *Pinus patula* and average seasonal rainfall (June to September) of 30-years interval from 1983 to 2012 (P<0.05).

Furthermore, cross-dating and the development of tree-ring chronologies for the study species were done successfully. Of the 15 stem discs that were collected for the study species, 14 of the sampled discs were cross-dated within and between trees (93.3% success rate); while 1 discs could not be cross-dated, although these discs were later replaced by other samples and eventually cross-dated against the tree ring chronologies developed for the study species (Figure 5). As mentioned earlier, in this study, cross-dating of the study species was successful with a mean (GLK %) range of 85–100%, (P<0.001) and t-values ≥4.

Out of the 60 measured tree ring series for each of the study species, 59 were retained for chronology building because of the accurate visual crossdating and COFECHA verification. The remaining series had correlation values less than the critical correlation level and hence they were removed before chronology development. The statistical crossdating quality of the discs was controlled by COFECHA outputs.

Accordingly, the mean correlations against the master chronology were 0.672, which is greater than the critical correlation level of COFECHA, 0.4093 at 99% confidence interval (Table 2). This implies that the statistical crossdating for the species is highly significant.

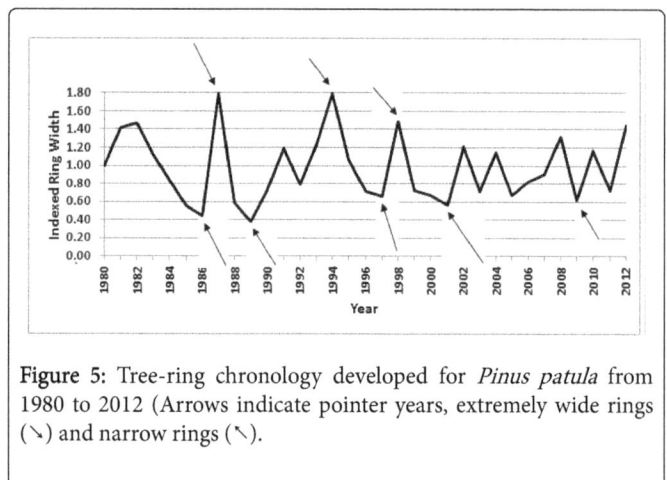

Figure 5: Tree-ring chronology developed for *Pinus patula* from 1980 to 2012 (Arrows indicate pointer years, extremely wide rings (↘) and narrow rings (↘).

COFECHA output	P. patula
Number of dated series	59
Master series	33 years (1980-2012)
Correlation with master	0.672
Average mean sensitivity	0.537
Auto-correlation (unfiltered)	0.6
No. of flags	1
No. of segments	67

Table 2: Summarized COFECHA output of the study species.

Climate–growth relationships

Though variations exist in the degree of correlation between the species, correlation statistics showed that species' growth in the study site was positively correlated with the climate variables treated. The species showed highest and significant values of correlation coefficient (r) with precipitation amounts of the rainy season (June to September) (Table 3).

Study species	Correlation coefficient (r) value			
	Max. Temp	Min Temp	Annual RF	Rainy Season
P. patula	0.24	0.1	0.27	0.42[*]

Table 3: Pearson correlation (r) values between the radial growth chronology of the study species and different climate variables using a 30-year common interval from 1983 to 2012. [*]High correlation during the rainy season.

The monthly correlation analysis resulted in both positive and negative relationships. The variables treated showed positive relationships during the months of January, August and September. Rainfall during the month of July had positive and significant correlation with the growth of the study species. Rainfall and minimum temperature during December had the highest negative relationships though the relationships were insignificant (Figure 6).

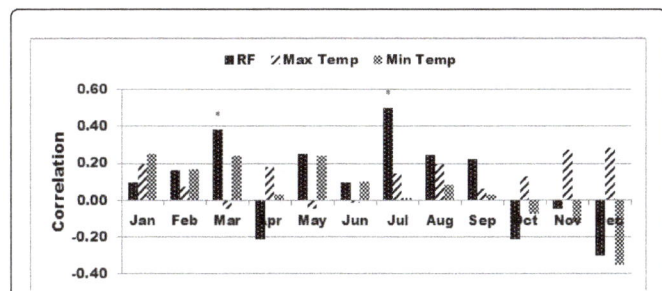

Figure 6: Correlations between indexed ring width of *Pinus patula* and average monthly climate variables (*P<0.05).

Furthermore, simple linear regression analyses indicated significant positive relationships between annual radial increments and major rainfall season for the study species. This showed that the climate of the rainy season (summer rainfall) is very significant to the growth of the study species. With the remaining climate variables analyze, the study species showed statistically insignificant relationship.

Looking at the scatter plots, we find that the variance from the regression line explained by the major climatic variables varied substantially for the species. In this regard, the coefficient of determination (R2) allows us to explain the existing relationships in terms of variations from the regression or trend line. Accordingly, taking the highest R2 outputs, one can realize that 17% of the variations in the ring widths of *P. patula* are explained by the amount of rainfall during the major rainfall season. And, when we take the lowest R2, we observe that 1% of the variations in the ring widths *P. patula* are explained by the amount of mean monthly maximum and minimum temperature (Figure 7).

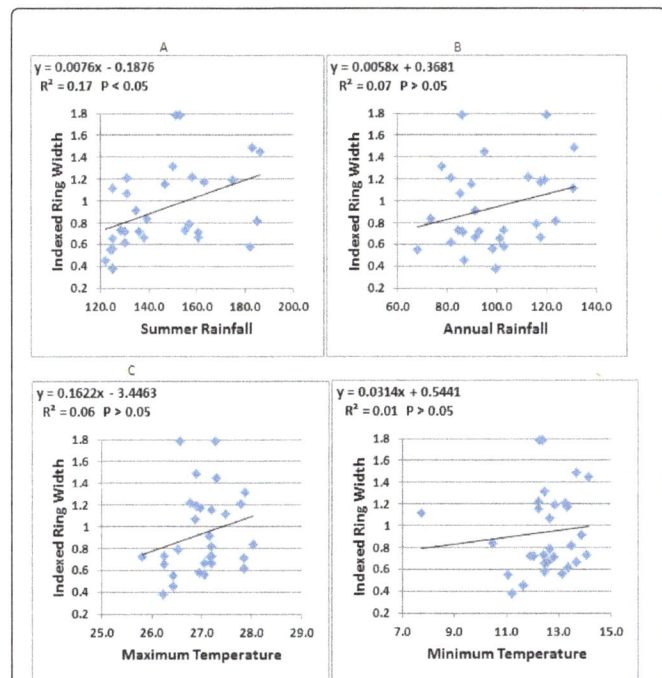

Figure 7: Simple linear regression between indexed ring widths of *Pinus patula* and the four climate variables analysed.

In general, both methods of analyses employed in this study (correlation and simple linear regression) resulted in similar outcomes in that they both showed high relationship between rainy season precipitation and radial growth of the species. In fact, the degree of relationship varied by variables treated.

Discussion

Tree-ring formation and cambial activity

The study species showed the formation of distinct rings. The variation in the distinctiveness of rings is mainly caused by differences in wood structure, which is species-specific [40]. For instance, the variations in the vessel size and distribution are a common feature in the wood structure of *P. patula*, as described for other tropical woods [31]. These wood anatomical features are indicators for periodic dormancy induced by seasonal precipitation patterns with a distinct dry season in tropical species [11,41].

Ring width patterns of individuals from the same tree species at a study site have been successfully cross-dated. This indicates that there is an external oscillating factor triggering tree growth [42] which is the seasonal precipitation. This is shown by many other dendrochronological studies on tree species from tropical Africa and from the tropics world-wide [14,30,40,42].

Although the study species form annual rings, crossdating was challenging. False rings and wedging rings occurred in the study species and are common problems in ring studies and in effective crossdating of tropical rainforest trees [2,43,44]. These irregular rings characters were successfully detected through differences in the anatomy of the ring and by checking the continuity of rings over the entire stem disc. False rings generally occurred just before the earlywood/latewood transition zone, suggesting that unusually low rainfall during earlywood growth might cause individuals to prematurely form tracheids with thicker cell walls, causing a false ring to appear before the formation of latewood cells. Wedging rings are believed to result from competition of species growing under poor light conditions [12].

The high incidence of such intra-annual growth zones and anomalous ring characters indicated the necessity to work with entire stem discs and makes the use of increment cores difficult [30,17,42]. Especially when working with new tropical species, as is the case in this study, it is absolutely necessary to start with entire stem discs for estimating dating accuracy and to learn anatomical differences between annual and non-annual growth zones. This was the main reason as to why this study preferred stem discs to increment cores, though it was ecologically destructive.

Growth periodicity in tropical areas is triggered when there is a dry season with less than 50 mm precipitation per month [11]. Decrease in precipitation at the beginning and during the dry season result in decreased soil water potential [14]. Species with different growth strategies respond differently to such climate events. At the most severe part of the dry period, when plant water potential reaches its permanent wilting point during the dry season, drought deciduous species shed their leaves and this is necessarily followed by a decrease in wood growth; and evergreen and deciduous species enter a period of cambial dormancy, which implies formation of growth boundaries. In agreement with this reasoning, both evergreen and deciduous species may show distinct growth rings. Our results also agree with previous studies, which showed that the study species form distinct annual growth rings [11,14,44].

Species' growth response to climate variability

Correlation and simple linear regression analyses revealed significant relationships between radial growth and precipitation and indicated that annual growth of the species is primarily influenced by water availability during the growing (rainy) season. This corroborates with the findings of several authors who explained species' sensitivity to water availability in tropical areas [2,14,28,39]. Precipitation, somehow, determines the amount of moisture available in the soil. If there is high rainfall, the moisture available in the soil will get increased and this in turn initiates wood cambial activity and growth of species. In contrast, dry conditions will slow down soil moisture and this leads to cambial dormancy and hence new growth ceases until the next wet season [45-49]. In fact, the amount of soil moisture is also dependent on soil type.

In the study area, variation in growth in the study species is at least partly determined by rainfall as the relationship between growth and rainfall is generally positive. Studies showed that tree growth is influenced by rainfall in climates with comparable precipitation levels of 1000-1700 mm. With a further rise in annual precipitation, however, it seems that growth is not limited by rainfall anymore, but by fluctuating patterns in photosynthetic radiation or air temperature [31,50,51].

Monthly correlation analysis showed that available moisture during the early and late summer months is proved to be a vital factor of xylem (cell) production because these months make up the bulk of the growing season when trees need moisture the most. Moreover, rainfall during the beginning of dry season (November to December) had negative relationship with the growth of the species signifying the importance of long term moisture availability for proper growth. The relationship between radial growth and maximum temperature during the rainy season is higher implying that exotic species tend to prefer warmer climatic conditions for their growth. These findings are comparable with the findings of other dendroclimatological studies in tropical areas [14,28,39].

Besides this, the statistical analyses between tree growth and climatic variables suggest that temperature has some degree of influence on the growth variation of the species under study. The response of tree growth to a change in temperature may differ in predictable ways. High latitude (temperate) tree growth may be temperature-limited and thus benefit from some degree of warming, as opposed to warm-adapted species [1,6,52]. In line with this reasoning, the study species reacted to temperature differently. That is, warmer temperature conditions had higher correlations with radial growth of *P. patula*. This implies that it benefited from high temperature amounts of the study site, as it is an exotic temperate species.

Conclusion

This study infers that the presence of anatomically distinct, annual growth rings, and high-quality cross dating, as shown by a high inter-series correlation and statistically significant correlations with climate makes the study species useful candidate for dendrochronological studies in tropical regions in general and Ethiopia and the study site in particular. Despite this, however, the scope of this study is somehow limited owing to some factors like deficiency of long term and persistent meteorological data, and scarcity of time and finance. Thus, this study raises issues that merit further investigation.

Acknowledgement

Above all, I would like to present my thankfulness, with utmost sincerity, to the Almighty God for the wonderful privileges He bestowed up on me and my family. Next, I would like to express my gratitude to Erlangen University (Germany) for the provision of the research grant. I also thank Dr. Aster Gebrekirstos, Dr. Solomon Zewdie and Dr. Motuma Tolera for their consistent advice and follow-up from the beginning to the completion of this research. My deepest thanks also go to the National Meteorological Agency of Ethiopia, Hawassa Branch Directorate and Hawassa University, WGCF-NR.

References

1. Sass U, Killmann W, Eckstein D (1995) Wood formation in two species of Dipterocarpaceae in peninsular Malaysia. IAWA J 16: 371-384.

2. Fritts HC (1976) Tree Rings and Climate. Academic Press, London.

3. Cook ER (1995) Temperature histories from tree rings and corals. Clim Dynam 11: 211-222.

4. Verheyden A, Kairo JG, Beeckman H, Koedam N (2004) Growth rings, growth ring formation and age determination in the mangrove Rhizophora mucronata. Ann Botany 94: 59-66.

5. Bauch J, Quiros L, Noldt G, Schmidt P (2006) Study on the wood anatomy, annual wood increment and intra-annual growth dynamics of Podocarpusoleifolius var. macrostachyus from Costa Rica. Appl Botany Food Qual 80: 19-24.

6. Brauning A, Burchardt I (2006) Detection of growth dynamics in tree species of a tropical mountain rain forest in southern Ecuador. TRACE - Tree Rings in Archaeology, Climatol Ecol 4: 127-131.

7. Hughes MK (2002) Dendrochronology in climatology-the state of the art. Dendrochronologia 20: 95-116.

8. Schweingruber FH (1988) Tree rings: basics and applications of dendrochronology. Kluwer Academic Publishers.

9. Briffa KR (2000) Annual climate variability in the Holocene: interpreting the message of ancient trees. Quatern Sci Rev 19: 87-105.

10. Smith D, Lewis D (2007) Dendrochronology. In: Encyclopaedia of Quaternary Science. Elsevier, Boston, pp: 459-465.

11. Worbes M (1999) Annual growth rings, rainfall dependent growth and long-term growth patterns of tropical trees from the Caparo Forest Reserve in Venezuela. Ecology 87: 391-403.

12. Worbes M (2002) One hundred years of tree ring research in the tropics: a brief history and an outlook to future challenges. Dendrochronologia 20: 217-231.

13. Wils TH, Robertson I, Eshetu Z, Sass-Klaassen U, Koprowski M (2009) Periodicity of growth rings in Juniperus procera from Ethiopia inferred from crossdating and radiocarbon dating. Dendrochronologia 27: 45-58.

14. Gebrekirstos A, Mitlohner R, Teketay D, Worbes M (2008) Climate-growth relationships of the dominant tree species from semi-arid savanna woodland in Ethiopia. Trees-Structure and Function 22: 631-641.

15. Sass-Klaassen U, Couralet C, Sahle Y, Sterck FJ (2008) Juniper from Ethiopia contains a large-scale precipitation signal. Int J Plant Sci 169: 1057-1065.

16. Wils TH, Robertson I, Eshetu Z, Touchan R, Sass-Klaassen U, et al. (2010) Crossdating Juniperus procera from North Gondar, Ethiopia. Trees 25: 71-82.

17. Krepkowski J, Brauning A, Gebrekirstos A, Strobl S (2011) Cambial growth dynamics and climatic control of different tree life forms in tropical mountain forest in Ethiopia. Trees 25: 59-70.

18. Couralet C, Sass-Klaassen U, Sterck F, Bekele T, Zuidema PA (2005) Combining dendrochronology and matrix modelling in demographic

studies: an evaluation for Juniperus procera in Ethiopia. Forest Ecol Manag 216: 317-330.

19. Tolera M, Menger D, Sass-Klaassen U, Sterck FJ, Copini P, et al. (2013) Resin secretory structures of *Boswellia papyrifera* and implications for frankincense yield. Annals of Botany 111: 61-68.

20. Tolera M, Sass-Klaassen U, Eshete A, Bongers F, Sterck FJ (2013) Frankincense tree recruitment failed over the past half century. Forest Ecol Manag 304: 65-72.

21. Hevia A, Sánchez-Salguero R, Linares JC, Olano JM, Camarero JJ, et al. (2016) Proceedings of the DENDROSYMPOSIUM 2015: May 20th-23rd, 2015 in Sevilla, Spain. German Research Centre for Geosciences 14: 74-77.

22. Wils THG, Sass-Klaassen UGW, Eshetu Z, Brauning A, Gebrekirstos A, et al. (2011) Dendrochronology in the dry tropics: the Ethiopian case. Trees 25: 345-354.

23. Cherubini P, Gartner BL, Tognetti R, Braker OU, Schoch W, et al. (2003) Identification, measurement and interpretation of tree rings in woody species from mediterranean climates. Biol Rev 78: 119-148.

24. Eshete G, Stahl G (1999) Tree rings as indicators of growth periodicity of acacias in the Rift Valley of Ethiopia. Forest Ecol Manag 116: 107-117.

25. Betru S (2006) Vegetation-environment interaction at small scale watershed: A study in Wondo Genet watershed vegetation. Environmental studies, Laboratory of Biosphere Information Sciences, Tokyo.

26. Gemechu T (2005) Prospects of Sustainable Natural Resource Management and Livelihood Development in Wondo Genet Area, Southern Ethiopia. MSc thesis, Addis Ababa University.

27. Kebede M, Kanninen M, Yirdaw E, Lemenih M (2012) Soil Seed Bank and Seedlings Bank Composition and Diversity of Wondo Genet Moist Afromontane Forest South Central Ethiopia. Botany 8: 170-180.

28. Couralet C (2010) Community dynamics, phenology and growth of tropical trees in the rain forest Reserve of Luki, Democratic Republic of Congo. Ghent University, Belgium.

29. Schweingruber FH, Borner A, Schulze ED (2007) Atlas of woody pant stems: evolution, structure and environmental modifications. Springer-Verlag, Berlin, p: 229.

30. Worbes M, Fichtler E (2010) Wood Anatomy and Tree-Ring Structure and Their Importance for Tropical Dendrochronology. In: junk WJ, Piedade MTF, Wittmann F, Schöngart J, Parolin P (eds.) Amazonian Floodplain Forests: Ecophysiology, Biodiversity and Sustainable Management. Ecological Studies 210: 329-346.

31. Worbes M (1989) Growth rings, increment and age of trees in inundation forests, savannas and a mountain forest in the Neo-tropics. IAWA J 10: 109-122.

32. Rinn F (2003) TSAP-Win, Software for tree-ring measurement analysis and presentation. Rinntech, Heidelberg, Germany.

33. Stokes MA (1996) An Introduction to Tree-Ring Dating. University of Arizona Press, Tucson, AZ.

34. Yamaguchi DK (1991) A simple method for cross-dating increment cores from living trees. Can J Forest Res 21: 414-416.

35. Enquist BJ, Leffler AJ (2001) Long-term tree ring chronologies from sympatric tropical dry-forest trees: individualistic responses to climatic variation. Tropical Ecol 17: 41-60.

36. Holmes RL (1983) Computer-assisted quality control in tree-ring dating and measurement. Tree-Ring Bulletin 43: 69-78.

37. Cook ER (1985) A time-series analysis approach to tree-ring standardization. University of Arizona.

38. Grissino-Mayer HD, Holmes RL, Fritts HC (1992) International tree-ring data bank program library: user's manual. Tucson: Laboratory of Tree-Ring Research, University of Arizona, p: 104.

39. Grissino-Mayer HD, Butler DR (1993) Effects of climate on growth of shortleaf pine (*Pinus echinata* Mill.) in northern Georgia: A dendroclimatic study. Southeastern Geographer 33: 65-81.

40. Brienen RJ, Zuidema PA (2005) Relating tree growth to rainfall in Bolivian rain forests: a test for six species using tree ring analysis. Oecologia 146: 1-12.

41. Schongart J, Orthmannw B, Hennenbergw KJ, Porembski S, Worbes M (2006) Climate–growth relationships of tropical tree species in West Africa and their potential for climate reconstruction. Global Change Biology 12: 1139-1150.

42. Worbes M (1995) How to measure growth dynamics in tropical trees-a review. IAWA J 16: 337-351.

43. Dunisch O, Bauch J, Gasparotto L (2002) Formation of increment zones and intraannual growth dynamics in the xylem of *Swietenia macrophylla*, *Carapa guianensis*, and *Cedrela odorata* (Meliaceae). IAWA J 23: 101-119.

44. Gebrekirstos A, Demel T, Masresha F, Mitlohner R (2006) Adaptation of five co-occurring tree and shrub species to water stress and its implication in restoration of degraded lands. Forest Ecol Manag 229: 259-267.

45. Powers RF (1981) Nutritional ecology of ponderosa pine (*Pinus ponderosa* Laws) and associated species. Ph.D. dissertation, University of California, Berkeley.

46. Carter GA, Miller JH, Davis DE, Patterson RM (1984) Effect of vegetative competition on the moisture and nutrient status of loblolly pine. Forest Resour 14: 1-9.

47. Oliver WW (1986) Growth of California red fir advance regeneration after over story removal and thinning. USDA Forest Resources Paper PSW, p: 180.

48. Byrne SV, Wentworth TR, Nusser SM (1987) A moisture strain index for loblolly pine. Forest Resour 17: 23-26.

49. Allen HL, Dougherty PM, Campbell RG (1990) Manipulation of water and nutrients-practice and opportunity in southern U.S. pine forests. Forest Ecol Manag 30: 437-453.

50. Clark DA (1994) Climate-induced annual variation in canopy tree growth in a Costa-Rican tropical rain-forest. Ecology 82: 865-872.

51. Stahle DW, Mushove PT, Cleaveland MK, Roig FA, Haynes GA (1999) Management implications of annual growth rings in Pterocarpus angolensis from Zimbabwe. Forest Ecol Manag 124: 217-229.

52. Way DA, Oren R (2010) Differential responses to changes in growth temperature between trees from different functional groups and biomes: a review and synthesis of data. Tree Physiol 30: 669-688.

16

Extreme Point Rainfall Events Analysis of Gorakhpur under Climate Change Scenario

Kailash Chand Pandey*

Climatologist in G.E.A.G. India, Gorakhpur Environmental Action Group, Gorakhpur, India

Abstract

The rainfall is an important parameter for the wellbeing of around 45 lac people of the district. However, certain extreme rainfall events occurring in different seasons cause disastrous situation over some parts of the district. In view of this, we have scanned the daily rainfall data of Gorakhpur to find out their extreme point rainfall events (24 hrs. rainfall) and examined whether there is any change in the number and the intensity of such events during past four decades. This analysis reveals that their number has gone up considerably after 1980 with an alarming rise in intensity thereafter. It is further noticed that after 2000 district is affected with a heavy downpour. The highest records were established over this part of the country on different time scales. It is conjectured that these events may be associated with the global and the regional warming under the climate change scenario. In the event of their continuation, there would be severe impact on societal and environmental issues warranting appropriate precautionary measures in near future to safe guard the interest of vast population of this district.

Keywords: Climate change; Precipitation; Extreme Rainfall; Temperature; Intensity

Introduction

Location and physical geography of district

Gorakhpur is one the most populated districts of Eastern Uttar Pradesh situated. It is situated between 26°13′N and 27°29′N latitude and 83°05′E and 83°56′E longitude having long stretches of fertile alluvial plains split apart by perennial flow of gangetic[1] river system. District Gorakhpur shares common boundary with district Azamgarh on south, Basti on west and district Deoria on east. It shares international border with Nepal on north (Figure 1).

Perched close to Himalayan mountain range[2] and 1.65 per cent land under forest cover[3], beside large number of lakes and water bodies, Gorakhpurdistrict finds itself nestled in unique microcosm of ecosystem. Three major seasons relay in succession of each other render range to agro-climatic conditions and to agriculture production and food economy[4] of country (Figure 2).

District finds itself nestled in unique microcosm of ecosystem. Three major seasons relay in succession of each other render range to agro-climatic conditions and to agriculture production and food economy[5] of country.

Gorakhpur city

Gorakhpur is head quarter of district and is also main center of commerce and trade in district. Population of Gorakhpur city is 692,519[6] and is rapidly increasing. Percentage of urban population to total population of district is 18.78[7] and changing fast with large floating population come and go every day in search of job in city. Pace of urbanization is fast eroding natural ecosystem in peri-urban areas and wetland[8] within city. It exacerbates micro-climate of city/region

Figure 1: Location of Gorakhpur city.

---

[2]Lower Himalayan mountain range is approx. 150 km.
[3]www.mofe.nic.in
[4]651719 metric tons of food grain production in Gorakhpur district
[5]651719 metric tons of food grain production in Gorakhpur district
[6]Census 2011
[7]Census 2011
[8]Ramgarh Tal lake is 18 kms.

*****Corresponding author:** Kailash Chand Pandey, Climatologist in G.E.A.G. India, Gorakhpur Environmental Action Group, Gorakhpur, India
E-mail: pand992000@hotmail.com

---

[1]Rapti, Rohin and ChhotiGandak river

due to heat-island, increasing aerosol and changes in natural drainage line of river.

Gorakhpur city is situated 78 meter above mean sea level, which is not very high from level of river bed. It does not allow low lying areas of city to drain properly, causing water to stand for 2-3 months in a year.

Location, physical geography of district, perennial flow of river systems, plentiful of forest cover, low relief and large number of water bodies render a unique micro-climate to city. Rapid pace of urban expansion however is gradually rasping natural ecosystem around city by either filling low-lying areas with solid waste or building constructions on it. It perhaps is giving birth to some new ecosystem and building climate risks in city (Figure 3).

Gorakhpur is fortunate to enjoy the heavy rainfall spells in the monsoon season due to tropical weather system. The summer monsoon

Figure 2: Himalaya mountain region.

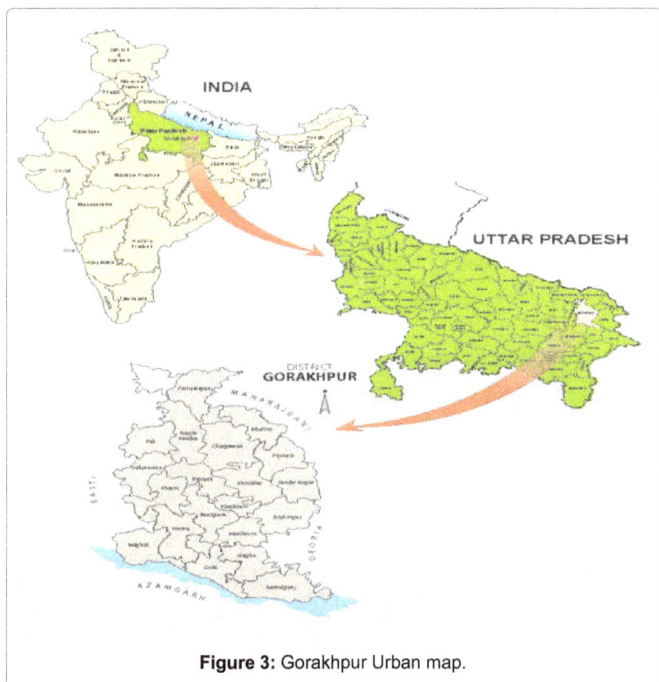

Figure 3: Gorakhpur Urban map.

season (June-Sep) is the main rainy season contributing 72-78% of the annual rainfall. Although, the contribution from other seasons, viz. the winter and pre-monsoon season and the post-monsoon season rainfall are not very significant, they are quite important for the other crops also. Main weather systems which bring rainfall to the region are monsoon low pressure area, depressions, thunderstorms, western disturbances etc. The orography of this place also influences the intensity and distribution of rainfall [1].

In view of the paramount importance of the rainfall from economic, societal and scientific points, extensive work has been carried out over the years on its various facets like trends, disaster events, spatio temporal variability, seasonal contribution etc. Goswami et al. [2] used grid point data at 100 km resolution [3] and demonstrated a significant increasing trend in the frequency and the magnitude extreme monsoon rain events in central north east India over the past 50 years. The information of the peak rainfall intensities at the station is instrumental for the planning of urban development, disaster management and for the study the environmental aspects pertaining to water runoffs in the vicinity of station. Therefore, present analysis is carried out using the station data and all the seasons are considered.

Criterion for Extreme Point Rainfall Event

The rainfall of 100 mm/day may be an extreme for the Gorakhpur region whereas it may not be a significant amount for the northeast region or along the west coast of India during summer monsoon. Even in summer monsoon season, west coast of India gets heavy rainfall spells in the first fortnight of June while the northern part of the country is devoid of the rainfall. Therefore for this write up the magnitude of extreme point rainfall event is taken as a fixed threshold for this place. Considering the climatological data, the magnitude of the extreme point rainfall event at the station is defined as its highest 24-hour rainfall reported in a particular month during the entire period of the data availability. This definition is adopted in order to examine whether there was any change in the number and intensity of the extreme point rainfall event in the recent decades.

Data and Methodology

Gorakhpur station data availability of at least 43 years up to June 2013 is considered. Only the cases with the minimum rainfall of 100 mm/day aretaken into account to give weightage to the high rainfall values.

The instances of Extreme Point Rainfall Events at the station are classified chronologically according to the decades. The high rainfall events occurred at the station after 1980 are compared with those of the earlier period to assess whether the previous Extreme Point Rainfall Events are exceeded in recent decades. Subsequently, the extreme rainfall events occurred on different time scales are also discussed in this paper.

Results

In order to compare the extreme point rainfall events in different periods. Four time slots are considered i.e. I. Period up to 1980, II. (1981-1990), III. (1991-2000), IV. (2001 to June 2013). Accordingly the outcome of the analysis is briefly presented below.

Period up to 1980

The magnitude of extreme point rainfall events recorded at the station, the dates of occurrence of these events and the data length of the station is shown in Table 1. The bold digits in the Table 1 indicate

Date	Rainfall in mm	Data Length
16-6-70	126.8	43 year
19-8-71	85	
25-8-72	104.4	
11.8.73	67.8	
28.6.74	54	
19.7.75	66.5	
14.9.76	45.8	
11.8.77	76.1	
18.7.78	123.8	
15.7.79	85.1	
08.6.80	175.8	

Table 1: Period- (1970-1980).

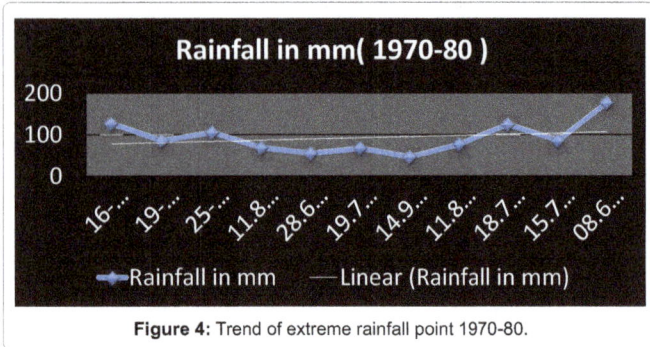

Figure 4: Trend of extreme rainfall point 1970-80.

Date	Rainfall in mm
31.7.81	108
05.6.82	131.1
19.7.83	90.5
09.6.84	259
16.9.85	137.6
30.9.86	84.2
11.8.87	175.9
16.8.88	86.4
11.7.89	147.4
12.8.90	67.1

Table 2: Period- (1981-1990).

that the rainfall was the highest for all the month (all time record) In case the station has registered extreme point rainfall events for more than one month; only a case with the maximum rainfall is taken into account (Figure 4).

Period: 1981-1990

High rainfall instances reported during 1981-90 is shown in Table 2. Bold digits indicate the rainfall more than 250 mm. Some notable instance is given below.

Gorakhpur recorded extremely heavy rainfall of 259 mm on 09 June 1984 which was highest in 72 years. It was due to intense thundershowers activity (Figure 5).

Period: 1991-2000

An extreme point rainfall event for this period is depicted in Table 3. A typical case is highlighted below.

Gorakhpur experienced very heavy rainfall of 200 mm on 09 July 1998. It was associated with passage of depression over east U.P. embedded with monsoon circulation (Figure 6).

Period: 2001-2013

An extreme point rainfall event for this period is depicted in Table 4. A typical case is highlighted below.

Gorakhpur recorded 342.9 mm rainfall on 29 June 2013 crossing its previous all-time highest of 284.5 mm reported on 19 August 1912. City was hit miserably due to unprecedented deluge. It was mainly due to passage of low pressure area over east U.P. and embedded with monsoon circulation and monsoon trough was very close and north of Gorakhpur (Figure 7).

Rainfall events exceeding 100 mm/day

Gorakhpur station which reported the rainfall 100 mm per day has been identified 21 times during (1970-June 2013).

Surpassing of all-time records

Gorakhpur station recorded 342.9 mm rainfall on 29 June 2013. It was the record as the highest 24 hrs. Rainfall over this place during last 113 years and surpassed the all-time record of 284.5 mm on 19 Aug 1912. It is further noticed that four of top seven rain events have

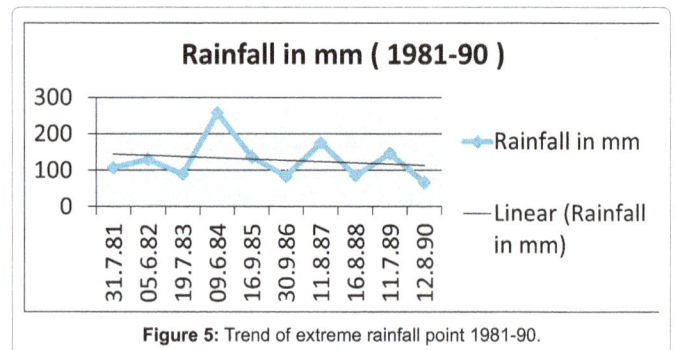

Figure 5: Trend of extreme rainfall point 1981-90.

Date	Rainfall in mm
01.7.91	93.5
13.7.92	75.1
25.8.93	107.7
23.7.94	75.5
17.7.95	86
12.8.96	82.5
12.9.97	74
09.7.98	200
17.8.99	68.9
07.8.2000	124

Table 3: Period- (1991-2000).

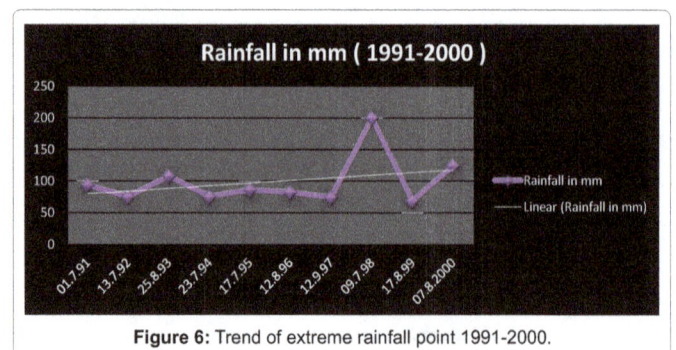

Figure 6: Trend of extreme rainfall point 1991-2000.

Date	Rainfall in mm
22.8.2001	158
04.7.2002	40
19.8.2003	63.5
28.8.2004	218
15.8.2005	91
27.8.2006	85.1
14.8.2007	74.4
13.7.2008	128
05.7.2009	129
12.7.2010	189
03.7.2011	132.09
17.9.2012	110.6
29.6.2013	342.9

Table 4: Period- (2001-June 2013).

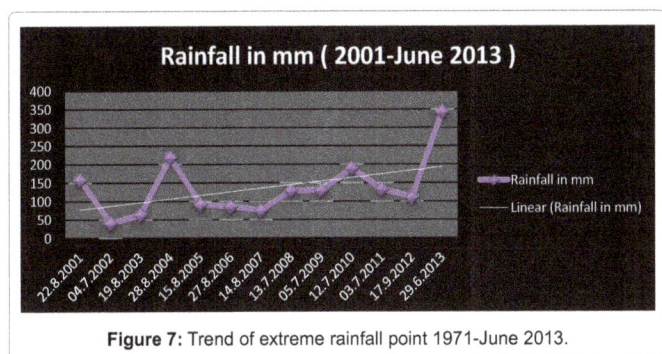

Figure 7: Trend of extreme rainfall point 1971-June 2013.

occurred after 2000 (Table 4) indicating the rise in intensity of extreme point rainfall events in the recent times.

High rainfall spells on different time scales

The cases of extreme rain events for 24 hrs. are described above. There are some instances of very high rainfall reported from 1970 on different time scales. They are described below.

Short duration record rainfall: On 09 July 1998 Gorakhpur station recorded 80 mm rainfall in one hour exceeding the previous record 73.0 mm (01 hour) held earlier on 11 August 1972.

Unseasonal heavy rainfall instances: Gorakhpur station reported 96.9 mm rainfall on 06 Oct 1970 getting more than 100 mm in same month i.e. 106.7 mm on 19 Oct 1987. It is an event of the four decade for the city. However, it was a curse to the city as it was harvesting season of kharif crops.

Rainiest year of Gorakhpur: On considering the data of 100 year (1901-2000). Annual mean rainfall of this place is 1238.9 mm. the two rainiest years over this place during last 112 year was 1989 and 2008 with actual recorded rainfall 2087 mm and 1991 mm respectively.

Discussion of Results

Rise in number and intensity of extreme point rainfall events in recent decades

The results figure out is that the majority (70-80%) have registered their extreme point rainfall events during 1981-90 and 2001-June 2013. During the 2001-June 2013 station reported the rainfall events surpassing the intensity of their previous highest rainfall. Some records were established on different time scales varying from hourly to annual

scales with the most of them noticed from 2000. The stations have experienced an alarming rise (40-50 %) in their intensity.

Possible cause of rise in extreme point rainfall events and their intensity in recent decades

It is well established fact that the global average surface temperature has increased during last 150 years and eleven years of the recent time (1995-2006) were among the warmest years. The global land surface has warmed at the rate of 0.07°C per decade during the past century [4]. From the late 1950s the rise is noticed in the lowest 8 km of the atmosphere. The details are available in the third assessment report [5] of the Intergovernmental Panel on Climate Change (IPCC).

The study over this place indicates that mean maximum temperature of Gorakhpur has increased by 0.88°C for the period (1901-2010). Whereas mean minimum temperature has decreased insignificantly. Changes in maximum and minimum temperature are more significant in post 1960 period. Rate of decrease in mean minimum temperature is not commensurate with increase in mean maximum temperature. It leads to overall warming of atmosphere [6].

The sea surface temperature of the oceanic region around India has also gone up by 0.6°C in 100 years and by about 0.15°C per decade from 1971 [7,8].

The rise in sea surface temperature more evaporation and the increase in the surface air temperature lead to deeper convection. Besides the warming of upper levels enhances the moisture holding capacity of the atmosphere. As such under this scenario, the weather systems like thunderstorms, the depressions etc., would have more potential for intense precipitation as compared to the cooler environment. Therefore, it is conjectured that the accelerated warming during last three decades and the warmest period of recent 11 years, could be the major cause for the increase in the extreme rainfall spells during past four decades with the sharp rise their intensity after the mid-1990s.

Although, the climate models have certain uncertainties and the atmospheric process are not well understood, it may be mentioned that some model projections show that current rise in incidence of hot summers is likely to continue in the northern hemisphere [4]. During next two decades, warming about 010-0.2°C per decade is expected to take place due to greenhouse gases emissions [5]. The extremes in the temperature [9] and the intensity of heavy rainfall events may increase in the future over the India region. In the light of this analysis, under climate change scenario, it is quite likely that the steep rise in the instances of extreme point rainfall events may continue in coming decades.

Conclusions

This analysis over the station Gorakhpur with a long data series shows that majority of them have reported their highest 24 hours rainfall during (2001-June 2013) with an alarming rise in intensity thereafter. Record rainfall events on different time scales (hourly to annual) have also taken place in recent decades. The instances of extreme point rainfall events have affected this place. These events may be associated with and regional warming signaling the effect of the climate changes over the region. Therefore if the trend of global warming continues, the extreme point rainfall events also may continue to occur in the future. They would pose serious problems in some parts of the district due to their adverse impact on socio economic issues like damage of property, infrastructure and life. Such spells may cause river silting in rapti and Rohini River as it originates from hills of Nepal. In view of these points it is imperative that proper care need to be exercised in near future for the work of planning, disasters management and the environmental protection for the sustainable development of the human beings over this region.

Acknowledgement

Author is grateful to the Rockefeller Foundation for the support to carry out this study. Thanks are also due to our president Dr. Shiraz AWajih G.E.A.G. for his regular encouragement and valuable support during the entire study. I would like to thanks to S.K. Nateshan for a critically reviewing of the manuscript and offering valuable comments.

References

1. May W (2004) Simulation of the variability and extremes of daily rainfall during the Indian summer monsoon for present and future times in a global time-slice Experiment, Climate Dynamics 22: 183-204.

2. Goswami BN, Venugopal V, Sengupta D, Madhusudan MS, Xavier PK (2006) Increasing trend of extreme rain events over India in a warming environment, Science 314: 1442-1445.

3. Francis PA, Gadgil S (2006) Intense rainfall events over the west coast of India. Met and AtmospPhys 94: 27-42.

4. Jones PD, Moberg A (2003) Hemispheric and large-scale surface air temperature variations : An extensive revision and update to 2001. J Climate 16: 206-223.

5. Houghton JT, Ding Y, Griggs DJ, Noguer M, Linden PJ, et al. (2001) Climate Change 2001: The Scientific Basis. Cambridge Univ Press, Third Assessment Report, IPCC.

6. Pandey K (2013) Prognostic Study of Climate Behavior in Gorakhpur.

7. Kothawale DR, Rupa Kumar K (2002) Tropospheric temperature variation over India and links with the Indian summer monsoon:1971-2000. Mausam 53; 289-308.

8. Kothawale DR, Rupa Kumar K (2005) On the recent changes in surface Temperature trends over India. Geophysical Res Lett.

9. Ajaymohan RS and Rao SA (2008) Indian Ocean dipole modulates the number of extreme rainfall events over India in a warming environment.Jr Met Soc Japan. 86:245-252.

17

An Assessment of Effective Temperature, Relative Strain Index and Dew Point Temperature Over Southwest Nigeria

Abuloye AP[1], Nevo AO[1], Eludoyin OM[2], Popoola KS[1] and Awotoye OO[1]

[1]*Institute of Ecology and Environmental Studies, Obafemi Awolowo University, Ile-Ife, Nigeria*

[2]*Adekunle Ajasin University, Akungba-Akoko, Ondo State, Nigeria*

*Corresponding author:** Abuloye AP, Institute of Ecology and Environmental Studies, Obafemi Awolowo University, Ile-Ife, Nigeria
E-mail: padeolu1@gmail.com

Abstract

This study examined the variations in some bioclimatic characteristics of the south-western Nigeria and residents' coping strategies to heat stress. Climate (temperature, dew point temperature, relative strain index and effective temperature) data for 10 meteorological stations between 1961 and 2013, and responses of residents of the region were examined. The study showed that temperature has increased by about 1°C between 1961-1990 and 1991-2010 periods at most stations, and that the region close to the Atlantic coast are under heat stress conditions, which has increased in the 1991-2010 period. Coping and mitigation strategies of the vulnerable settlements are largely personal adjustments; mainly change of cloth/wear and sleeping outdoor at night–in rural settlements while main technological adjustment was through use of fan and air conditioners.

Keywords: Temperature change; Dewpoint temperature; Effective temperature; Relative strain index; Thermal climate

Introduction

The study of human comfort has generated interests across the world, because comfort often forms the basis for planning for housing, healthcare and recreation facilities [1,2]. Knowledge of the thermal comfort ensures adequate warning for future extreme climate scenarios and for adequate preparation of humans and their livelihood against extreme climate conditions [3].

The developing countries have been reported to be characterized by increasing population but relatively poor social infrastructural facilities to cope with extreme climate effects. While many studies have been carried out in temperate region and in developed countries, only a few studies have been done in many developing countries, including Nigeria [4,5].

Human thermal comfort is an expression of feeling of satisfaction with the prevailing weather condition [6,7]. Extreme climate conditions outside the comfort zone are conditions of heat and cold stresses. Stressful climatic conditions are known to decrease work productivity, and increase (heat-related) mortality or morbidity [3,8-10]. Thermal stresses in tropical environment are often associated with increased heat and cold stress, and are typically described in terms of too high or too low humidity, temperatures and in excess of certain thermal thresholds [5,11-13]. Thermal thresholds are usually in terms of unitary (temperature, humidity and other heat related climatic elements) and integrative indices [14].

Abuloye [15] modified a list of human thermal comfort indices by De Freitas and Grigorieva [14], and listed 82 integrative indices, out of which only three (Effective Temperature (ET), Relative Strain (RS) and Temperature Humidity Indices (THI)) have been extensively reported on Nigeria [3,4]. These previous studies have shown that the results from ET and THI are not significantly different for Nigeria, and that

ET is more suitable for the country. In addition, the studies indicated that, since Nigeria is a large country (with a population of more than 140 million spread over an area of 923 800 km^2), each of the geographical regions deserve to be studied for variability in the thermal climate.

The south-western part of Nigeria is known as the most populated and industrialised within the country, especially as it is the location for the second most populated (Lagos), and second and third largest city in Africa (Ibadan and Ogbomosho, respectively). Studies on the south-western region have focused on the temporal changes in the temperature and relative humidity [12,16-20], and existing studies with regional focus is scarce to find.

In addition, each of these studies has focused on one settlement or another in the region, and relatively dated. Furthermore, published studies informing about the perception of residents of the region (and Nigeria, in general) to thermal comfort have only focused on the people in tertiary schools [21], and the perceptions of the people in either the rural or urban areas in this region are less reported. Studies on adaptation of humans to stressful thermal climatic conditions have nevertheless suggested that adaptive methods (which may be anticipatory or reactive; behavioural, physiological or psychological) can vary over space, and can inform about the coping capability of the people with climate change [22].

Subsequently, information presented in this study is aimed at enhancing policy issues on peoples' preparedness to extreme climatic conditions in the region. Specific objectives are to determine the spatial and temporal variations in the thermal conditions over the south-western Nigeria; and examine the adaptive methods of the residents of two purposively selected settlements for coping with extreme weather conditions in the region.

This study is presented as a case study for regional study of thermal comfort in developing countries. Indices evaluated for thermal conditions in this study are Effective Temperature (ET), Relative Strain

Index (RSI) and Dew Point temperature (Td), based on previous studies that showed that they are relevant to tropical climates. Although the Td is not a comfort index, it is a measure of humidity, and this also has significant implication for thermal stress [23]. The ET and RSI are also not state-of-the-art indicators because they do not take into consideration the effects of radiation and wind velocity, which are heat elements [7].

However, studies on the regional climate with respect to the indices and humidity are few, and the results from the traditional indices will provide a control for the understanding of the more state-of-the-art indices in the area.

Study Area

The southwest of Nigeria lies between 2°3'-6° E and 6°2'-8°4' N (Figure 1), and is characterised by the tropical rainforest in the south and tropical guinea savanna in the north. The climate of region is strongly influenced by the moisture-laden tropical maritime (mT) and dust-laden tropical continental (cT) air masses [24].

The southern sub-region is characterized by a general low relief (0-200 m above sea level), and rises gently northwards to the area of crystalline rocks where inselbergs (981 m) rise abruptly above the surrounding plains [25]. In the urban centres, construction of roads, buildings, factories, manufacturing plants, bridges and culverts, farmlands and others have reduced drainage channels and erosion passages and or diverted the natural courses of others [26].

The population of the region, as at 2006 was over 36 million, and was projected to be about 183 million in 2015, and about 433 million by 2050. The total land area is about 166,361 km [27].

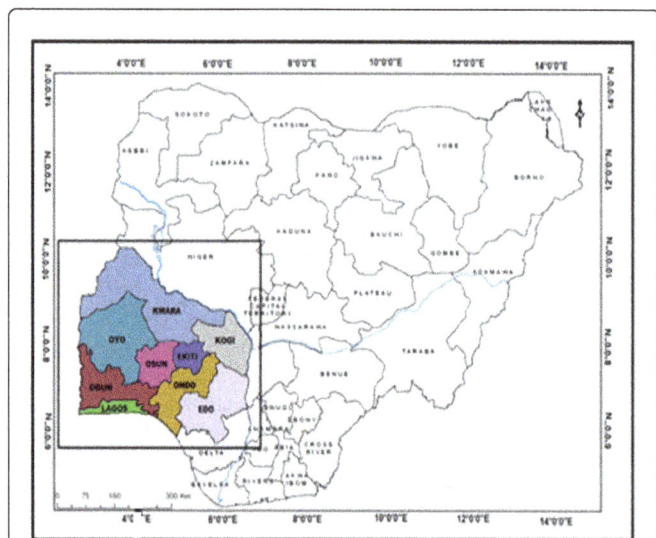

Figure 1: Map of Nigeria, showing the location of the states that make up the southwest region.

Previous studies showed evidence of regional difference in the climate of Nigeria, as typical of large countries, including India, United States of America and China [13,21,28].

Data

Data used for this study include the monthly temperature and relative humidity records for all the 10 synoptic stations (with the records from 1960 onwards) in the region (Figure 2) [29]. The data were obtained from the office of the Nigerian Meteorological Agency's office at Oshodi, Lagos. Data were examined for spurious values and evidence of non-climatic heterogeneity and instrumental errors as advised by the World Meteorological Organisation (1989).

The thermal comfort indices (ET, RSI and Td) were derived from the temperature and corresponding relative humidity data using equations i–iii [11,28] (for ET); Giles (for RSI); Wolkoff and Kjargaard [10] (for RSI)):

$$ET = T - 0.4(T - 10) \times (1 - \frac{RH}{100}) \qquad (1)$$

$$RSI = \left(\frac{T - 21}{58 - e}\right) \qquad (2a)$$

$$e = \frac{(RH \times Vp)}{100} \qquad (2b)$$

$$Vp = 6.11 \times 10^{7.5\frac{T}{237.3} + T} \qquad (2c)$$

$$T_d = T - \left(\frac{100 - RH}{5}\right) \qquad (3)$$

Where ET: Effective Temperature, H_{rh}: Relative Humidity, RSI: Relative Strain Index, Vp: Vapour Pressure (hPa), T_d: Dewpoint Temperature, T: Air Temperature, RH: Relative Humidity, e: Actual Vapour Pressure.

The mean and variations of the elements and indices were computed for 1961-1990, 1991-2010 and 1961-2010, and descriptive maps were produced with the moving average interpolation statistics in geographic information systems, for the periods. For the generation of comfort map for the study area, comfortable region for Td, RSI and ET were 17.5–23°C [23], 0.1–0.2 [30] and 18.9–25.6°C [11,31]. Areas outside these thresholds were mapped as discomfort zone in the study.

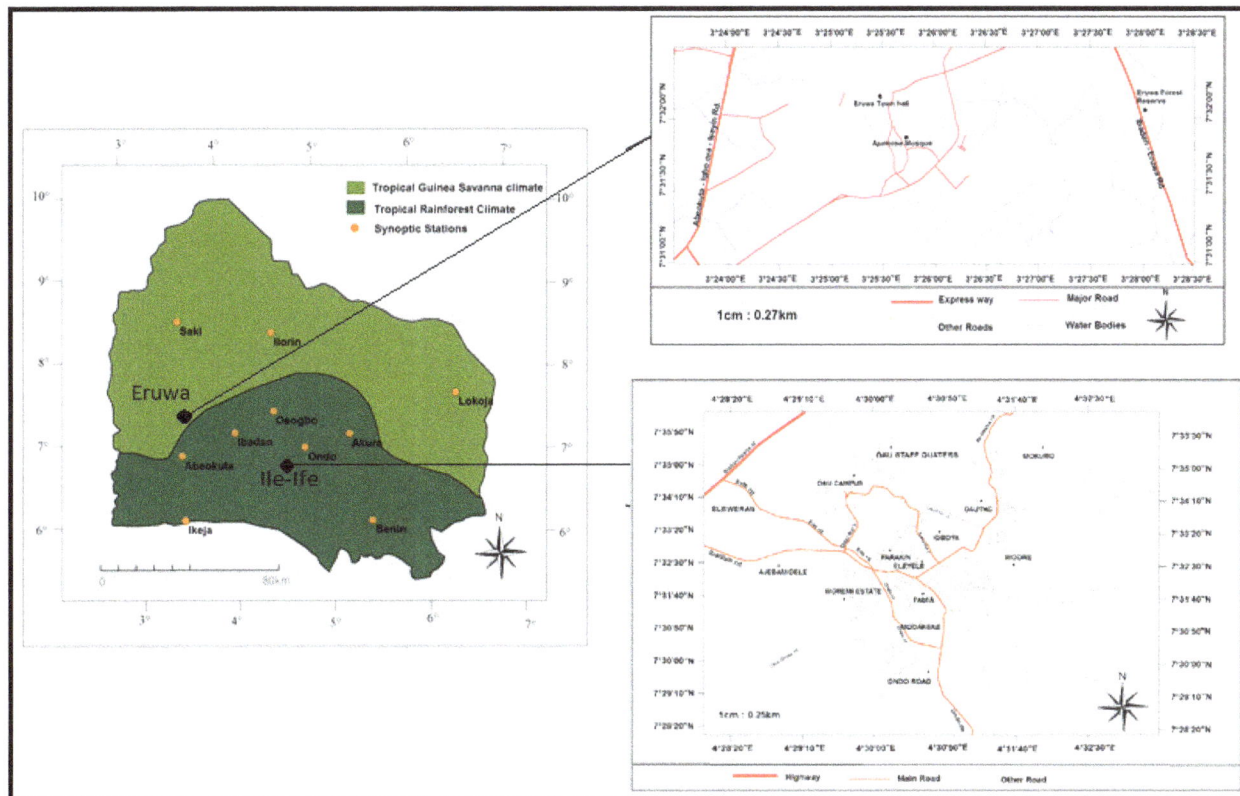

Figure 2: Selected meteorological stations and settlements for the perception study.

In addition, responses of residents of two purposively (based on convenience and relatively small size, and significant dichotomy across different landuse areas) selected settlements (Ile-Ife, a medium-size University town, and Eruwa, a semi-rural settlement in the region) were elicited with a semi structured questionnaire (Appendix). These settlements also represent locations in the two (tropical rainforest and guinea savanna) main ecological zones in the regions (Figure 2). In each settlement, occupants of every third building in a street) were sampled in identified residential, industrial (e.g. sawmills and a metal scrapping and smelting firm in Ile-Ife) and commercial (markets) areas in each settlement. In all, 501 (321 and 180, in Ile-Ife and Eruwa, respectively) copies of questionnaire, which represents the responses of about 0.5% of the entire population of the two communities were returned after two weeks of administration by ten field personnel. Most of the copies of questionnaire for the residential areas were administered in the evening and weekends, because most respondents were not often in the house in the other periods (Table 1).

Results

Regional pattern of temperature and thermal variables

Mean variations, trends and change in 1960-1990 and 1991-2013 means: The summary statistics for the 50 (1961–2010) year mean monthly temperatures (minimum, average and maximum) and those of ET, RSI and Td at all the selected stations are presented in Tables 2 and 3, respectively. Minimum temperature over the region varied between 21.1 and 22.9°C while maximum temperature varied between 30.9 and 32.2°C; mean temperature was from 26.2 to 28.1°C. Dewpoint temperature values that varied from 21.3 to 24.4°C while RSI values ranged from 0.09 to 0.15 (no unit). The ET values were between 24.6 and 26.7°C. The temporal variations in the temperature variables were generally below 10%. Furthermore, whilst Akure (a medium-size administrative city in the region) exhibited decreasing trend in all the temperature variables, other stations exhibited increasing trends in minimum (except Ibadan and Ikeja) and maximum (except Ilorin and Lokoja) over the years. Correspondingly, relative humidity has increased in Akure, and Saki. Comparison of the 1961–1990 and 1991–2010 values of the temperature variables indicated significant (p ≤ 0.05) increase (though with small 30-year average in most cases) at most stations (except Ilorin for mean and minimum temperature, Benin for maximum, Lokoja and Osogbo, for the mean temperature). Relative humidity also significantly reduced at most stations, except at Saki, Lokoja and Akure (p ≥ 0.05).

Variable	Option	Percentage (Frequency)	
		Ile-Ife	Eruwa
Age (years)	<18	1.9 (6)	-
	18-30	33.7 (106)	-
	31-60	51.1 (161)	76.1 (137)
	>60	13.3 (42)	23.9 (43)
Gender	Male	49.8 (160)	48.9 (88)
	Female	50.2 (161)	51.1 (92)
Height (meter)	<1.2	2.0 (6)	-
	1.2-1.79	46.4 (140)	15.9 (28)
	>1.8	51.7 (156)	84.1 (148)
Body type	Slim	13.1 (42)	8.9 (16)
	Moderate	60.4 (194)	52.2 (94)
	Fat	26.5 (85)	38.9 (70)
Marital status	Single	38.2 (120)	-
	Married	61.8 (194)	100 (180)
Minimum educational qualification	No formal education	1.9 (6)	-
	Primary	1.2 (4)	4.4 (8)
	Secondary/technical	42.7 (137)	31.7 (57)
	Tertiary	54.2 (174)	63.9 (115)
Job category	Student	18.7 (60)	-
	Self Employed	28.7 (92)	21.1 (38)
	Government Employee	35.5 (114)	34.4 (62)
	Private Employee	17.1 (55)	44.4 (80)
No of dependant(s)	None	27.9 (88)	-
	<5	62.9 (198)	97.2 (175)
	>5	9.2 (29)	2.8 (5)
Monthly income (in Naira (N))	<20,000	21.6 (69)	-
	20,000-50,000	51.1 (164)	40.6 (73)
	>50,000	27.4 (88)	59.4 (107)
Approximate No of work hour (Hours)	<5	8.7 (28)	2.2 (94)
	5-8	65.4 (210)	92.8 (167)
	>8	25.9 (83)	5.0 (9)
Category of occupants	Outdoor under shade	35.4 (107)	18.9 (34)
	Outdoor without shade	9.3 (28)	5.0 (9)
	Indoor	55.3 (167)	76.1 (137)

Table 1: Demographic and socio-economic characteristics of the respondents.

The patterns of temporal change in RS, Td and ET were different across the stations. Whereas the Td trend at most of the stations showed significant increase between the 1961–1990 and 1991–2010 values (except at Akure, Benin and Ilorin, where it significantly decreased, or Ondo and Saki, where changes in Td was not significant), RSI significantly reduced at most of the stations (except Saki, where it has significantly increased and other stations, where such variation was not significant). Significant 1960–1990 and 1991–2013 change occurred in the values of ET at all but the Ibadan station, and such trend of change was negative for most stations (except Abeokuta, Ikeja and Saki) (Table 3).

	Station	Mean (Min-Max) (°C)	CV (%)	Trend (a ± b(x)) (R2 in parenthesis)	1961-1990 (mean) (°C)	1991-2010 (mean) (°C)	Variation	
							ANOVA	
							P-value	F-value
Minimum temperature	Abeokuta	23.6 (15.7-26.5)	5.8	0.01x+21.3 (0.1)	23.2	23.9	0.002	609.8
	Akure	21.4 (13.2-23.8)	6.7	-0.001x+21.8 (0.01)	21.6	21.4	0.001	39.9
	Benin	22.9 (18.4-26.3)	4.3	0.003x+22.2(0.2)	23.1	23.5	0.002	252
	Ibadan	22.3 (16.4-29.8)	5.2	-0.003x+21.5 (0.2)	22.6	22.8	0.001	14.7
	Ikeja	23.3 (20-29)	5.7	0.005x+22 (0.3)	23.5	24.1	0.002	399.1
	Ilorin	21.3 (14.5-25)	7.8	0.002x+20.7 (0.05)	21.7	21.7	0.535	0.7
	Lokoja	22.8 (14.1-27.7)	9	0.001x+22.4 (0.01)	22.7	23.3	0.002	129.8
	Ondo	22.2 (18.5-25.1)	4.1	0.002x+21.8 (0.07)	22.3	22.5	0.001	57.9
	Osogbo	21.1 (13.5-24.3)	7.5	0.003x+21.1 (0.001)	21.5	21	0.001	122.9
	Saki	22.7 (19.5-25.4)	4.8	0.002x+21.7 (0.03)	22.6	22.8	0.003	44.8
	Summary	22.4 (13.5-29.8)	6		22.48	22.7	0.06	167.2
Maximum temperature	Abeokuta	32.7 (28.0-38.1)	8	0.003x+31.7 (0.03)	32.7	32.9	0.002	9.44
	Akure	31.0 (26.5-36.5)	8	-0.01x+32.9 (0.03)	31.4	30.9	0.002	92.68
	Benin	31.4 (27.0-37.0)	6	0.002x+31 (0.02)	31.7	31.7	0.749	0.102
	Ibadan	31.3 (23.7-38.0)	8	0.002x+30.8 (0.02)	31.6	31.9	0.003	18.17
	Ikeja	30.9 (26.9-35.3)	6	0.002x+30.5 (0.02)	31	31.4	0	66.33
	Ilorin	32.2 (27.5-37.9)	8	-0.001x+32.6 (0.01)	32.2	32	0.031	4.666
	Lokoja	33.1 (28.0-39.1)	7	-0.001x+33.1 (0.001)	33.7	33.1	0.002	110.9
	Ondo	30.9 (26.1-34.2)	8	0.03x+30.0 (0.03)	30.9	31.4	0.003	57.87
	Osogbo	31.2 (25.0-37.2)	8	0.001x+30.9 (0.01)	31.3	31.5	0.001	11.26
	Saki	31.6 (25.0-37.2)	7	0.002x+30.8 (0.004)	31.5	31.7	0.013	6.165
	Summary	31.6 (25-38.1)	7		31.8	31.9	0.08	37.8
Mean temperature	Abeokuta	28.1 (23.9-32.2)	5.9	0.004x+26.5(0.05)	27.9	28.4	0.002	138.9
	Akure	26.2 (22.4-30.1)	5.3	-0.03x+27.3 (0.04)	26.5	26.1	0.001	128.6
	Benin	27.2 (24.3-31.3)	4.9	0.002x+26.5 (0.07)	27.4	27.6	0.003	28.6
	Ibadan	26.8 (23.4-31.4)	5.9	0.03x+26.1 (0.1)	27.1	27.3	0.003	21.46
	Ikeja	27.2 (21.6-31.3)	5.4	0.003x+26.2 (0.1)	27.2	27.7	0	205.2
	Ilorin	26.8 (22.3-30.7)	6.1	0.001x+26.2 (0.01)	26.9	26.8	0.041	4.185
	Lokoja	28.0 (23.9-32.9)	6.2	0.001x+26.2 (0.003)	28.2	28.2	0.571	0.32
	Ondo	26.5 (23.5-38.9)	5.2	0.002x+25.9 (0.1)	26.6	26.9	0.002	100.9
	Osogbo	26.2 (23.1-30.2)	5.5	0.001x+26.0 (0.01)	26.4	26.3	0.119	2.44
	Saki	27.1 (24.8-30.4)	5.2	0.002x+26.2 (0.01)	27	27.2	0.001	20.42
	Summary	27 (22.3-38.9)	5.5		27.1	27.3	0.07	65.1

Table 2: Mean, range, variation, trend (change of values over time) and temporal variations in temperature variables over selected station in southwest Nigeria (1961-2010).

	Station	Mean (Min-Max) (°C)	CV (%)	Trend (a ± b(x)) (R2 in parenthesis)	1961-1990 (mean) (°C)	1991-2010 (mean) (°C)	Variation	
							ANOVA	
							P-value	F-value
Effective temperature Index	Abeokuta	26.7 (20.1-29.9)	5.4	0.01x+25.4 (0.04)	26.6	26.9	0.002	94.2
	Akure	24.6 (17.6-28.0)	1	-0.01x+26.5(0.03)	25.1	24.2	0.004	209.9
	Benin	26.0 (19.2-30.4)	0.8	-0.003x+25.3 (0.1)	26.2	26.5	0.001	33.8
	Ibadan	25.5 (22.0-29.8)	5	0.002x+24.9 (0.1)	25.8	25.8	0.252	1.32
	Ikeja	25.9 (21.2-29.5)	4.8	0.001x+25.2 (0.004)	25.8	26.4	0.003	392.5
	Ilorin	24.9 (16.1-28.8)	6.9	-0.001x+25.1 (0.004)	25	24.7	0.005	39.49
	Lokoja	25.6 (18.6-29.6)	12	-0.001x+26.0 (0.01)	26.1	25.3	0.004	99.79
	Ondo	25.3 (19.4-36.2)	4.4	-0.002x+24.9 (0.1)	25.5	25.7	0.002	54.5
	Osogbo	25.0 (21.1-28.5)	4.5	-0.001x+24.8 (0.01)	25.2	25	0.001	42.23
	Saki	25.3 (21.8-27.6)	4.1	0.002x+24.5 (0.02)	25.3	25.4	0.001	13.52
	Summary	25.5 (16.1-36.2)	6.6		25.7	25.6	0.03	98.1
Relative Strain Index (no unit)	Abeokuta	0.12 (0.05-0.19)	23	1.0-0.03x (0.04)	0.12	0.13	0.002	120.1
	Akure	0.09 (-0.4-0.17)	53	0.1-0.01x (0.004)	0.09	0.09	0.004	65.3
	Benin	0.15 (-0.41-0.18)	37	0.09 +0.1x (0.04)	0.11	0.11	0.2	25.1
	Ibadan	0.1 (0.04-0.18)	27	0.1+0.1x (0.06)	0.11	0.11	0.3	21.5
	Ikeja	0.11 (0.01-0.18)	24	0.1-0.02x (0.1)	0.11	0.12	0.001	205.2
	Ilorin	0.1 (-0.39-0.17)	25	0.1+0.08x (0.01)	0.1	0.1	0.2	17.4
	Lokoja	0.09 (-0.39-0.20)	28	0.1-0.03x (0.001)	0.12	0.11	0.001	71.8
	Ondo	0.11 (0.04-0.31)	28	0.08+0.25x (0.06)	0.1	0.1	0.4	89.7
	Osogbo	0.1 (0.04-0.16)	36	0.1+0.38x (0.01)	0.09	0.09	0.119	2.43
	Saki	0.11 (0.07-0.16)	60	0.1+0.04x (0.01)	0.1	0.11	0.01	20.5
	Summary	0.1 (-0.39-0.31)	34		0.11	0.11	0.02	63.9
Dewpoint temperature (°C)	Abeokuta	24.4 (6.9-27.8)	8	0.003x+23.2 (0.02)	24.3	24.5	0.003	27.3
	Akure	21.3 (6.2-25.4)	19	-0.01x+25.9 (0.06)	22.4	20.7	0.001	296.5
	Benin	23.8 (2.2-29.3)	10	-0.002x+23.2 (0.02)	23.9	24.3	0.002	35.9
	Ibadan	22.9 (15.7-27.6)	6.5	0.002x+22.5 (0.02)	23.2	23.1	0.006	22.9
	Ikeja	23.5 (14.6-27.4)	6.1	0.001x+23.1 (0.03)	23.1	23.9	0.004	559.9
	Ilorin	21.5 (3.2-26.6)	13	-0.002x+22.1 (0.01)	21.5	21	0.002	45.8
	Lokoja	22.0 (7.4-27.6)	19	0.003x+22.9 (0.01)	22.6	21.3	0.003	123.8
	Ondo	23.0 (5.7-34.3)	6.5	0.001x+22.7 (0.02)	23.2	23.2	0.965	0.326
	Osogbo	22.6 (14.7-26.6)	6.3	0.001x+22.4 (0.01)	22.9	22.5	0.001	135.7
	Saki	22.0 (11.5-25.6)	12	0.002x+21.1(0.004)	22	22.1	0.721	0.396
	Summary	22.7 (2.2-34.3)	11		22.6	22.5	0.189	135.7

Table 3: Mean, range, variation, trend (change of values over time) and temporal variations in effective temperature, relative strain index and dewpoint temperature over selected station in southwest Nigeria (1961-2010).

Descriptive thermal maps for the region: The results of the moving average interpolation of the thermal climate variables investigated in this study are presented as Figures 3-6. Minimum temperature appears to decrease as one move away from the coastal areas, whereas maximum temperature was slightly higher around Lokoja than the other parts of the region. Except for the area around Abeokuta, where both minimum and maximum temperatures were higher than most part of the surroundings, maximum temperature tend to vary inversely with minimum temperature over the region (Figure 3). Both Td and ET were also higher around Abeokuta–Ikeja axis and Benin in the south than the settlements in the other part of the southwestern Nigeria while the RSI interpolation showed lower values in the inner settlements (Osogbo–Ondo-Akure –Ilorin axis) of the region than the other parts (Figure 3d-3f). Furthermore, whereas the values of minimum and mean temperature around the Ikeja and Lokoja axes have generally increased by 1°C, in 1991-2010 period, when compared to the 1961-1990 means, there was a decrease in the maximum temperature, also, by about 1°C around Lokoja in 1991-2010 period (Figure 4). In addition, Td and ET interpolations over the study area exhibited an increase of at about 1°C, over the coastal region of the Atlantic Ocean, and a converse decrease by 1°N in the northern sub-region. The RSI unit was also lower around the northern part in the 1991-2010 periods (Figure 5).

Assessment of the Td, RSI and ET over the study area with the recommended thresholds of 17.5–23°C, 0.1–0.2 and 18.9–25.6°C, respectively, showed that areas further into the interior of the region are more thermally comfortable than the coastal region, especially with respect to Td and ET but RSI results describes the entire region as thermally comfortable, however (Figure 6).

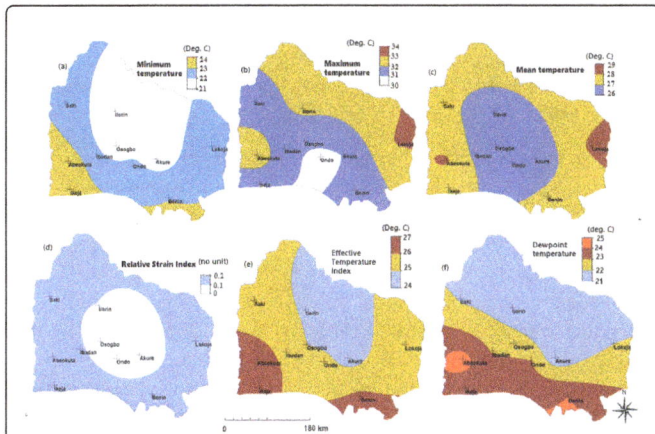

Figure 3: 50 (1961–2010)–year Average temperature (minimum, maximum and mean), relative strain index, effective temperature and dewpoint temperature over the southwest Nigeria. The cross indicates the location of the meteorological station whose data were interpolated.

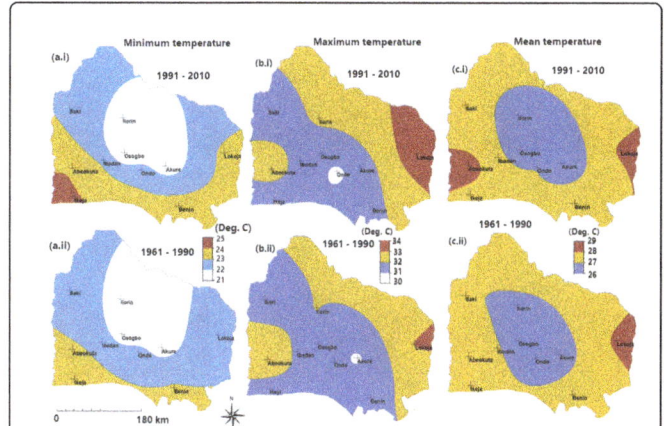

Figure 4: Variation in the minimum, mean and maximum temperature over the southwest Nigeria in 1961-1990 and 1991–2010 years periods. The cross indicates the location of the meteorological station whose data were interpolated.

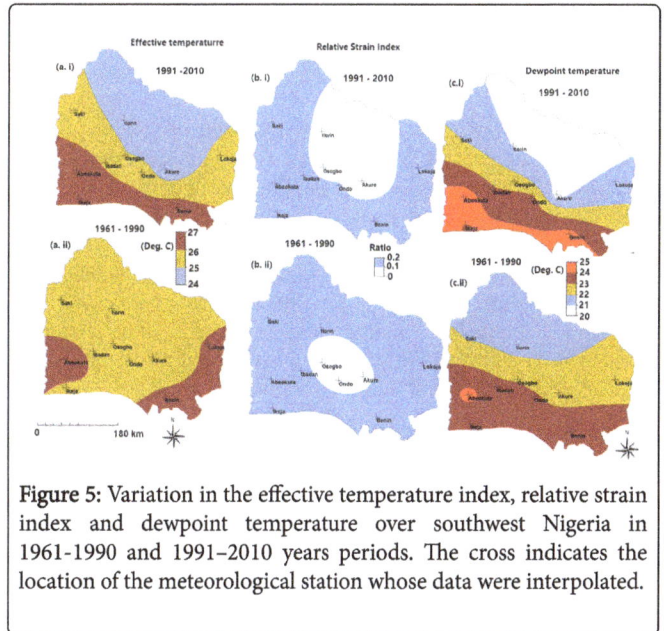

Figure 5: Variation in the effective temperature index, relative strain index and dewpoint temperature over southwest Nigeria in 1961-1990 and 1991–2010 years periods. The cross indicates the location of the meteorological station whose data were interpolated.

When compared across the 1961-1990 and 1991-2010 periods, the results of the interpolation indicate that whereas thermal discomfort has aggravated in more areas in the western sub-region of the study area, the thermal condition in eastern region has improved in the 1991-2010, than the preceding 30-year period (Figure 6).

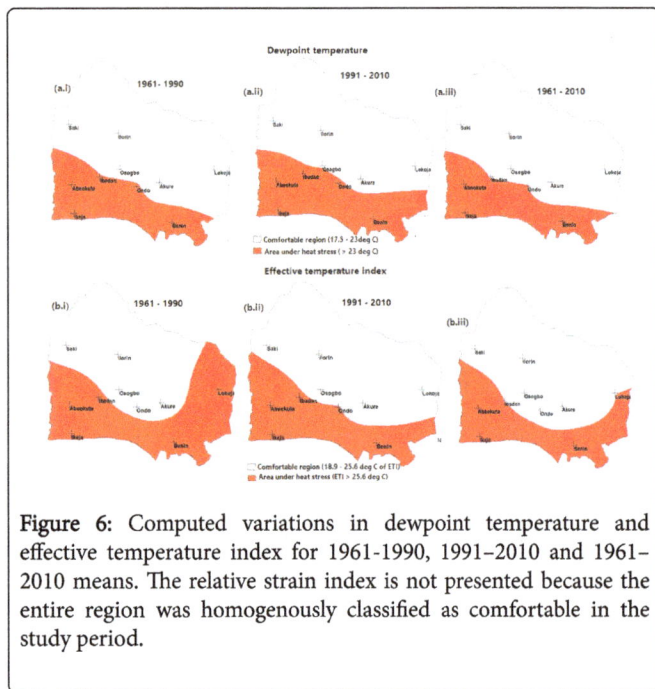

Figure 6: Computed variations in dewpoint temperature and effective temperature index for 1961-1990, 1991–2010 and 1961–2010 means. The relative strain index is not presented because the entire region was homogenously classified as comfortable in the study period.

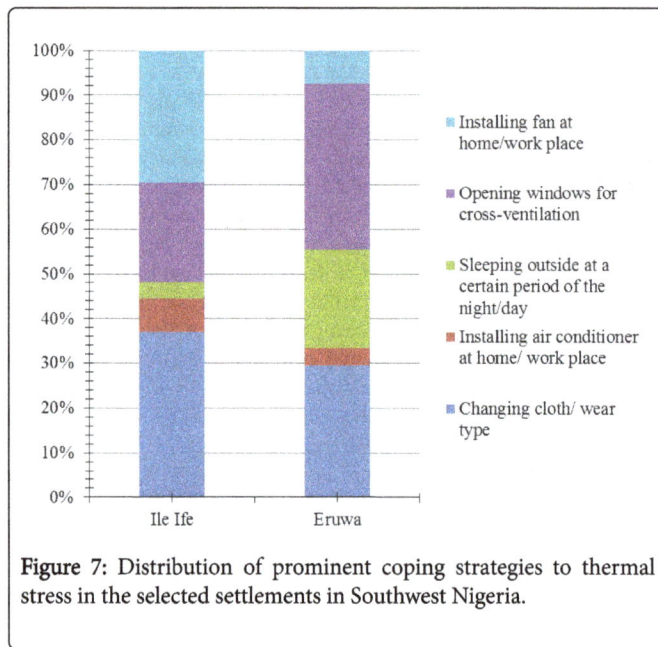

Figure 7: Distribution of prominent coping strategies to thermal stress in the selected settlements in Southwest Nigeria.

Perception of residents of selected settlements on comfortable weather condition

Characteristics of respondents: Over 94% of the respondents in both settlements have had at least primary education and were able to understand the content of the questionnaire and its purpose. The entire sampled population at Eruwa were aged above 30 years and majority of the respondents from Ile Ife were at least 18 years. About 81% of the respondents were employed and made at least N20, 000 ($100) monthly while more than 70% have dependants, and worked for a minimum of 5 h daily. In addition, majority of the respondents work indoor (55.3% and 76.1% respectively in Ile Ife and Eruwa) and under shade, if outdoor (35.4 and 18.9, respectively) (Table 1).

Coping strategies to extreme weather conditions in the southwest Nigeria: Concerns were raised over heat stress in the study area. At least 80% of the respondents at both the settlements of Eruwa and Ile-Ife considered the time between 0800 and 1200 Nigerian Local Standard Time (LST) as comfortable, and the periods between 1200 and 1800 LST as thermally uncomfortable. The period between 1800 and 2100 LST was also considered as comfortable by about 70% of the respondents. There was no significant (p ≤ 0.05) difference in the perceptions of the different categories of residents (those outdoor, undershade, outdoor without shade, and indoor) at the periods.

Coping strategies, however, varied among the different the majority of people in the two locations (Ile-Ife and Eruwa). Whilst the first three coping preferences at Ile-Ife were cloth/wear type, (ii) installation of fan and (iii) installation of air conditioners at home and offices, the most important preferences at Eruwa (in order of importance) were (ii) opening of ventilation for cross ventilation, (i) changing changing of cloth or wear type, and (iii) sleeping outside the house (outdoor) at certain times of the night or day (Figure 7).

Discussion

A number of recent studies have focused on the thermal comfort of Nigeria, probably because of the increased awareness on climate change and its effects on the people and their livelihoods [32-36]. Nigeria is a typical developing country where people have been noted to be vulnerable to the effects of extreme climatic conditions because of poor economic, political and technological responses [2,37,38]. Studies have noted that many sub-Saharan countries are vulnerable to conditions of livelihoods (e.g., food insecurity, water scarcity and environmental degradation) due to the intensive effects of climate-related causes, including decortication, desertification and temperature increase [39].

In Nigeria, our previous studies [4,21], among others have indicated that human comfort in the southwest Nigeria may be under heat stress in the recent years, but we could not determine the certainty of the stretch of the vulnerable area in the region, since the studies-like majority of other previous studies from other sources-have been based on smaller, country-wide scale. In the present larger, regional-scale study, the southern (coastal) settlements- consisting of the areas around Benin, Abeokuta, Ikeja, Ondo and Ibadan, which are mainly traditional large cities in the southwest Nigeria, are characterized by heat stress, as typical of many large cities, globally [40-42]. A major difference between the countries in developing and developed economies is however the reaction to the increasing thermal discomfort. Whilst in the southwest Nigeria, the common mitigating/coping strategies to heat stress are personal adjustments (changing personal variables, such as clothing, moving to a different location, etc.), and few instances of technological or environmental adjustments (e.g., opening or closing windows or shades, turning on of fans and/or air conditioners, especially among economically buoyant populations), approaches for coping strategies in many developed countries have developed into policy issues, early warning system and speedy responses. Metcalf et al. [43], for example, noted the development of urban greening as a social movement in New York whereas the few capital cities (e.g., Ikeja and Akure) that promote urban greening in Nigeria do it for aesthetic justification. Again, while studies have reported the adoption of early warning system on impending heat stress in north-eastern North America [3], cases of heat stress are still rarely reported in Nigerian hospitals. In our study area, majority of respondents argued that effects of climatic discomfort are 'not-life-threatening', and do not require reporting in hospitals. Nonetheless, heat stress has been linked with violence, heat stroke, rashes and meningitis among people [9,44,45].

Furthermore, this study has shown an increase in temperature by about 1°C in the southwest Nigeria between 1960–1990 and 1991–2013. Temperature increase in the southwest Nigeria may be associated with increased urbanization, transportation and industrial activities in the region, especially with its role as the commercial and industrial hub of the country. Recent studies have reported massive industrial plans for the area, but these plans are often unassociated with a parallel policy on impact assessment and mitigation [46,47]. For example, Oketola and Osibanjo noted that rapid and haphazard industrialization has taken place in part of the southwest Nigeria without environmental considerations, and that pollution abatement technologies are largely absent. Earlier, Hettige et al. [48] has also indicated that many developing countries (including Nigeria) lack the necessary information to set priorities, strategies, and action plans on environmental protection issues. Studies have also linked urbanization and associated activities with development of urban heat islands in some major cities in the region [20,32]. Also, studies have shown that a number of heat-related mortality cases have been reported in Nigeria and neighboring countries [49,50]. Similarly, concerns for sustainable development have increased, globally with the post Kyoto Protocol assessments that indicate the increased culpability of developing countries to contribute to global warming.

Conclusion

The study showed that there was significant increase in temperature and selected indices of thermal comfort at most studied stations in 1991-2013 over 1960-1990, and that while the thermal variables have increased, relative humidity exhibited significant decrease. It also showed that resident's coping strategies were mainly of personal adjustment, with few cases of environmental adjustments in urban settlements, or among economically buoyant population. Subsequently, there is need for policy re-drive in the region towards sustainable urbanization and development, and with concerns for the impact of many developmental actions on the humans.

References

1. De Freitas C (2003) Tourism climatology: evaluating environmental information for decision making and business planning in the recreation and tourism sector. Int J Biometeorol 48: 45-54.
2. World Health Organisation (WHO) (2011) Regional consultation on health of the urban, Proceedings of the 2010 Regional consultation of Mumbai, India, Regional Office for South East Asia, UNFPA, p: 82.
3. Kalkstein LS, Smoyer KE (1993) The impact of climate change on human health: some international implications. Experientia 49: 969-979.
4. Eludoyin OM, Adelekan IO (2013) The physiological climate of Nigeria. Int J Biometeorol 57: 241-264.
5. Eludoyin OM, Adelekan IO, Webster R, Eludoyin AO (2014) Air temperature, relative humidity, climate regionalization and thermal comfort of Nigeria. Int J Climatol 34: 2000-2018.
6. Lee DH (1953) Physiological climatology as a field of study, Annals of the Association of American Geographers 43: 127-137.
7. Terjung W (1967) The geographical application of some selected physio-climatic indices to Africa. Int J Biometeorol 11: 5-19.
8. Havenith G, Holmer I, Parsons KC (2002) Personal factors in thermal comfort assessment; clothing properties and metabolic heat production. In Energy Building 34: 581-591.
9. Alcamo J, Moreno JM, Novaky B, Bindi M, Corobov R, et al. (2007) Europe, climate change 2007: impacts, adaptation and vulnerability. Contribution of Working Group II to the Fourth Assessment Report of the Intergovernmental Panel on Climate Change. Cambridge University Press.
10. Wolkoff P, Kjargaard SK (2007) The dichotomy of relative humidity on indoor air quality. Environ Int 33: 850-857.
11. Ayoade JO (1978) Spatial and seasonal patterns of physiologic comfort in Nigeria. Arch Met Geoph Biokl B 26: 319-337.
12. Olaniran OJ (1982) The physiological climate of Ilorin, Nigeria. Arch Met Geoph Biocl 31: 287-299.
13. Candido C, Dear RJ, Lamberts R, Bitterncourt L (2010) Air movement acceptability limits and thermal comfort in Brazil's hot humid climate zone. Build Environ 45: 222-229.
14. De Freitas CR, Grigorieva EA (2015) A comprehensive catalogue and classification of human thermal climate indices. Int J Biometeorol 59: 109-120.
15. Abuloye AP (2015) Assessment of variations in thermal climate and response to thermal stress in some parts of the southwest Nigeria, Unpublished M.Sc. thesis, Obafemi Awolowo University, Ile-Ife, Nigeria.
16. Omogbai BE (1985) Aspects of Urban Climate of Benin City, Nigeria. Atmos Clim Sci 4: 241-252.
17. Adebayo YR (1987) A note on the effect of urbanization on temperature in Ibadan. J Climatol 7: 185-192.
18. Adebayo YR (1991) Heat island in a humid tropical city and its relationship with potential evaporation. Theor Appl Climatol 43: 12-26.
19. Aina SA (1989) Aspects of the Urban Climate of Oshogbo, Unpublished M.Sc Dissertation, University of Ibadan, Nigeria.
20. Akinbode OM, Eludoyin AO, Fasae OA (2008) Temperature and relative humidity distributions in a medium-size administrative town in Southwest Nigeria. J Environ Manag 87: 95-105.
21. Eludoyin OM (2015) Assessment of daytime physiologic comfort, its perception and coping strategies among people in tertiary institutions in Nigeria. Weather and Climate Extremes 10: 70-84.
22. De Dear R, Brager G, Cooper D (1997) Developing an Adaptive Model of Thermal Comfort and Preference. Final Report of ASHRAE RP-884.
23. Lawrence MG (2005) The relationship between relative humidity and the dewpoint temperature in moist air: A simple conversion and applications. Bul American Meteorol Soci 86: 225-233.
24. Omogbai BE (2010) Rain days and their predictability in south-western region of Nigeria J Hum Ecol 31: 185-195.
25. Ileoje NP (2001) A New Geography of Nigeria, New Revised Edition. Longman Publishers: Ibadan, Nigeria.
26. Oyinloye RO, Oloukoi J (2012) Spatio-temporal assessment and mapping of the land use/land cover dynamics in the central forest belt of Southwestern Nigeria. Res J Environ Earth Sci 4: 720-730.
27. National Population commission (NPC) (2006) National population census, Abuja, Nigeria.
28. Gregorczuk M, Cena K (1967) Distribution of effective temperature over the surface of the earth. Int J Biometeorol 11: 145-149.
29. Aderinto SA (2006) Implication of Automatic Weather Observing Stations in Nigerian Meteorological Agency, Unpublished IMO report.
30. Balafoutis CJ, Makrogiannis TJ (2003) Hourly discomfort conditions in the city of Thessaloniki (North Greece) estimated by the relative strain index (RSI), pp: 1-5.
31. Kyle WJ (1994) The human bioclimate of Hong Kong. In: Brazdil R, Kolar M (eds.) Contemporary Climatology, Masaryk University, Brno, Czech Republic pp: 345-350.
32. Ayanlade A, Jegede OO (2015) Evaluation of the intensity of the daytime surface urban heat island: how can remote sensing help? Int J Image Data Fusion 6: 348-365.

33. Ayanlade A (2016) Seasonality in the daytime and night-time intensity of land surface temperature in a tropical city area. Sci Total Environ 557: 415-424.

34. Mirrahimi S, Mohamed MF, Haw LC, Ibrahim NLN, Yusoff WFM, et al. (2016) The effect of building envelope on the thermal comfort and energy saving for high-rise buildings in hot-humid climate. Renewable and Sustainable Energy Reviews 53: 1508-1519.

35. Igbawua T, Zhang J, Chang Q, Yao F (2016) Vegetation dynamics in relation with climate over Nigeria from 1982 to 2011. Environ Earth Sci 75: 1-16.

36. Ilesanmi AO (2016) Doctoral research on architecture in Nigeria: Exploring domains, extending boundaries. Front Architect Res 5: 134-142.

37. United Nations International Strategy for Disaster Reduction Regional Office for Africa (UNISDR) (2012) Disaster Reduction in Africa, Special Issue on drought.

38. Pezzoli A, Santos Dávila JL, d'Elia E (2016) Climate and Human Health: Relations, Projections, and Future Implementations. Climate 4: 18.

39. Lawson ET (2016) Negotiating stakeholder participation in the Ghana national climate change policy. Int J Clim Change Strat Manag 8.

40. Algeciras JAR, Consuegra LG, Matzarakis A (2016) Spatial-temporal study on the effects of urban street configurations on human thermal comfort in the world heritage city of Camagüey-Cuba. Build Environ 101: 85-101.

41. Yang YE, Wi S, Ray PA, Brown CM, Khalil AF (2016) The future nexus of the Brahmaputra River Basin: Climate, water, energy and food trajectories. Glob Environ Change 37: 16-30.

42. Baklanov A, Molina LT, Gauss M (2016) Megacities, air quality and climate. Atmos Environ 126: 235-249.

43. Metcalf SS, Svendsen ES, Knigge L, Wang H, Palmer HD, et al. (2016) Urban Greening as a Social Movement, In Urban Sustainability: Policy and Praxis pp: 233-248.

44. Mohammed I, Nasidi A, Alkali AS, Garbati MA, Ajayi-Obe EK, et al. (2000) A severe epidemic of meningococcal meningitis in Nigeria. Tran R Soc Trop Med Hyg 94: 265-270.

45. Lin T, Ho T, Wang Y (2011) Mortality risk associated with temperature and prolonged temperature extremes in elderly populations in Taiwan. Environ Res 111: 1156-1163.

46. Oketola AA, Osibanjo O (2009) Estimating sectoral pollution load in Lagos by Industrial Pollution Projection System (IPPS): Employment versus output. Toxicol Environ Chem 91: 799-818.

47. Ajayi DD (2007) Recent trends and patterns in Nigeria's industrial development. Afr Dev 32.

48. Hettige H, Martin P, Singh M, Wheeler D (1994) IPPS-The Industrial Pollution Projection System, World Bank, Policy Research Working Paper.

49. Greenwood B (2006) Editorial: 100 years of epidemic meningitis in West Africa-has anything changed? Trop Med Int Health 11: 773-780.

50. Sawa BA and Buhari B (2011) Temperature variability and outbreak of meningitis and measles in Zaria, northern Nigeria. Res J Applied Sci Eng Technol 3: 399-402.

18

Assessment of Drought Recurrence in Somaliland: Causes, Impacts and Mitigations

Abdulkadir G*

Food and Agriculture Organization of the United Nations, Koddbur, Hargeisa, Somaliland

*Corresponding author: Abdulkadir G, Food and Agriculture Organization of the United Nations, Koddbur, Hargeisa, Somaliland
E-mail: Abdulkadir.Gure@fao.org

Abstract

This paper presents a comprehensive review and analysis of the available climatological data and information on droughts to examine the major causes of droughts recurrence in Somaliland by analyzing the drought occurrence in the past decades with special focus on drought categories and its impact on the livelihoods and sustainable development of Somaliland. The primary data used for this study was collected from the rainfall stations across Somaliland as well as climate data retrieved from CHIRPS gridded rainfall dataset. However, the main findings of the present study were; Somaliland is characterized by drought, which is known to have the most far-reaching impacts of all natural disasters. This obvious challenge is most likely to aggravate due to slow progress in drought risk management, increased population and massive land degradation. The study also found that after a large scale failure of the rains during the 2016 Deyr season have led to severe drought conditions across Somaliland, resulting in extensive growing season failures and record low vegetation. The most seriously affected areas in this current drought are the eastern regions. On the other hand, based on the available climatological data from the past, it clearly shows that Somaliland is likely to face extreme and widespread droughts in the coming years as climate change is anticipated to increase the intensity and frequency of drought. As a result, there is a clear need for increased and integrated efforts in drought mitigation to lessen the negative impacts of recurrent droughts.

Keywords: Variability coefficient; Precipitation; Standard deviation

Introduction

Based on Palmer [1], drought is defined as permanent (from the beginning times of droughts until the end of the duration) and unusual deficit of moisture and is classified into four interrelated categories. The first type is called Meteorological drought and is defined by climatic variables (precipitation and humidity) and the duration of the dry period. Hydrological drought which is the second type of drought is associated with the effects of periods of rainfall shortfalls on the water levels of rivers, reservoirs and lakes and aquifers.

The third type is Agricultural drought which occurs when there is not enough water available for a particular crop to grow in a particular time. It is important to mention that Agricultural drought is normally evident after meteorological drought but before a hydrological drought. In general, drought is a complex phenomenon, which varies every time in terms of its onset, intensity, duration and geographical coverage and causes a serious hydrological imbalance [2].

In Horn of Africa, drought and its consequences (degradation of environmental and natural resources), continues persistent largely due to climate changes, increased human population, inadequate institutional capacities, civil unrest and high poverty levels in the region [3]. In Somaliland, which lies in an arid and semi-arid environment, is frequently experienced recurrent episodes of drought which has become serious natural hazards.

It affects large proportion of the population in a number of ways such as causing loss of life, crop failures, food shortages which might lead to malnutrition, health problems and mass migration. The most seriously affected areas are eastern regions namely Sanaag, Sool and Togdheer. These regions face frequent reduction of water and moisture. However, in the current drought (2015-2016), most of the regions of Somaliland suffering the worst drought in several decades which represents the most severe food security emergency.

Two consecutive seasons (Gu and Deyr) of significantly below average rainfall in Somaliland have resulted in failed crop production, depletion of grazing resources and significant livestock mortality. Pastoralists and agro-pastoralists who occupy the vast majority in Somaliland are hardly meet basic water requirements during the current drought and the problem will most likely get worse due to the climate change.

In this study, the main purpose is merely to investigate the major causes and impacts of drought recurrence in Somaliland and its outcome is expected to add up to the understanding the causes of the deepening marginalization and vulnerability of the Somaliland pastoralists due to the recurrent droughts and facilitate the mitigation of future droughts.

Study Area

The geographical coverage of the study area is the self-declared state of Somaliland that broke away from Somalia in 1991, declaring independence using the borders of the former British Somaliland (Figure 1). To date, Somaliland has not been recognized by the international community as well as the federal government of Somalia.

However, since its establishment, Somaliland has been enjoying peace, working political system and survived much of the chaos and violence that overwhelmed the rest of Somalia.

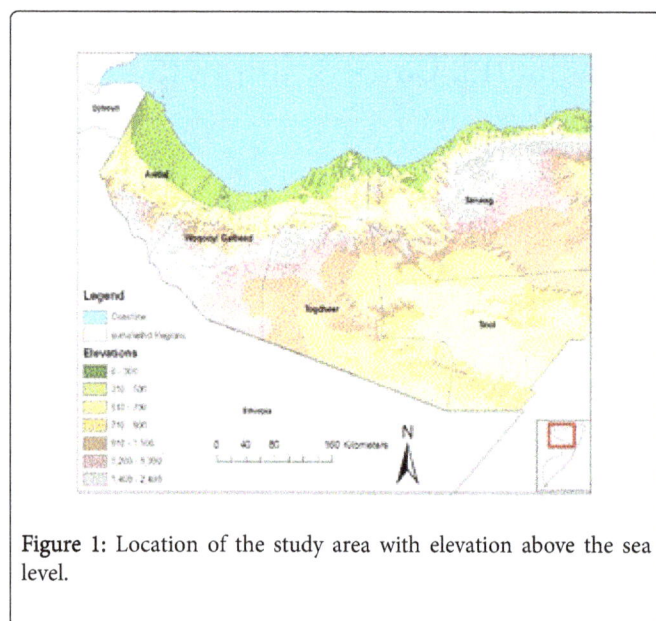

Figure 1: Location of the study area with elevation above the sea level.

Pastoralism has been the major livelihood of Somaliland whereby populations with their livestock follow seasonal migration patterns depending upon rainfall and pasture availability. Nevertheless, for the

past decades, there has been a dramatic changes in the socio-economy of the pastoral population as a result of the recurring droughts and massive change of land use, which in turn had negative impacts on availability of fodder, and thus adversely affecting livestock production.

Climate

Generally, Somaliland has an arid to semi-arid climate, with four seasons. The first main rainy season of Gu occurs between April and June, when around 50-60% of rain falls. The second rainy season is called Deyr (from August to November) and accounts 20-30% of total rainfall. The two dry seasons in are Jilaal and Hagga, which occur between December and March and July and August, respectively (Table 1).

The amount of rainfall received annually reduces further to the north except for areas around Sheikh, Hargeisa and Borama that receive between 500 mm and 600 mm per year [4]. While the area around Erigavo receives up to 400 mm annually [5]. The northern coastline is characterized by low rains of less than 100 mm per year. The rest of Somaliland receives an annual rainfall of 200-300 mm [4]. In the higher altitudes of the mountains and plateau areas temperatures vary considerably with the seasons, with a mean annual temperature of 20-24°C, while the coastal region has mean annual temperatures of 28-32°C.

Rainfall Station	Jan	Feb	Mar	Apr	May	Jun	Jul	Aug	Sep	Oct	Nov	Dec	Annual
Aburin	4.3	12	29.3	77.5	65.2	38.4	50.6	68.3	66.4	31.6	11.1	3.2	457.8
Borama	4.3	19.6	41.9	96.3	59.3	31.3	68	103.7	78.3	20.6	10.3	3.8	537.2
Burao	2.9	2.5	11.1	71.9	65.2	7.8	4.7	3.8	7.9	88.5	61.7	8.2	336.3
Caynabo	2.9	4.4	11.9	48.7	55.3	17.1	9.5	15.2	33.2	37.9	12.7	3.2	251.9
Dilla	5	13.9	36.4	75.3	65.2	47.7	72.7	94.9	72.7	25.3	11.9	3.2	524
Eeerigavo	7.8	6.3	18.2	34.3	51.4	30.6	6.3	25.3	64.8	7.1	4.7	1.3	258.2
Elfweyne	3.6	4.4	11.9	34.3	45.5	18.5	7.9	17.7	37.2	21.3	7.1	1.9	211.2
Gebilley	1.4	6.3	30	57.6	61.3	50.5	71.2	84.7	62.5	19	9.5	1.9	455.8
Hargeisa	2.9	10.6	26.1	80.8	65.2	34.2	42.7	58.2	64.8	30.8	9.5	1.9	427.8
Odweyne	3.6	8.2	21.3	59.8	59.3	28.5	26.9	40.5	50.6	35.6	11.1	4.43	349.6
Quljeed	6.4	19	39.5	84.1	63.2	39.8	71.2	102.5	75.9	25.3	13.4	3.8	544.1
Sheikh	4.3	10.8	27.7	69.7	67.2	33.4	28.5	46.8	64	56.9	19.8	11.4	440.4

Table 1: Somaliland long-term average rainfall for selected stations.

Soils

Soil types of Somaliland closely follow its geomorphology and are characterized by poor structure, high permeability, low moisture retention ability and inadequate internal drainage [6]. Moreover, soil erosion has been a major challenge as a result of land clearing, cutting trees for charcoal production and overgrazing of livestock. Gully erosion is a also another major challenge to both rain fed farming land and grazing land and has made large parts of farming land in Somaliland unproductive and is spreading at an alarming rate [6].

Land Use

In Somaliland, land is mainly used for livestock production and rearing or mix farming (crop and livestock production). Most of the regions in Somaliland are dry and cannot support rain fed agriculture except for small pockets of land in the areas around Hargeisa, Gebiley and Borama that receive amounts of rainfall that can support rainfall dependent agriculture. The eastern regions of Togdheer, Sool, and Sanaag are almost exclusively relying on livestock raising [6]. The coastal grasslands are used for extensive livestock grazing especially in the dry season as water is more available in these areas than in the wood land [6].

Methods

Data

The primary data used for this study was collected from the meteorological stations of Somaliland managed by the Department of Meteorology and Food Security of the Ministry of Agriculture which is technically and financially supported by the FAO-SWALIM. It was also used CHIRPS gridded rainfall dataset produced by the Climate Hazards Group at the University of California, Santa Barbara.

The Climate Hazards Group Infrared Precipitation with Station data (CHIRPS v.2) is a global daily, pentadal and monthly precipitation product explicitly designed for monitoring agricultural drought and global environmental change over land [7]. The CHIRPS is a gridded rainfall time series that brings together high-resolution average rainfall data gleaned from satellites and weather station data to provide a comprehensive picture of rainfall patterns from 1981 to the near-present. In addition, there were a secondary data used for this study collected from previous reports and literature (Figure 2).

It was also carried out statistical and situational analysis by holistically evaluating the available information on drought and rainfall data to compile a picture of the current drought situation in Somaliland.

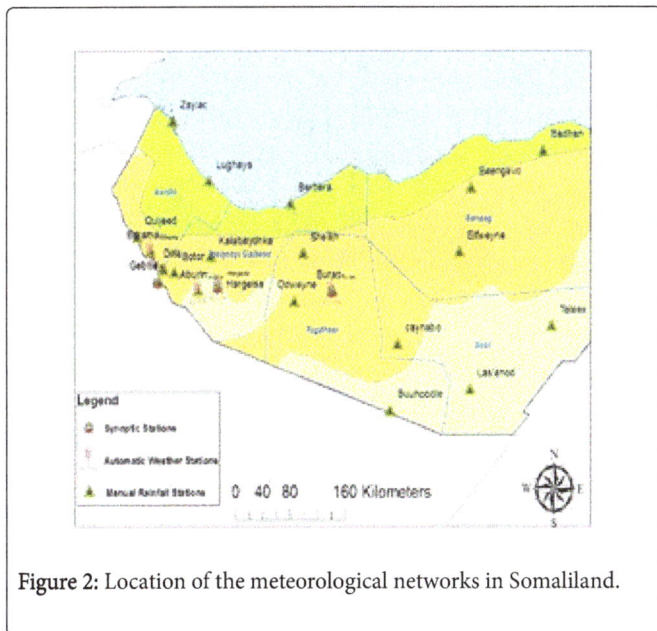

Figure 2: Location of the meteorological networks in Somaliland.

Standardized Precipitation Index (SPI)

Standardized Precipitation Index (SPI) is used to indicate meteorological drought, which is mainly caused by a deficiency of precipitation. McKee et al. [8] suggested that the SPI ranges corresponding to different severity levels of drought (Table 2). The SPI calculation for any location is based on the long-term precipitation record for a desired period. A positive SPI value indicates greater than mean precipitation while a negative value indicates less than mean precipitation.

The SPI has been used to analyse drought severity in different parts of the world. It has been used in Australia [9], in Karnataka, India [10], in Mexico [11], in Iran [12] and among other places in the world and is calculated as follows:

$$Spi = \frac{X - \bar{X}}{\sigma} \ (1)$$

Where: σ=Standard deviation

X=Precipitation

$\bar{X}$=Mean precipitation.

No.	SPI	Climate Situation
1	>2	Intensely moist
2	1.5 to −1.49	Very moist
3	1 to −1.49	Moderately moist
4	0 to −0.99	Low moisture
5	0 to −0.99	Mild drought
6	−1 to −1.49	Moderate drought
7	−1.5 to −1.99	Severe drought
8	<2	Intense drought

Table 2: SPI index values and the drought intensity.

GeoCLIM

The GeoCLIM program is part of a set of agro- climatic analysis products developed by the FEWS NET and United States Geological Survey (USGS). This is a spatial analysis tool designed for climatological analysis of historical rainfall and temperature and evapotranspiration data.

It can be used to blend station information with satellite data to create improved datasets and analyse seasonal trends and historical climate data. Moreover, the GeoCLIM is also used to examine drought for a selected region by calculating the standardized precipitation index (SPI). In this paper, it was also used GIS (geographical information system) tool to produce a wide range of different maps of the study area.

Results

For the past decades, Somaliland has seen an increase in hazardous events such as droughts, flash floods, massive land degradation, and invasion of alien species. However, drought remains the only major disaster causing huge damages to its populations and the economy. Currently, drought conditions still prevail in most parts of Somaliland particularly in the eastern regions which had a poor rainfall performance in the previous Deyr and Gu rainy seasons. In some areas of Sanaag, Togdheer and Sool regions, drought conditions have persisted even longer and spread to western regions of Somaliland.

Therefore, in order to understand why intense dry and below normal rainy seasons in 2016 turn into crises in 2017, we need to look beyond the shortage of rain. In this paper, it was analyzed the performance of the 2016 Deyr and Gu rainfall activities as well as the historical drought disasters in Somaliland to find out the actual causes of the current drought.

It is evident from the analysis that throughout its recent history, Somaliland has experienced 15 major droughts during the period between 1960 and 2017 as indicated in Table 3. The droughts that occurred in 1973–1974, 1984, 1991, 2010/2011 and 2016/2017 were most intense and widespread.

Disaster	Date
Drought	1964
Drought	1973/1974
Drought	1979/1980
Drought	1984
Drought	1987
Drought	1991
Drought	1994
Famine	36220

Drought	36526
Drought	37043
Drought	37226
Drought	2003/2004
Drought	2006/2007
Drought	2010/2011
Drought	2016/2017

Table 3: Historical drought disasters in Somaliland.

However, the analysis of droughts during in the specified period indicated that droughts have intensified in terms of their frequency, severity and coverage over the last two decades.

Figure 3 shows calculated CDI (Compiled Drought Index) time series for Hargeisa station for monthly time units. CDI is a Microsoft Excel-based software package that provides an easy-to-use means of calculating indices for monitoring drought. It was developed by FAO-SWALIM.

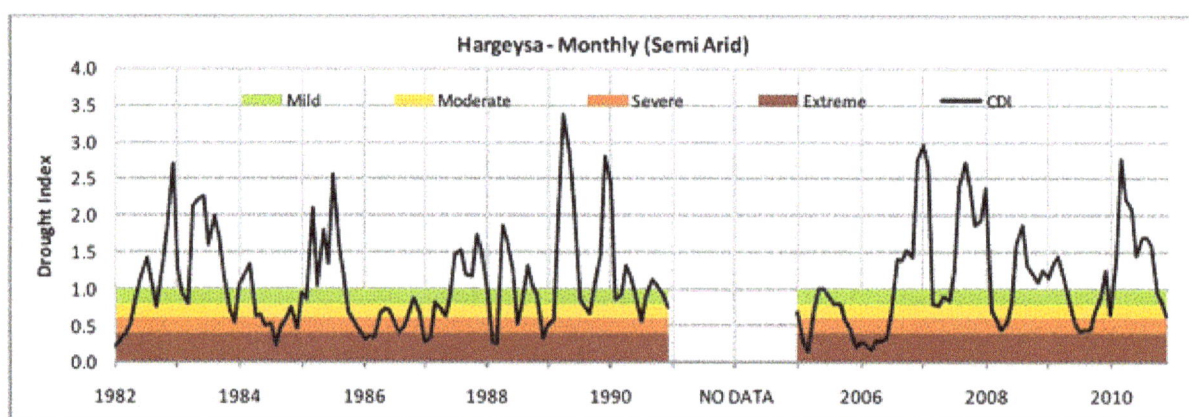

Figure 3: Calculated CDI time series for Hargeisa station for monthly time units.

Severities of meteorological drought for Somaliland were also calculated using (SPI). Based on the historical rainfall data, it was easy to analyze whether the probability of rainfall is less than or equal to a certain amount. The calculation of SPI values for the Gu and Deyr rainy seasons were carried out using CHIRPS gridded rainfall dataset to get a general view of the alternation of moist and dry periods. Figures 4-7 depict the comparison of SPI and rainfall deviations for Somaliland during 2015/2016 Deyr and Gu rainy seasons.

During the 2015 Gu and Deyr rainy seasons, the values of SPI in Somaliland are more stretched between 1.5 to -1.5 as showed in Figures 4 and 5. On the other hand, during the 2016 Gu and Deyr rainy seasons, the values of SPI in Somaliland are more stretched between 2.0 to -1.5 and 1.0 to -1.0 respectively as can be seen in Figures 6 and 7. In the agro-pastoral areas of Somaliland, the SPI values tend to be higher >1.5, for the events of excess rainfall.

It also interests to note that the SPI value was lower in eastern parts of Somaliland than the other parts of Somaliland during the analysed

period accept 2015 Deyr rainy season where large portion of Sanaag and few pocks in Togdheer regions have recorded higher positive value of SPI. With increasing the rain in the Deyr 2015, the drought value decreased in the western and southern parts. With decreasing rain in Gu (2016), the drought increased in the 2016 Deyr in the eastern regions.

On the other hand, rainfall data of 10 years of for four different cities in the eastern and western parts of Somaliland were also analysed specifically for drought investigation, which may be used for long term planning of irrigation system in the area.

During 10 years period, different categories of drought year was experienced for the four cities. Table 4 shows the characteristics of drought in 12 month time scale from 2007 to 2016 for four different cities in Somaliland namely Borama and Gebiley in the west and Odweyne and Erigavo in the East.

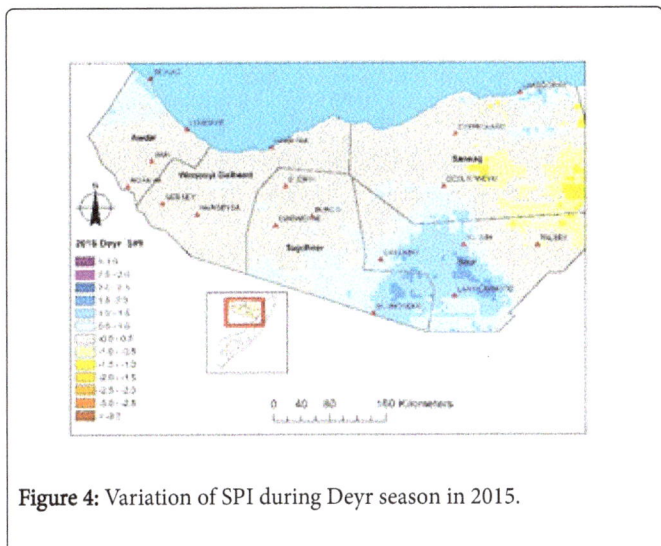

Figure 4: Variation of SPI during Deyr season in 2015.

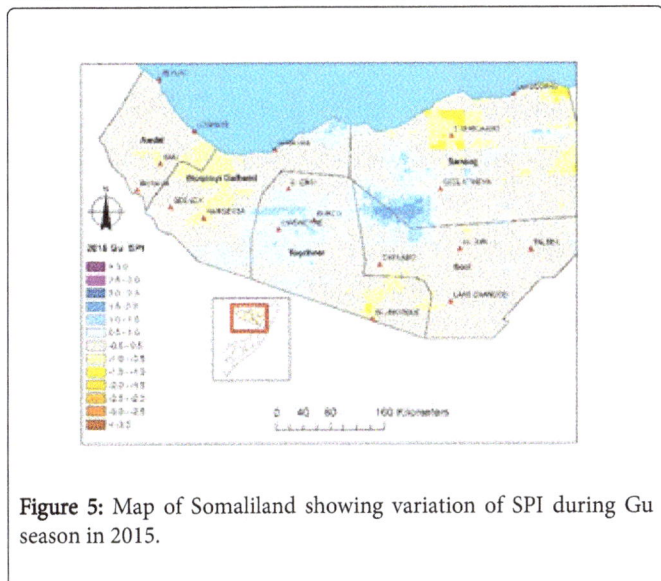

Figure 5: Map of Somaliland showing variation of SPI during Gu season in 2015.

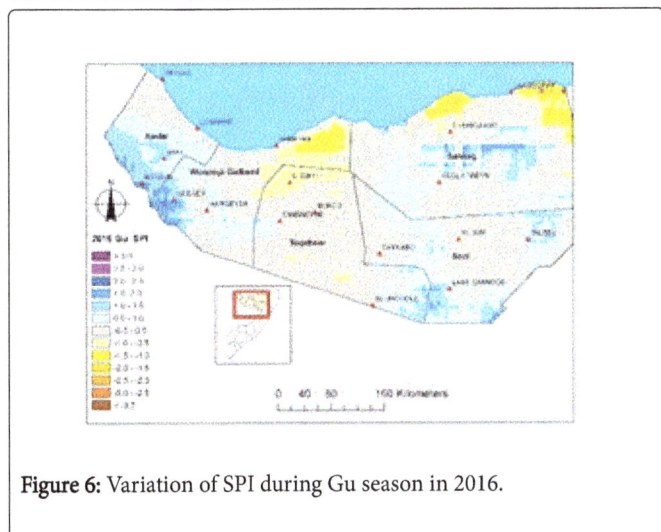

Figure 6: Variation of SPI during Gu season in 2016.

Plotting a time series of years against SPI gives a good indication of the drought history of a particular station.

Result show that the most severe drought values had occurred in 2009 for Gebiley with SPI values of -1.92 and with total rainfall of 293.5 mm and for Borama in 2014 and 2015 with SPI values of -1.60 and -1.94 and with total rainfall of 340 mm and 299 mm respectively (Figure 8a and 8b).

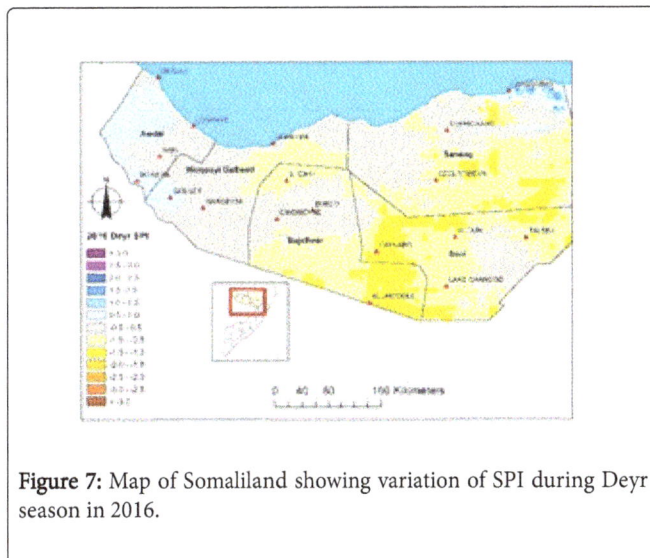

Figure 7: Map of Somaliland showing variation of SPI during Deyr season in 2016.

Whereas severe drought was also seen in 2008, 2015 and 2016 in Odweyne district in Togdheer region with SPI values of -1.50, -1.93 and-1.55 and with total rainfall of 144 mm, 85.5 mm and 137 mm respectively (Figure 8c).

In Erigavo, a moderate drought was observed during 2012 and 2016 with total rainfall of 177.4 mm and 187.5 mm respectively (Figure 8d). Moderate drought was also observed in Borama in 2009, 2011 and 2016 with total rainfall of 414.5 mm, 405.5 mm and 370.5 mm respectively.

Similarly, moderate drought was also observed in Odweyne in 2007, 2009 and 2016 with total rainfall of 205 mm, 189.5 mm and 332 mm respectively.

Is also interesting to note that there was no severe or moderate drought occurred in the year of 2013. But it has only been a low moisture year for Gebiley and Borama with SPI values of 0.07 and 0.71 with total rainfall of 461.5 mm and 624.8 mm respectively.

While the same year of 2013, the eastern cities of Odweyne and Erigavo has experienced a year with intense moist with SPI values of 2.02 and 2.86 (with total rainfall of 627 mm and 443 mm).

Seasonal (Gu and Deyr) rainfall distribution and coefficient of rainfall variation

Figures 9 and 10 show the average seasonal (Gu and Deyr) rainfall distribution in Somaliland for the period between 1981 and 2016. During the Gu rainy season, much of the rains are concentrated in the agro-pastoral areas of Awdal and Waqoyi-Galbed regions as well as parts of Togdheer region particularly the areas around sheikh with an average rainfall of between 157-223 mm (Figure 9).

City	Year	Annual rainfall total (mm)	Drought severity index (SPI)	Drought category	City	Year	Annual rainfall total (mm)	Drought severity index (SPI)	Drought category
Gebiley	2007	697.8	2.86	Intensely moist	Odweyne	2007	205	-1.05	Moderate drought
	2008	452	-0.04	Borderline drought		2008	144	-1.5	Severe drought
	2009	293.5	-1.92	Severe drought		2009	189.5	-1.17	Moderate drought
	2010	536.5	0.95	Low moisture		2010	259.5	-0.66	Borderline drought
	2011	434.5	-0.25	Borderline drought		2011	434.5	0.62	Low moisture
	2012	511.5	0.66	Low moisture		2012	332	-0.13	Moderate drought
	2013	461.5	0.07	Low moisture		2013	627	2.02	Intensely moist
	2014	411	-0.53	Borderline drought		2014	281	-0.5	Borderline drought
	2015	392	-0.75	Borderline drought		2015	85.5	-1.93	Severe drought
	2016	572	1.37	Moderately moist		2016	137	-1.55	Severe drought
Borama	2007	432.6	-0.85	Borderline drought	Erigavo	2007	277.5	0.3	Low moisture
	2008	493.5	-0.36	Borderline drought		2008	296.5	0.59	Low moisture
	2009	414.5	-1	Moderate drought		2009	283	0.38	Low moisture
	2010	781.1	1.98	Very moist		2010	350	1.42	Moderately moist
	2011	405.5	-1.07	Moderate drought		2011	218	-0.62	Borderline drought
	2012	456.8	-0.65	Borderline drought		2012	177.4	-1.25	Moderate drought
	2013	624.8	0.71	Low moisture		2013	443	2.86	Intensely moist
	2014	340	-1.6	Severe drought		2014	349	1.41	Moderately moist
	2015	299	-1.94	Severe drought		2015	241.5	-0.26	Borderline drought
	2016	370.5	-1.35	Moderate drought		2016	187.5	-1.1	Moderate drought

Table 4: Variation of SPI for four cities in Somaliland during the period between 2007 and 2016.

In contrast, low average seasonal distribution was seen in the areas along the coast of Somaliland and few pockets in the eastern and southern part of Sool and Sanaag regions respectively.

Meanwhile, during the Deyr season, much of the rains are concentrated in the areas around sheikh and southern parts of Togdher and Sool region with an average rainfall of between 69-104 mm. Similarly, during the Deyr season, the eastern parts of Somaliland shows low average seasonal distribution compared to western parts particularly the northern parts of Sanaag region and parts of eastern Sool region (Figure 10).

In Somaliland, climate variability and change on the other hand greatly influence social and natural environments, with subsequent impacts on natural resources. Figures 11 and 12 show the coefficient of rainfall variability over Somaliland during the Deyr and Gu rainfall season.

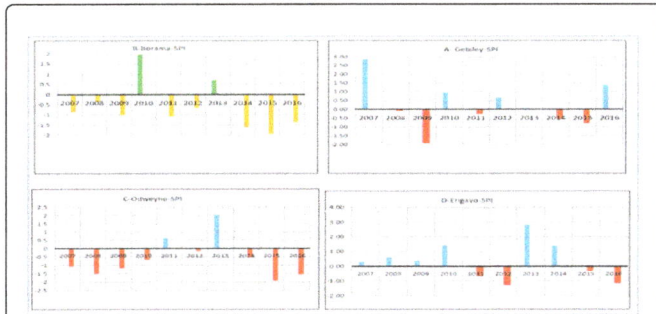

Figure 8: Variation of SPI values for Four Cities in Somaliland during the period between 2007 and 2016.

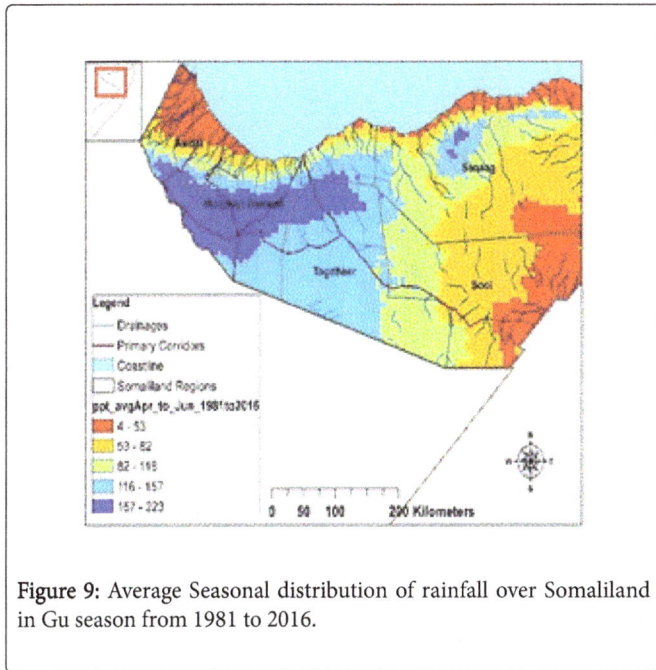

Figure 9: Average Seasonal distribution of rainfall over Somaliland in Gu season from 1981 to 2016.

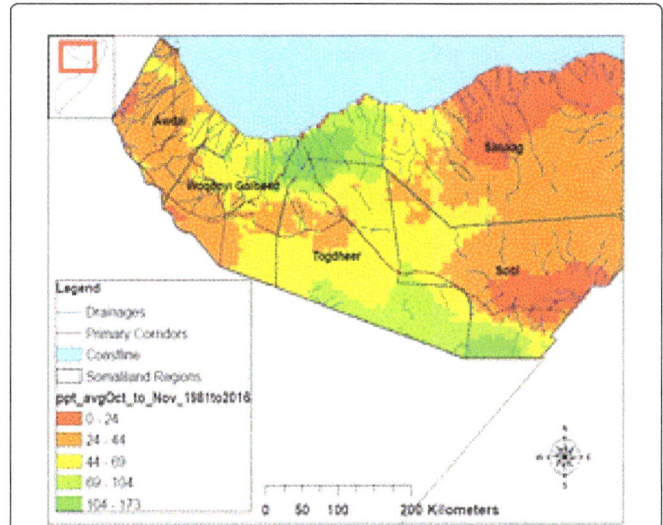

Figure 10: Average Seasonal distribution of rainfall over Somaliland in Gu season from 1981 to 2016.

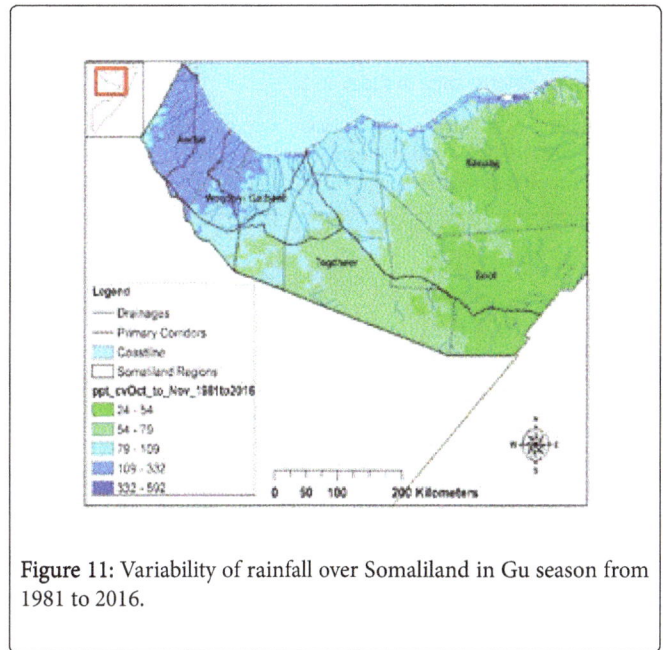

Figure 11: Variability of rainfall over Somaliland in Gu season from 1981 to 2016.

During the Deyr season, the analysis of coefficient of variation from 1981 to 2016, there was high variability in Awdal region and some parts of Waqoyi Galbed region as well as the coastal areas of Somaliland (Figure 11).

The higher coefficient of rainfall variability implies that the precipitation is likely to be irregular in these areas; heavy rainfall at certain periods and low or no rainfall in some other years. Similarly, during the Gu season, coefficient of variation from the same period (1981 to 2016) was high in Awdal region, particularly in the coastal areas (Figure 12).

2016 Seasonal rainfall performance

The 2016 Deyr rainfall performance has been mixed across Somaliland. In the eastern half of Somaliland (except for sheikh), the performance of the rainfall was poor as it was recorded below average rains during the Deyr season.

Most of these pastoral regions have also experienced rainfall deficits in the previous Gu rains. The poor rains of 2016 Deyr have also impacted the major agro-pastoral zones of Guban and West Golis in Awadal and Waqoyi galbed regions.

The areas with rainfall deficits include Odweyne (8 mm), Erigavo (12 mm), Aynabo (24 mm) and El-Afwayne (0 mm) as well as their surroundings (Figure 13).

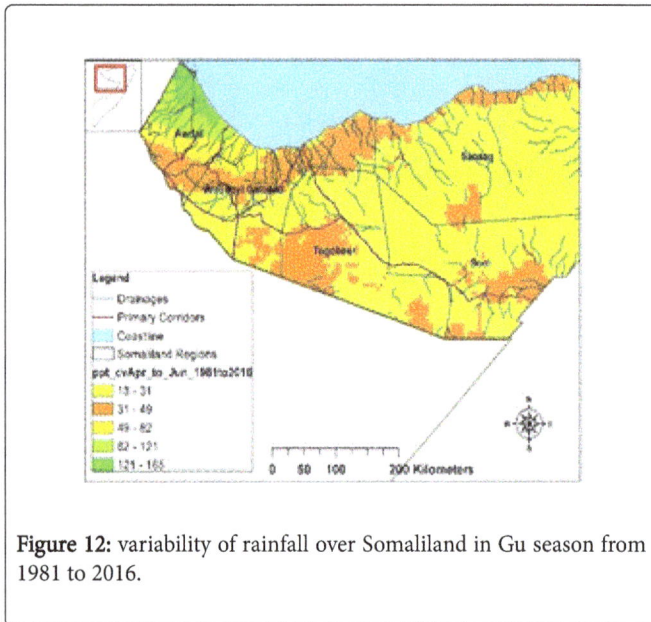

Figure 12: variability of rainfall over Somaliland in Gu season from 1981 to 2016.

This situation of long-term persistent dryness has put local livelihoods under considerable stress, especially in the agro pastoral areas of western Somaliland where considerable number of migrants from eastern regions come to these areas.

Insufficient rain during Deyr season has also led significantly to large moisture deficits and abnormal dryness, which have negatively affected cropping activities in agro pastoral areas of Somaliland.

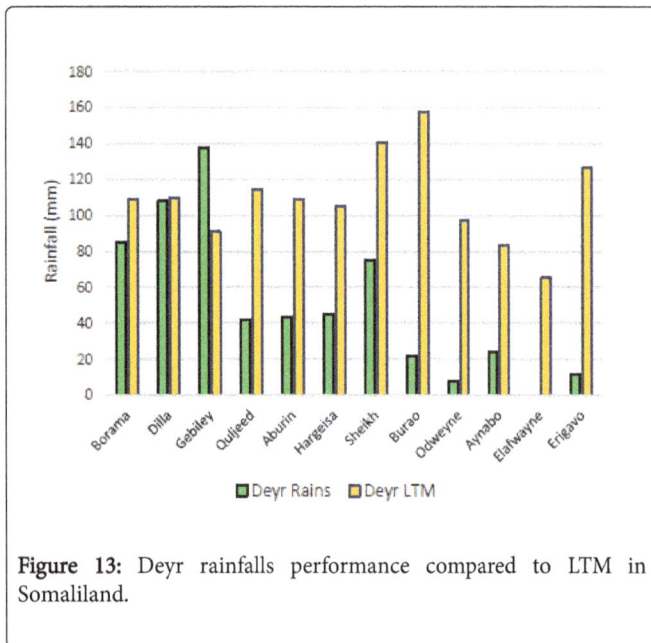

Figure 13: Deyr rainfalls performance compared to LTM in Somaliland.

On the other hand during the 2016 Gu season, good rains were received in parts of Somaliland particularly the western regions. Although the spatial distribution of rainfall was quite good, the accumulative rainfall amounts for some areas were remained depressed following prolonged dry period in preceding months.

Most of stations in the west of Somaliland recorded significant rains while the eastern regions have received little rains. The agro-pastoral

areas of Guban and West Golis, which have been dry for a long period of time, received some moderate to heavy rains which caused flash floods.

Dilla, Aburin and Gebiley have recorded the highest amounts of rainfall in Somaliland during the 2106 Gu rainy season.

While other nearby areas such as Hargeisa recorded significant amounts of rainfall. Moreover, the rains were well distributed across Somaliland during the 2016 Gu rainy season. The stations that recorded highest amounts of rainfall include Gebiley (310 mm), Dila (306 mm), Aburin (292 mm) and Hargeisa (271.5 mm).

Some stations are however recorded below average rains during the 2016 Gu rainy season. These stations include; El-Afwayne (25.5 mm), Aynabo (75 mm), Sheikh (89 mm) and Odweyne (119 mm), as seen in Figure 14.

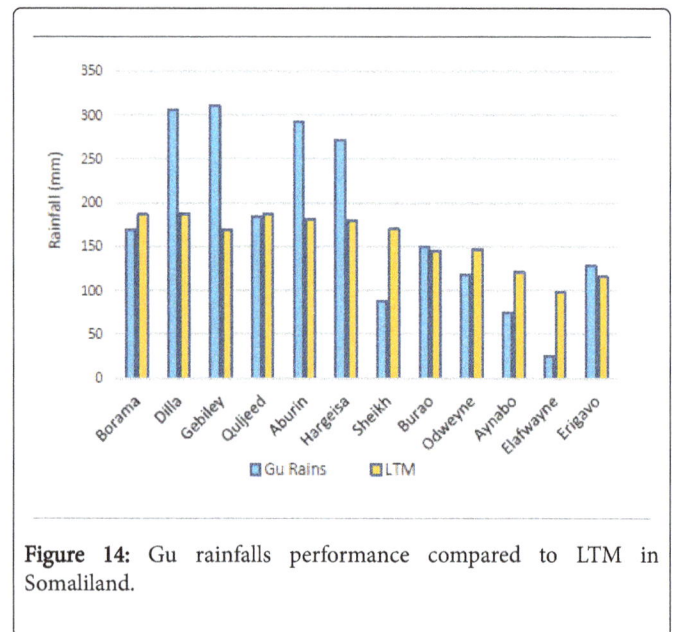

Figure 14: Gu rainfalls performance compared to LTM in Somaliland.

It is also important to mention that the March-May (2017) rainfall for the Greater Horn of Africa, which is also the main rainy season for Somaliland as it covering most of the Gu rainfall season, forecasts on below average rainfall for Somaliland (Figure 15).

The regional consensus climate outlook for the March to May 2017 season indicates an increased likelihood of below normal to near normal rainfall over Somaliland (yellow zones in Figure 15). Increased probabilities of near to above normal rainfall are indicated over parts of Galmudug and Puntland (the blue zones in Figure 15).

The UK-Met Office forecasts indicate below to near average rainfall for most of parts of Somaliland in broad agreement with the GHACOF forecast. Whereas combining a number of models into a single "ensemble" forecast (WMO) allows a better picture.

Overall, most likely outcome for 2017 Gu rainy season is for below average rainfall for Somaliland. The unfavourable rainfall forecast is further compounded by increased expectations of above-average temperatures, which could further delay crop development, negatively affecting yields and harvests (Figures 16-18).

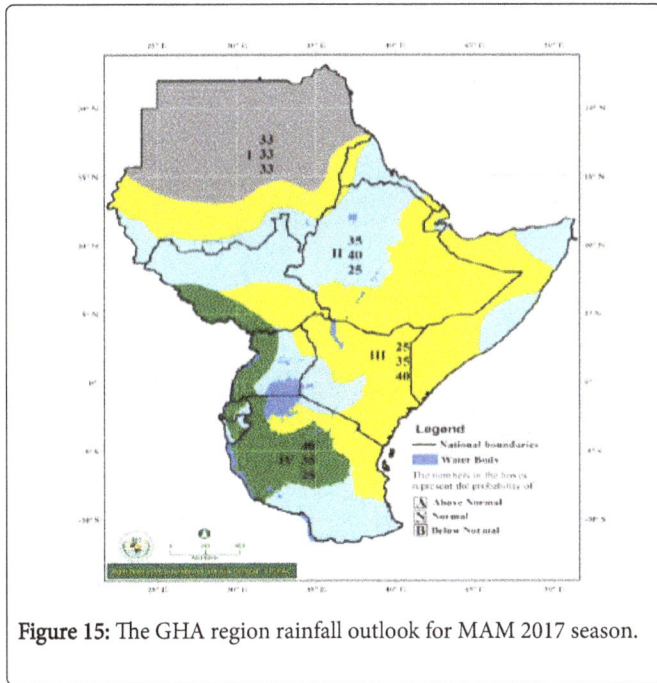

Figure 15: The GHA region rainfall outlook for MAM 2017 season.

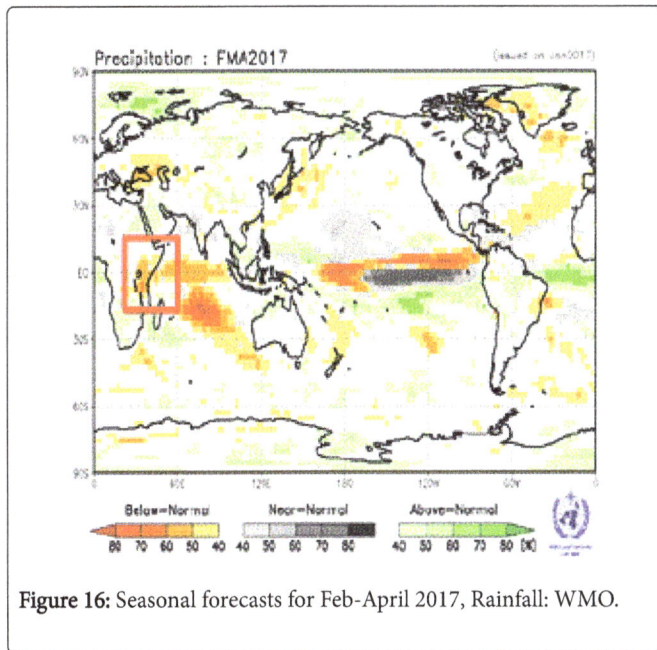

Figure 16: Seasonal forecasts for Feb-April 2017, Rainfall: WMO.

Discussion

Extreme weather events have become more frequent and more intense since the middle of the 20th century worldwide and are expected to become even more pronounced throughout the 21st century due to climate change [13]. Particularly, the number and length of warm weather spells and heat waves have increased globally. In Africa, drought is a part of natural climatic variability on the continent, which is quite high at intra-annual, decadal and century timescales [14]. In general, drought is one of the most complex natural phenomena, that is difficult to quantify and manage, and has multiple and extreme social and economic consequences. The magnitude of these impacts is determined by the level of development, population density and structure, demands on water, government institutional capacity.

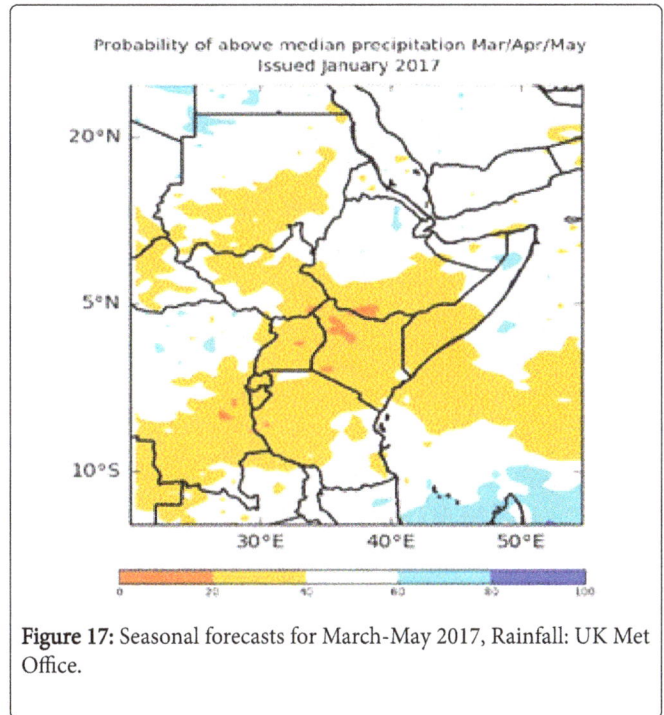

Figure 17: Seasonal forecasts for March-May 2017, Rainfall: UK Met Office.

Many studies attempted to investigate the natural causes that could be associated with droughts in Africa [15-21]. Some of them also focused on anthropogenic factors, such as climate change, aerosol emissions, land use practices and resulting land-atmosphere interactions, contributing to drought inducing mechanisms [16,17,22-24]. The review of these studies revealed that there are a number of factors contributing to inducing drought conditions. However, despite regional differences in the factors causing droughts in a specific region, El Niño–Southern Oscillation (ENSO) and SSTs are regarded major influencing factors across the continent.

Somaliland is, however, vulnerable to drought recurrence, mainly because of its geographical location, the fragile environment and the variable climate. The main underlying causes of most droughts in Somaliland can be related to changing weather patterns manifested through the excessive build-up of heat on the earth's surface, meteorological changes which result in a reduction of rainfall, and reduced cloud cover, all of which results in greater evaporation rates. Furthermore, the resultant effects of drought are aggravated by human activities such as deforestation, overgrazing and poor cropping methods, which diminish water retention of the soil, and inappropriate soil conservation techniques, which lead to soil degradation. Frequent drought occurs in Somaliland as it locates in arid regions where the rainfall usually performs poorly and is accompanied by relatively high evaporation rates for prolonged periods. However, in most cases, drought is caused by a deficiency of either precipitation or an inadequacy of inland water resources supplies for a prolonged period. Since most of the surface water resources in Somaliland are usually sustained by rainfall, inadequate rainfall is usually the major cause of drought.

Droughts have disastrous impacts on Somaliland communities. Therefore, in order to analyze the severities of meteorological drought, the SPI values of different years and seasonal were analyzed in this study with actual rainfall and rainfall deviation from normal particularly drought prone areas. The intensity of the drought event can be classified according to the magnitude of the negative SPI values such that the larger the negative SPI values the more serious the event would be. A drought event occurs when the SPI is continuously reaches an intensity of -1.0 or less. The event ends when the SPI becomes positive. Each drought event, therefore, has a duration defined by its beginning and end, and intensity for each month that the event continues. In Somaliland, the droughts that occurred in 1973–1974, 1984, 1991, 2010/2011 and 2016/2017 were most intense and widespread.

Analysis of 10 years rainfall data for four different cities in Somaliland was performed specifically to discover how drought was recurring during these periods. During the analyzed periods, different categories of drought year were experienced for the four cities. The most severe drought had occurred in 2009 for Gebiley and 2014 and 2015 for Borama. Whereas severe drought was seen in 2008, 2015 and 2016 in Odweyne district. In Erigavo, a moderate drought was observed during 2012 and 2016. While another moderate drought was also observed in Borama in 2009, 2011 and 2016. Similarly, moderate drought was also spotted in Odweyne in 2007, 2009 and 2016.

From the rainfall analysis using SPI, it was found that even the very small amount of rainfall that was inadequate to maintain enough soil moisture resulted in SPI values of around −1.5 which otherwise would represent extreme dryness with the values around −2.0 and below. Similarly, excess rainfall events in Somaliland had a SPI value of around 1.5. It is obvious that a reduction in precipitation with respect to the normal precipitation amount is the primary driver of drought, resulting in a successive shortage of water for different natural and human needs. Deficit of rainfall over a period of time at a certain location could lead to various degrees of drought conditions, affecting water resources, agriculture and socioeconomic activities. As a result, the rainfall patterns, particularly rain failure or erratic rainfall are frequently the cause of recurrent droughts in Somaliland. Environmental degradation is accelerated during periods of drought due to poor land use activities such as cutting down of trees for fuel; wood and charcoal burning for income and overgrazing.

In 2015/2016 drought, many people had to sell what little they had, making them even more vulnerable to the next drought and would also mean that fewer people have enough animals to be able to donate to others in need. Meanwhile, livelihoods have suffered very much from consecutive droughts in Somaliland and most people do not perceive any future in pastoralism as many pastoralists switched to charcoal production to compensate for economic losses from pastoralism.

For many years, mobile pastoral communities in Somaliland have been coping with changing environmental conditions, and as a result they have a long established capacity for adaptation. However, changes in their environments in recent years have weakened their adaptive strategy, which is now increasing their vulnerability. Normally, when dangerous hazards, like droughts occur, the whole agro-pastoral production system collapses with disastrous consequences for the affected populations. Enormous financial resources are then required for humanitarian aid and even more to recover the production systems and livelihoods of the affected communities. Within the Somali context, local communities, diaspora groups, local organizations and local authorities are typically the first responders in crises [25].

Currently, there are many challenges that face Somaliland when it comes drought mitigation. The main challenge is to convince policy

and other decision makers that investments in mitigation are more cost effective than post-impact assistance or relief programmes. The process of dealing with drought during the disaster management would only be effective when an appropriate knowledge and required technology is readily available to improve preparedness and mitigation impacts.

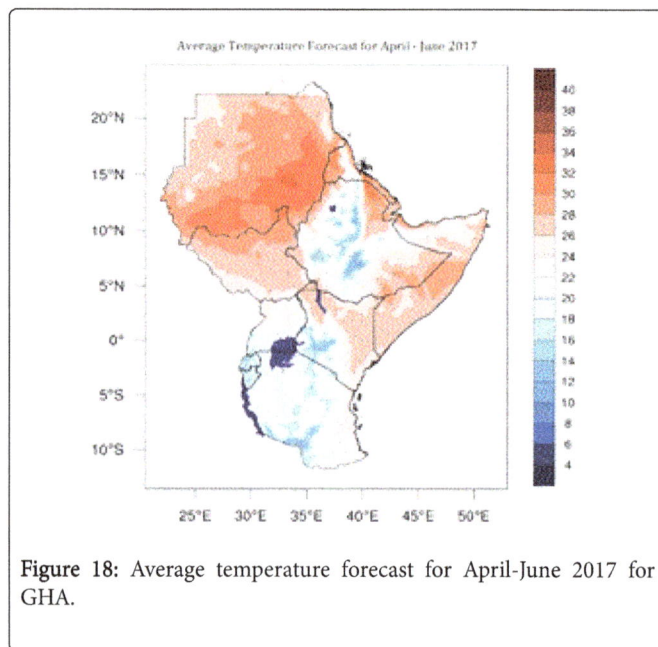

Figure 18: Average temperature forecast for April-June 2017 for GHA.

Structural mitigation measures is required in order to lessen the drought recurrence in Somaliland such as provision of an appropriate crops to farmers and engineering projects such as sand dams to harvest rain water during the rainy season. It is also required non- structural mitigations such as policies, knowledge development and functioning practices assumed to limit the adverse impacts of natural hazards. On the other hand, access to climate information is important for effective preparedness and response to adverse weather trends and disasters by the vulnerable communities in Somaliland.

2016 Deyr were well below average and pessimistic forecasts for the next rainfall season raise the possibility of a third poor rainfall season for Somaliland that would lead to a situation similar to the 2010-2011 humanitarian crisis. Therefore, early action is key to prevent the situation from getting as bad as in 2011 as the humanitarian response was too slow. It is also important to underline that the forecasted performance of the 2017 Gu rainfall is expected to have mixed implications for food security, livestock production and productivity, water, and health in different parts of Somaliland. Thus, in order reduce the impacts of the forecasted rainfall in Somaliland, there is need to strengthen disaster risk reduction strategies including response capacities, coordination, resource mobilization, communication and advocacy at national level (Figure 19). On the other hand, drought predictions based on the global climate models simulations show varying results and therefore remain uncertain for Somaliland. The available evidence from the past clearly shows that the Somaliland is very likely to face extreme and widespread droughts in the future. The vulnerability is likely to increase due to fast growing populations, increasing water demands and degradation of land and environmental resources.

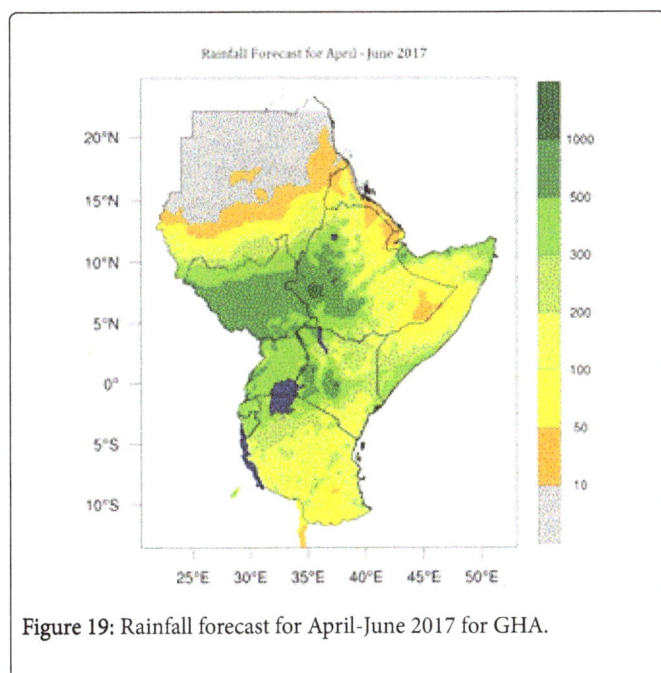

Figure 19: Rainfall forecast for April-June 2017 for GHA.

Recommendations

The ministry of Agriculture of Somaliland and NERAD, as the responsible institutions for provision of met services and Early Warning Systems (EWS), should provide farmers and pastoralists simple and understandable (Somali language), timely and reliable weather forecast information to facilitate early intervention by the government and its partners.

In order to strengthen the resilience of the people, Governments of Somaliland is strongly advised to increase investment in community infrastructure and social services with a focus on improving rural communities by providing training to farmers, pastoralists in special techniques such as soil and water conservation, water harvesting that can play a major role in drought disaster mitigation.

More importantly, there is an urgent and dire need to progress on various fronts of drought mitigation such as early warning and forecasting, long-term planning and capacity building. The global climate models simulations show varying results and therefore remain uncertain for Somaliland. The available evidence from the past clearly shows that the Somaliland is very likely to face extreme and widespread droughts in the future. The vulnerability is likely to increase due to fast growing populations, increasing water demands and degradation of land and environmental resources.

Conclusion

Conclusions of this paper is that Somaliland's aridity and drought recurrence have been increasing significantly during the past decades and this attributed to widespread drying particularly the eastern regions. The vulnerability to droughts is high in Somaliland due to poverty and large dependency on rain fed agriculture. Furthermore, water scarcity caused by aridity and recurrent droughts has been threatening food security by reducing agricultural productivity. The drought trends are also projected to continuously increase in the next coming years due to the climate change. However, if proper and

detailed study of various rainfall data is analyzed, the severity of droughts can be known and therefore various measures can be taken to cope up with the problems of drought. The study has also noted that the capacity of Somaliland people to face these recurrent droughts compounded by shortcomings in government's capacity to install long-term drought mitigation measures have been weakened as the population resilience mechanism reduced. As a result, a contingency planning and management of droughts is essentially required.

Acknowledgement

I am grateful to the many people who have contributed directly or indirectly to the preparation of this publication. Special thanks go to Mr. Ali Ismail from FAO-SWALIM as well as Dr Philip Amondi from IGAD Climate Prediction and Application Center (ICPAC).

Disclaimer

The opinions and recommendations expressed in this study are those of the author and do not necessarily reflect the views or policies of FAO. The author and FAO do not assume any responsibility for the geographic borders and names used for the purpose of mapping in this publication.

References

1. Palmer WC (1965) Meteorological Drought, Research Paper No. 45. U.S. Department of Commerce Weather Bureau: Washington, DC, USA.

2. Heim RR (2002) A review of twentieth-century drought indices used in the United States. Bull Amer Meteorol Soci 83: 1149-1165.

3. IFRC (2011) Drought in the Horn of Africa: Preventing the next disaster. International Federation of the Red Cross and Red Crescent Societies, Geneva, Switzerland.

4. Muchiri PW (2007) Climate of Somalia. Technical Report No W-01, FAO-SWALIM, Nairobi, Kenya.

5. Oduor AR, Gadain HM (2007) Potential of Rainwater Harvesting in Somalia, A Planning, Design, Implementation and Monitoring Framework, Technical Report NoW-09. FAO- SWALIM, Nairobi, Kenya.

6. Saleem U (2016) Territorial diagnostic report of the land resources of Somaliland. FAO-SWALIM, Nairobi, Kenya.

7. Funk C, Peterson P, Landsfeld M, Pedreros D, Verdin J, et al. (2015) The climate hazards infrared precipitation with stations-a new environmental record for monitoring extremes. Sci Data 2: 150066.

8. McKee BT, Doesken JN, Kleist J (1993) The relationship of drought frequency and duration to time scales, Proceedings of ninth Conference on Applied Climatology, Boston. 179-184.

9. Abawi Y, Dutta S, Ritchie J (2003) Potential use of climate forecasts in water resources management. In Stone R and Patridge I (eds.) Science for drought.

10. Sharma A, Dadhwal VK, Jeganathan C, Tolpekin V (2009) Drought monitoring using standardized precipitation index: a case study for the state of Karnataka, India.

11. Giddings L, Soto M, Rutherford BM, Maarouf A (2005) Standardized precipitation index Zones for Mexico. Atmosfera pp: 33-56.

12. Akhatari R, Bandarabadi SR, Saghafian B (2008) Spatio-temporal pattern of drought in north-eastern Iran. Option Mediterraneennes. Options Series A 80: 71-77.

13. Intergovernmental Panel on Climate Change (IPCC) (2012) Managing the risks of extreme events and disasters to advance climate change adaptation p: 582.

14. Nicholson SE (2000) The nature of rainfall variability over Africa on time scales of decades to millenia, Global Planet Chan 26: 137-158.

15. Caminade C, Terray L (2010) Twentieth century Sahel rainfall variability as simulated by ARPEGE AGCM, and future changes. Clim Dynam 35: 75-94.

16. Dai A (2011) Drought under global warming: a review, Wiley Interdisciplinary Reviews. Clim Change 2: 45-65.

17. Dai A (2013) Increasing drought under global warming in observations and models. Nat Clim Change 3: 52-58.

18. Dutra E, Magnusson L, Wetterhall F, Cloke HL, Balsamo G, et al. (2011) The 2010-2011 drought in the Horn of Africa in ECMWF reanalysis and seasonal forecast products. Int J Climatol 33: 1720-1729.

19. Giannini A, Biasutti M, Verstraete MM (2008) A climate modelbased review of drought in the Sahel: Desertification, the regreening and climate change. Global Planet Change 64: 119-128.

20. Giannini A, Salack S, Lodoun T, Ali A, Gaye AT, et al. (2013) A unifying view of climate change in the Sahel linking intra-seasonal, interannual and longer time scales. Environ Res Lett 8: 024010.

21. Hastenrath S, Polzin D, Mutai C (2005) Diagnosing the 2005 Drought in Equatorial East Africa. J Climate 20: 4628-4637.

22. Hwang YT, Frierson DMW, Kang SM (2013) Anthropogenic sulfate aerosol and the southward shift of tropical precipitation in the late 20th century. Geophys Res Lett 40: 1-6.

23. Lebel T, Cappelaere B, Galle S, Hanan N, Kergoat L, et al. (2009) AMMA-CATCH studies in the Sahelian region of West- Africa: An overview. J Hydrol 375: 3-13.

24. Zeng N (2003) Drought in the Sahel. Science 302: 999-1000.

25. Darcy J, Bonard P, Dini S (2012) Real Time Evaluation of the Humanitarian Response to the Horn of Africa Drought Crisis: Somalia 2011-2012. NY, USA: IASC.

19

Effect of Global warming on Indian Agriculture

Ruchita S[*] and Rohit S

Department of Sciences, Pandit Deendayal Petroleum University, Gandhinagar, Gujarat, India

[*]**Corresponding author:** Ruchita S, Petroleum University, Gandhinagar, Raisan, Gujarat-382007, India E-mail: ruchitapshah05@gmail.com

Abstract

Agriculture is the backbone of Indian economy which in turn relies on the monsoon season. Rising global temperature is not only causing climate change but also contributing to the irregular rainfall patterns. Uneven rainfall patterns, increased temperature, elevated CO_2 content in the atmosphere are important climatic parameters which affects the crop production. Research studies indicate that weathering parameters influence strongly (67%) compared to other factors like soil and nutrient management (33%) during the cropping season. The Intergovernmental Panel on Climate Change (IPCC) projected that the global mean surface temperature will likely rise and may result into uneven climatic changes. This rising temperature may affect crop yield at large scale. It has been reported over 20th century that rising temperature plays an important role towards global warming as compared to precipitation. Researchers have confirmed that crop yield falls by 3% to 5% for every 1°F increase in the temperature. In India, crop production may be divided into two seasons: Kharif (influenced by south-west monsoon) and rabi (mostly influenced by north-east monsoon). Present study shows that the crop production is dependent on temperature. Temperature vs. crop production shows a funnel shape for all the seasons. For the lower temperature both the properties are almost linearly correlated. In rabi, at the beginning production show a negative trend with temperature which slowly converts to the positive trend. In kharif that negative trend is not visible. At higher temperatures production increases for both the seasons but with large scattering. The findings may be helpful to study the effect of climate change on the crop production.

Keywords: Global warming; Climate change; Indian agriculture; Crop yield

Introduction

Mankind is in need of an equitable standard of living like adequate food, water, energy, safe shelter and a healthy environment for present as well as future generations. But casual acts of human race, such as emission of greenhouse gases by burning fossil fuels and deforestation has increased the earth's average surface temperature, which is defined as global warming. It is proved that the warming on the earth's surface over last 50 years is mostly due to the anthropogenic activities [1]. Further, it is predicted that the global mean surface temperature will likely be in the range of 0.3-0.7°C for the period 2016-2035 [2]. This rise in temperature may cause various changes such as sea level rise, melting of snow sheets and change in rainfall pattern. Hence, global warming can be considered as the major affecting parameter in changing the earth's climate.

Warming of the climate system is observed all over the world. Recent climate changes have shown its impact on natural as well as human systems. Any significant change in climate may affect agriculture at larger scale. Various factors such as increase in temperature change in rainfall pattern, increase of CO_2 content in atmosphere, frequency and intensity of extreme weather events may have significant impact in agriculture sector. It is predicted that increase in temperature will show overall negative effects on agriculture in the world [3]. Generally agricultural productivity in developing countries is expected to decline by 9-21% because of global warming [4]. In case of India, almost 70% of the population depends on agriculture for their livelihood. 23% of India's Gross National Product (GNP) representing agriculture sector alone, which plays a major role in the country's development and shall continue to hold an important place in the national economy [5]. 2.4% decrement in wheat yield was reported in China due to rising temperature over the past two decades [6]. Increasing global mean surface temperature is very likely to lead changes in precipitation [7]. It is globally accepted that precipitation is a leading factor affecting especially rain fed crop yield [8]. Too much precipitation can cause disease infestation in crops, while too little can be detrimental to crop yields; especially dry periods occur during critical development stages [7]. Carbon dioxide is one of the significant parameter for plant growth. IPCC projected that atmospheric concentration of CO_2 will increase from 368 μmol/mol to 540-970 μmol/mol in 2100 [3]. Research studies observed that a small increase in temperature (2-40°C) had larger effect than elevated CO_2 on grain quality [9,10]. Rising trend of global warming is considered to be more striking than precipitation over the 20th century [7].

Global Scenario

Climate change could be one of the affecting parameter all over the world. It is predicted by IPCC that many of the observed changes due to climate change are unprecedented [11]. Global sea level rise is projected to be between 0.17-0.41 m in the year 2050 [12]. It is observed that the rate of rising sea level has been larger than the mean rate during the previous two millennia, till the mid-19th century [13]. IPCC reported that changes in precipitation will be non-uniform and its extreme events over most of the mid-latitude and wet tropical regions will become more intense and frequent [14]. Recent finding of increasing trends in extreme precipitation leads to imply greater risks of flooding at regional scale [15]. Since 1850, last three decades has been consecutively warmer than any other decade on the Earth's surface. Heat wave frequency has increased since the middle of the 20th century in large part of Asia [2]. Moreover concentration of CO_2

and other greenhouse gases leads to increase the temperature. IPCC report states that the amount and rate of warming expected for the 21st century depends on the total amount of greenhouse gases that mankind emit [16]. These observed changes are responsible for varying the climate at different parts of the Earth and sometimes it may result into extreme weather events.

In 2005, hurricane Katrina strike U.S., which is considered to be one of the most powerful storms in last 100 years. It struck the Gulf Coast region and reports estimated that the greatest farm production loss takes place due to this disaster. Prior to Katrina, in the same year, mid-west portions had experienced significant crop losses due to prolonged drought [17]. Such climatic events affect the ecosystem worldwide. Climate change may increase or decrease the crop yield depending on the latitude of the area and irrigation application. Increasing temperature and varying precipitation may decrease the crop productivity in future [18]. Temperature could be an impactful parameter which affects crop yield all over the world. Hence, studies related to effects of temperature on crop yield may help agriculture sector in a better way to plan and enhance the economy in the future.

Indian Scenario

Like other countries, India has also started experiencing extreme weather events which lead to change the climate. As mentioned earlier, global warming is one of the major affecting parameter to change the climate. In India, it is observed that the annual mean temperature has increased at the rate of 0.42°C [19]. Indian agriculture system is based upon south-west and north-east monsoon. Almost 80% of the total precipitation comes from south-west monsoon in India. Any fluctuations and uncertainties in long range rainfall pattern may affect the agriculture sector and also lead to increase the frequency of droughts and floods at regional scale [20]. A significant increasing trend in rainfall was reported along the west coast, north Andhra Pradesh and North West India [21,22], and while significant decreasing trend was observed over parts of Gujarat, Madhya Pradesh and adjoining area, Kerala and northeast India [23]. North western region of India gets affected by western disturbances at small scale as such disturbances have impact only on rabi production [24] only for not more than 20-30 days.

Not only monsoon, but temperature has also shown its effect on agriculture. Extreme maximum and minimum temperature showed an increasing trend in the southern part whereas decreasing trend in the northern part of India [20]. Research studies show that with the increase in temperature, crop productivity is likely to decrease in future [18]. Hence, there is a need to study the dependency of temperature on crop productivity, stability, yield and quality to uplift the country's economy.

Effect of increasing temperature on Indian agriculture

Research studies shows that rise in global surface temperature would affect Indian agriculture. Several climatic factors which affect agriculture productivity are heat waves, high temperature [25,26], heavy and prolonged precipitation [27-29] and excess cold. These factors have positive as well as negative effects on crop production. Almost every year India faces several weather events due to changes in such climatic parameters in various regions which reduces crop yield. Varied nature of such weather events tends to affect the crop growth cycle and plant physiological processes [30]. In India, about 17% of the years during 1901-2010 were reported as drought years, which result

into severe impacts on agriculture, water resources, food security, economy and social life in the country [31]. The variation in temperature and precipitation above threshold value may affect photosynthesis and transpiration process in crops [32]. Excess rainfall and flood may leads to physical damage of the crops [27]. Studies predicted that changing trends in temperature and precipitation will continue to have significant impact on agriculture [7]. A small rise in temperature (1-2°C), especially in the seasonally dry tropical regions [1] would decrease crop yield [33].

Indian agriculture is divided into two main seasons: Kharif and Rabi based on the monsoon. It is reported that overall temperature rise is likely to be much higher during winter (rabi) rather than in rainy season (kharif) [34]. Moreover, it is predicted that the mean temperature in India will rise by 0.4-2.0°C in Kharif and 1.1-4.5°C in Rabi by 2070 [5]. Decline in agricultural productivity leads to increase food prices at state as well as at country level [35]. Hence, temperature could be one of the significant affecting factor which results into greater instability in agriculture of India.

Materials and Methods

The data sets for the annual mean temperature of India over the period 1990-2013 was taken from India Meteorological Department (IMD). Development in the agriculture and allied sectors of India are of interest to a wide spectrum of people across the world. The Directorate of Economics and Statistics of the Department of Agriculture and Cooperation, Government of India publishes "Agricultural Statistics at a Glance 2014" that presents comprehensive information on this sector [36]. All the data sets of food grain production including rabi and kharif were taken from this report. The report contains production data from 1990-91 to 2012-13. In India, kharif crops are sown at the beginning of south-west monsoon (i.e., June to September) and harvested during autumn season (i.e., September to October). Such crops are highly dependent on the timings as well as amount of rainfall. Millets (Bajra, Jowar), Cotton, Soya bean, Sugarcane, Turmeric, Rice, Maize, Moong (Pulses), Groundnut, Red Chilies are several major Kharif crops in India. Rabi crops are sown after north-east monsoon (i.e. October to February) and harvested during spring season (i.e. February to April). The farmers in India are mainly dependent on this monsoon for growing the crops. Such crops need cool climate during growth period but warm climate during the germination of seed and maturation. Wheat, Barley, Gram, Linseed, Mustard, Masoor, Peas are several major Rabi crops in India.

Result and Discussion

Trend of temperature in India

The average annual temperature shows an increasing trend over India (Figure 1). Hence the effect of warming is clearly visible from 1990 to 2013. Extreme high temperature (25.2°C) was noticed in the year 1996. High temperature may help some of the crops to grow faster, whereas some of them may get negatively affected [34]. Research confirms that every rise of 1°C temperature throughout the growing period, even after considering carbon fertilization will decline 4-5 million tons of wheat production in India [34]. Rice yield will decline by 10% for each 1°C increase in minimum temperature during the growing season [37]. Hence an uneven pattern of temperature may affect crop yield as well as economy of the country.

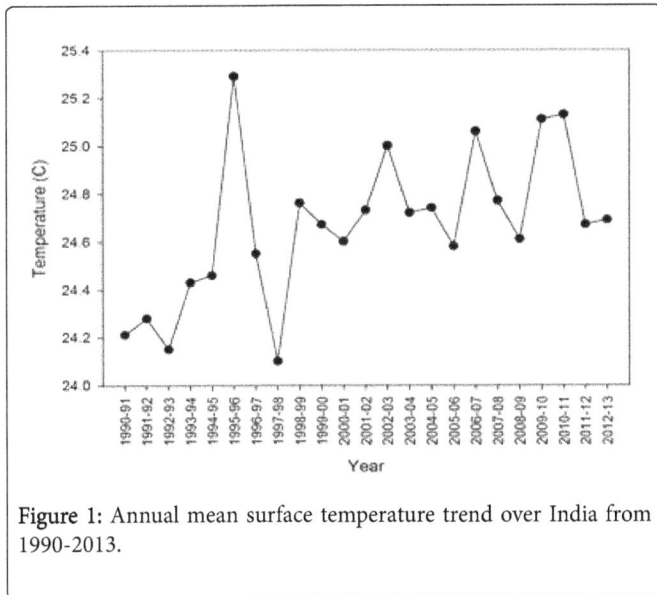

Figure 1: Annual mean surface temperature trend over India from 1990-2013.

Production of overall food grains in India

India is agriculture dependent country which produces varieties of food grains. The lines in the plots are drawn in such a way that almost all the points get cover within the structure. Hence, funnel like structure demonstrate the dependence of temperature on overall crop production in India (Figure 2).

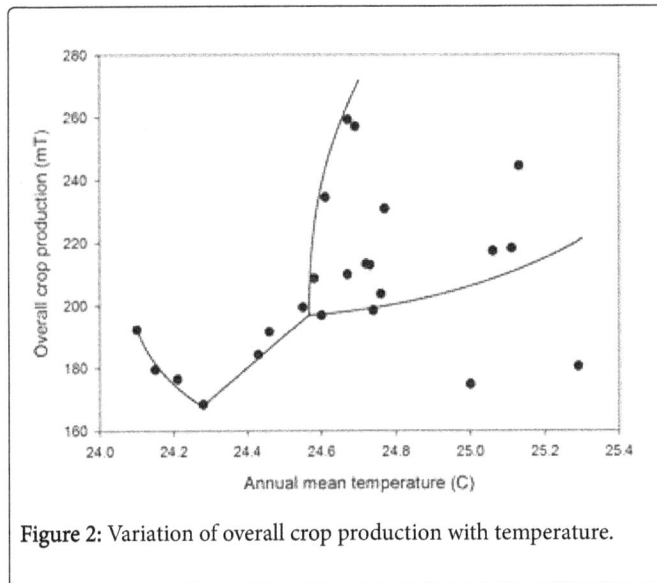

Figure 2: Variation of overall crop production with temperature.

Initially in low temperature range, tail portion show that crop production decreases with increase in temperature. After then, linear trend is observed which show that crop production increases with temperature. As temperature raises further, proportion of scattering increases. Thus, prediction at high temperature becomes difficult. This uneven pattern of temperature may affect crop yield in the country. Research study estimated that by 2020, food grain requirement would be almost 30-50% more than the current demand [38]. Hence, temperature could be one of the significant parameter which helps to visualize the crop response.

Effect of temperature rise on Kharif crops in India

The dependency of temperature on kharif crops shows funnel like structure in Figure 3, which is as similar as in Figure 2. At initial stage, tail portion is not observed which signifies that kharif production may not get affected at low temperature. After that, at high temperature, scattering in the data points was observed. Hence, this makes the prediction difficult at high temperature range. The production for the year 1991-1992, 1995-1996, 2002-2003 and 2008-2009 has not been placed within the funnel like structure as during these years, several weather events took place in India.

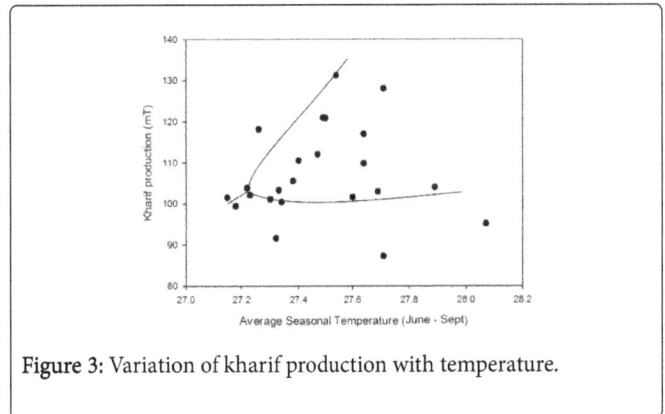

Figure 3: Variation of kharif production with temperature.

During the year 1991-1992, production fell by 5.3% compared to the previous year due to unpredictable behavior of south-west monsoon. Several depressions arose during the year 1995 which caused heavy to very heavy rainfall over Bihar plateau and Gangetic West Bengal. Such situation leads to flood and damage to crops was also reported [39]. The year 2002 was declared as one of the severe drought in India [30]. Due to this drought, production especially kharif crops got badly affected at large extent [40]. Overall deficit of 23% rainfall during the south west monsoon, in the year 2009-2010, adversely affected kharif production [34]. Hence due to such weather events, kharif production gets affected in India.

Effect of temperature rise on rice production in India

Rice is one of the major kharif crops in India. During north-east monsoon, almost two-third of total rice production takes place [41]. Several other studies reported that increase in temperature beyond critical limits may contribute to reduce rice yield in future [42,43]. Thus variation in north–east monsoon and rise in temperature may have impact on rice production.

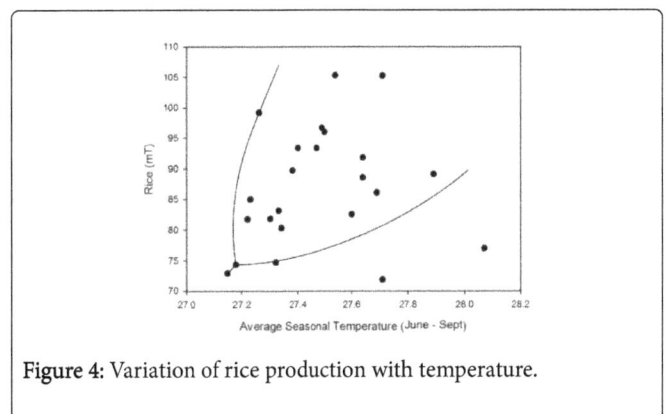

Figure 4: Variation of rice production with temperature.

Rice production also show funnel like structure in Figure 4, which is same as shown in Figure 3. As mentioned earlier in Effect of temperature rise on Kharif crops in India, tail portion is not observed which signifies that rice production does not get affected at low temperature. It may also be possible that if we observe below this temperature range then we may get the tail portion. Same as kharif, increase in production takes place along with temperature rise but with high scatter. The reason for deviation in data points is already discussed in Effect of temperature rise on Kharif crops in India.

Effect of temperature rise on Rabi production in India

Rabi production show funnel-like structure in Figure 5, which is similar as shown in Figure 3. Tail portion is observed which shows that rabi production gets affected at low temperature, which is same as Figure 2. After that rise in temperature shows linear relation with rabi production. Further rise in temperature shows high scattering. Same as for all, prediction becomes difficult at high temperature range. The production in the year 1995-1996 and 2006-2007 are not compiling within the funnel like structure due to extreme weather events took place in India.

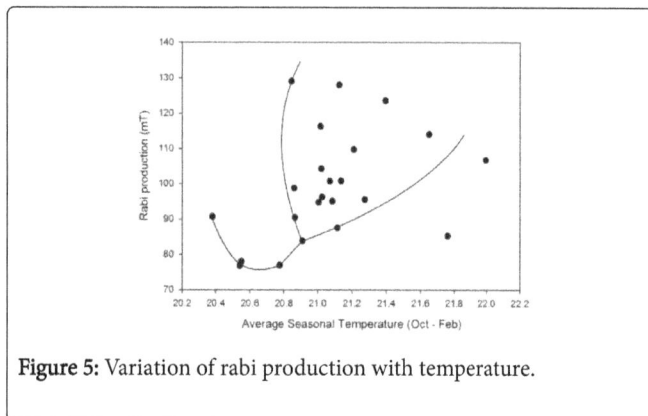

Figure 5: Variation of rabi production with temperature.

During the year 1995, cyclonic storm at various places has reported loss of huge crop yield. Hence, overall rabi production in this year was affected [39]. Moreover, the year 2007 was declared as flood year because series of floods hit India and hence production gets affected [30].

Effect of temperature rise on wheat production in India

Wheat is one of the major rabi crops in India. India is the second largest producer of wheat in the world after China. It is reported since past few years that the productivity of wheat is declining gradually even in Punjab and Haryana-the grainary of the Nation.

Wheat production also show funnel like structure in Figure 6, which is same as kharif production (Figure 5). At initial temperature range, production decreases with increase in temperature. Same as discussed earlier in Effect of temperature rise on Rabi production in India that scattering in wheat production is observed along with further rise in temperature.

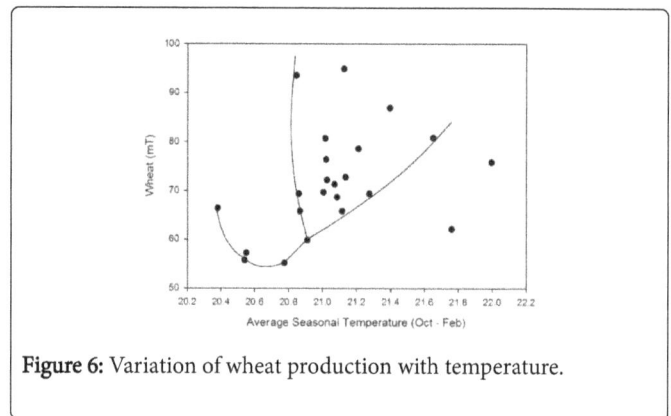

Figure 6: Variation of wheat production with temperature.

The reason for points which are deviating from funnel structure is already explained in Effect of temperature rise on Rabi production in India.

Conclusion

Present study shows that the crop production depends on temperature. Funnel–like structure is observed for overall production (including rabi and kharif) which signifies their dependency on temperature. At low temperature, tail portion was observed in rabi (wheat) production whereas not in kharif (rice). This shows that rabi production has affected comparatively more than kharif at lower temperature. At high temperature range, both types of production shows increasing trend. Moreover, in case of high temperature, it has been observed that scattering in production gets increases. Our study confirms the report of IPCC which states that crop production will get affected at high temperature. Hence, temperature can be one of the significant parameter in crop production studies. At high temperature, prediction of crop production may become difficult as the data points got more scatter. If anyhow, such predictions can be improved further then it may help farmers to make their field planning better, identification of appropriate crop type in particular field, estimation of crop yield and requirement of water for irrigation. In this way, damage to the crops can be minimized and better enhancement in the crop yield can be achieved. Hence, government needs to adopt such predictions and accordingly reframe their plans and policies which may help agriculture sector to uplift and hence can strengthen our economy. Predictions can be improved further by doing long term analysis as the present study contains data of only 23 years. Present study may be limited to the monsoon dominated region. Similar studies may be done for other regions as well to gain the confidence.

References

1. Intergovernmental Panel on Climate Change (IPCC) (2007) Climate Change 2007: Impacts, Adaptation and Vulnerability. Contribution of Working Group II to the Fourth Assessment Report of the Intergovernmental Panel on Climate Change, Cambridge University Press, Cambridge, UK.

2. Intergovernmental Panel on Climate Change (IPCC) (2014) Climate Change 2014: Synthesis Report. Contribution of Working Group I, II and III to the Fifth Assessment Report of the Intergovernmental Panel on Climate Change. IPCC, Geneva, Switzerland.

3. Intergovernmental Panel on Climate Change (IPCC) (2001) Climate Change: Synthesis Report 2001. Summary for Policy makers. Wembley, UK.

4. Cline W (2008) Global warming and agriculture: Impact estimates by country. Peterson Institute for International Economics, Washington, DC, USA.

5. Khan SA, Kumar S, Hussain MZ, Kalra N (2009) Climate Change, Climate Variability and Indian Agriculture: Impacts Vulnerability and Adaptation Strategies. Climate Change Crops pp: 19-38.

6. You L, Mark RW, Fang C, Wood S (2005) Impact of global warming on Chinese wheat productivity. Environment and Production Technology Division. International Food Policy Research Institute.

7. Neenu S, Biswas AK, Rao AS (2013) Impact of climatic factors on crop production-a review. Agri Rev 34: 97-106.

8. Izaurrale RC, Rosenberg NJ, Brown RA, Thomson AM (2003) Integrated assessment of Hardley Center (HadCM2) climate change impacts on agricultural productivity and irrigation water supply in the conterminous United States Part II. Regional agricultural production in 2030 and 2095. Agr Forest Meterol 117: 97-122.

9. Tester RF, Morrison WR, Ellis RH, Piggott JR, Batts GR, et al. (1995) Effects of elevated growth temperature and carbon dioxide levels on some physicochemicalproperties of wheat starch. J Cereal Sci 22: 63-71.

10. Williams M, Shewry PR, Lawlor DW, Harwood JL (1995) The effects of elevated temperature and atmospheric carbon dioxide concentration on the quality of grain lipids in wheat (Triticum aestivum L.) grown at two levels of nitrogen application. Plant Cell Environ 18: 999-1009.

11. Huber DG, Gulledge J (2011) Extreme Weather and Climate change: Understanding the link and managing the risk. Center for Climate and Energy Solutions.

12. Brown S, Lincke D, Nicholls RJ, Hinkel J (2015) The impacts of sea-level rise on European coasts in a 2°C world. Results and analysis of task 6.5 prepared as part of IMPACT2C: quantifying project impacts under 2°C warming. Southampton, GB, University of Southampton.

13. Kemp AC, Horton BP, Donnelly JP, Mann ME, Vermeer M, et al. (2011) Climate related sea-level variations over the past two millennia. Proceedings of the National Academy of Sciences of United States of America 108.

14. Kitoh A, Endo H (2016) Changes in precipitation extremes projected by a 20-km mesh global atmospheric model. Weather Climate Extremes 11: 41-52.

15. Kundzewicz ZW, Matczak P (2015) Hydrological extremes and security. Proc. IAHS.

16. Cicerone RJ, Nurse P (2014) Climate Change-Evidence and Causes. The Royal Society.

17. Schnepf R, Chite RM (2005) U.S. Agriculture after Hurricane Katrina: Status and Issues. The Library of Congress Congressional Research Service.

18. Kang Y, Khan S, Ma X (2009) Climate change impacts on crop yield, crop water productivity and food security-A review. Progress Natural Sci 19: 1665-1674.

19. Arora M, Goel NK, Singh P (2009) Evaluation of temperature trends over India. Hydrol Sci J 50.

20. Jain SK, Kumar V (2012) Trend analysis of rainfall and temperature data for India. Current Sci 102: 37-49.

21. Koteswaram P, Alvi SMA (1969) Trends and periodicities in rainfall at west coast stations in India. Curr Sci 38: 229-231.

22. Jagannathan P, Parthasarathy B (1973) Trends and Periodicities of Rainfall Over India. Indian Institute of Tropical Meteorology 101.

23. Krishnakumar KN, Rao GSLHVP, Gopakumar CS (2009) Rainfall trends in twentieth century over Kerala, India. Atmos Environ 43: 1940-1944.

24. Chand R, Singh C (2015) Movements of western disturbance and associated cloud convection. J Indian Geophy Union 19: 62-70.

25. Ciais P, Reichstein M, Viovy N, Granier A, Ogee J, et al. (2005) Europewide reduction in primary productivity caused by the heat and drought in 2003. Nature 437: 529-533.

26. Van der Velde M, Tubiello FN, Vrieling A, Bouraoui F (2012) Impacts of extreme weather on wheat and maize in France: evaluating regional crop simulations against observed data. Clim Change 113: 751-765.

27. Rosenzweig C, Tubiello FN, Goldberg R, Mills E, Bloomfield J (2002) Increased crop damage in the U.S. from excess precipitation under climate change. Glob Environ Change 12: 197-202.

28. Pathak H, Byjesh K, Chakrabarti B, Aggarwal PK (2011) Potential and cost of carbon sequestration in Indian agriculture: Estimates from longterm field experiments. Field Crops Res 120: 102-111.

29. Thakur P, Kumar S, Malik JA, Berger JD, Nayyar H (2010) Cold stress effects on reproductive development in grain crops: An overview. Environ Exp Bot 67: 429-443.

30. Mahdi SS, Dhekale BS, Choudhury SR, Bangroo SA, Gupta SK (2015) On the climate risks in crop production and management in India. Aust J Crop Sci 9: 585-595.

31. Niranjankumar K, Rajeevan M, Pai DS, Srivastava AK, Preethi B (2013) On the observed variability of monsoon droughts over India. Weather and Climate Extremes 1: 42-50.

32. Porter JR, Semenov MA (2005) Crop responses to climatic variation. The Royal Society Publishing 360: 2021-2035.

33. Lakshmikumar TV, Barbosa H, Rao KK, Jothi EP (2012) Some Studies on the Frequency of Extreme Weather Events over India. J Agricul Sci Tech 14: 1343-1356.

34. Aggarwal PK, Kumar SN, Pathak H (2010) Impacts of climate change on growth and yield of rice and wheat in the Upper Ganga Basin. Indian Agricultural Research Institute (IARI).

35. Udmale PD, Ichikawa Y, Kiem AS, Panda SN (2014) Drought Impacts and Adaptation Strategies for Agriculture and Rural Livelihood in the Maharashtra State of India. The Open Agriculture Journal 8: 41-47.

36. The Directorate of Economic and Statistics in the Ministry of Agriculture, Government of India (2014) Agricultural Statistics At Glance.

37. Peng S, Huang J, Sheehy J, Laza R, Visperas R, et al. (2004) Rice yields decline with higher night temperature from global warming. P Natl Acad Sci USA 101: 9971-9975.

38. Paroda RS, Praduman K (2000) Food production and demand in South Asia. Agr Econ Res Rev 13: 1-24.

39. India Meteorological Department (IMD) (1995) Report on cyclonic disturbances over North Indian Ocean during 1995. Regional Specialized Meteorological Centre for Tropical Cyclones Over North Indian Ocean (Abridged report for circulation during the annual meeting of WMO/ESCAP Panel on Tropical Cyclones).

40. The Planning Commission, Government of India (1991-92) Economy and the plan: An overview.

41. Saravanakumar V (2015) Impact of Climate Change on Yield of Major Food Crops in TamilNadu, India. SANDEE Working Paper pp: 91-15.

42. Dash SK, Hunt JCR (2007) Variability of climate change in India. Curr Sci 93.

43. Geethalakshmi V, Lakshmanan A, Rajalakshmi D, Jagannathan R, Sridhar G, et al. (2011) Climate change impact assessment and adaptation strategies to sustain rice production in Cauvery basin of Tamil Nadu. Curr Sci 101.

20

Assessing the Performance of WRF Model in Simulating Rainfall over Western Uganda

Mugume I[1*], **Waiswa D**[1], **Mesquita MDS**[2,3], **Reuder J**[4], **Basalirwa C**[1], **Bamutaze Y**[1], **Twinomuhangi R**[1], **Tumwine F**[5], **Sansa Otim J**[5], **Jacob Ngailo T**[6] and **Ayesiga G**[7]

[1]*Department of Geography, Geo-Informatics and Climatic Sciences, Makerere University, Uganda*

[2]*Uni Research Climate, Bergen, Norway*

[3]*Bjerknes Centre for Climate Research, Bergen, Norway*

[4]*Geophysical Institute, University of Bergen, Norway*

[5]*Department of Networks, Makerere University, Kampala, Uganda*

[6]*Department of General Studies, Dar er Salaam Institute of Technology, Tanzania*

[7]*Uganda National Meteorological Authority, Uganda*

Corresponding author: Mugume I, Department of Geography, Geo Informatics and Climatic Sciences, Makerere University, Uganda
E-mail: imugume@caes.mak.ac.ug

Abstract

Skillful rainfall prediction is important to sectors such as agriculture, health and water resources. The study assessed the ability of the Weather Research and Forecasting model to simulate rainfall over Western Uganda for the period 21st April to 10th May 2013 and tested six cumulus parameterization schemes. The root mean square error, mean error and the sign test method are used to assess the ability of the schemes to simulate rainfall along with an adapted contingency table. Results show that the Grell-Fretas scheme is better at simulating rainfall compared to other schemes over the study period while the Betts-Miller-Janji'c and the Kain-Fritsch schemes overestimated rainfall. However all the schemes under predicted heavy rainfall events but the Betts-Miller-Janjic and the Kain-Fritsch schemes over predicted the light rainfall. The variation of altitude presented a noticeable change in predicted rainfall where an increase of 25% in altitude increased the probability of prediction by 6.5% which shows a key role played by altitude in convection.

Keywords: WRF model; Adapted contingency; Rainfall; Parameterization schemes

Introduction

Numerical Weather Prediction (NWP) models, such as the Weather Research and Forecasting (WRF) model have gained widespread attention in weather and climate prediction over the 21st Century. These models are objective [1] and produce simulations by solving atmospheric governing equations [2]. They have dynamical cores that represent atmospheric processes and physical schemes that resolve the physics in sub-grid scale process. Resolving sub-grid processes requires parameterization such as the cumulus parameterization schemes [3,4] and the microphysical schemes [4,5] which have a great influenced on the precipitation simulated and also play an important role in determining the vertical structure of temperature and moisture fields of the atmosphere [5].

The use of NWP models in precipitation forecasting is already established in many operational weather and climate prediction centers. This can be partly explained by the demand of improved precipitation prediction since precipitation affects many economic sectors such as agriculture [6,7], fisheries [8], transport and other economic activities [9,10]. Accurate precipitation monitoring and prediction is thus important for spatial and temporal variability analysis [10] as well as climate change studies [11]. However, the skill of NWP models regarding quantitative precipitation prediction is a challenge [12,13] which was also observed by Opijah et al. [14] over the Greater Horn of Africa and is attributed by Ducrocq et al. [15] to model's failure to accurately simulate dynamical and physical processes as well as initial conditions. Additionally NWP models normally have systematic biases [1,16] that can be attributed to unresolved sub–grid scale atmospheric processes. In spite of these biases, NWP are used for downscaling coarse resolution atmospheric fields to generate fine scale representation of atmospheric processes [17-19].

Over East Africa, precipitation is majorly in form of rainfall and is influenced by the Inter-Tropical Convergence Zone (ITCZ); the monsoon wind systems; the tropical cyclones; semi-permanent subtropical anticyclones and easterly waves [6,20]. Fortunately NWP models have the ability to simulate these systems as observed by Nishant et al. [21]. Other features that influence rainfall include the complex topography, vegetation and inland water bodies which modulate local rainfall [6,20,22]. The ITCZ passes over the equator twice a year making the region to have two major rainfall seasons, namely, March-May (MAM) and September–November (SON). While passing over East Africa, the ITCZ exhibits both Zonal and meridional flows as discussed by Mugume et al. [6] but unlike the zonal, the meridional component is shallow (i.e., 2-5 km) [21].

The amount of precipitation simulated by NWP models is the sum of convective and non-convective rainfall which is contributed by the cumulus scheme and the microphysical scheme respectively. Therefore to assess the ability of WRF model to simulate rainfall over Uganda, it is important to obtain a skilful combination of the cumulus and microphysical schemes [5] but studies over the tropics such as Mayor and Mesquita [5] have shown the stronger impact of the cumulus scheme compared to the microphysical schemes regarding convective precipitation events. The different parameterization schemes have biases that depend on region, season or weather but it is important to use a combination with lowest mean error [23]. The cumulus schemes have a couple of assumptions and are based on many variables such as entrainment, cloud height, moist static energy, convective available potential energy and vertical temperature profiles among others.

The study tested six cumulus parameterization schemes by subjecting the simulated results to statistical performance measures including an adapted contingency table. The effect of varying model elevation was also investigated and the study was aimed at answering the questions: (1) What is the most skilful cumulus scheme over Western Uganda? (2) Which scheme is good at simulating light/heavy rainfall events? and (3) What is the effect of varying model elevation on simulated precipitation?

Data Sources and Study Methods

Data

In the study, daily observed rainfall data from 21st April to 10th May 2013 for five stations of western Uganda (Table 1) was obtained from Uganda National Meteorological Authority (UNMA). This data was evaluated using a point-to-point evaluation method with a node interpolated from the prediction grid as explained by Mayor and Mesquita [5]. The model was initialized using boundary conditions obtained from the National Centers for Environmental Prediction (NCEP) final reanalysis [24] and the resolution of this data is $1° \times 1°$, covering the period of study (21st April to 10th May 2013).

Name	Longitude	Latitude	Altitude (m)
Bushenyi	30.167	-0.567	1590
Kabale	29.983	-1.25	1869
Kasese	30.1	0.183	691
Masindi	31.717	1.683	1147
Mbarara	30.683	-0.6	1420

Table 1: Geographical data of study locations.

The study case

The study was aimed at stimulating rainfall over western Uganda because, during the study period, the region experienced heavy rainfall. For example in Masindi, 124.9 mm of rainfall was recorded on 6th May 2013. Over Kasese, heavy rainfall of 40.1 mm pounded the region on 1st May 2013 and caused devastating flooding in Kasese district after bursting of the river banks of Nyamwamba and Mobuku. According to the Disaster Relief Emergency Fund of the International Federation of Red Cross and Red Crescent, eight people were confirmed dead and an estimated 9,663 people were displaced as a consequence of the floods. In addition, infrastructure such as houses and bridges were destroyed.

Experiment design

The study was conducted using the WRF version 3.8 model which consist of a dynamical core and physical parameterization schemes and it is recommended for high resolution simulations by Mayor and Mesquita [5]. This model is developed for regional simulation and weather prediction [25] and is popularly employed in NWP simulations by many scholars including Mercader et al. [26], Flaounas et al. [3], Krogsater et al. [27], Rajasekhar et al. [28] and many others.

For our study, the mercator map projection was used for projecting nested domain (Figure 1) with the parent domain extending from 18.3306°N to 23.0638°S and 0.42832°W and 55.4283°E at a horizontal resolution of 30 km while the nest domain extended from 6.04777°N to 5.1772°S and 25.9259°E to 39.8676°E at a horizontal resolution of 10 km.

The coarse domain was used for large scale simulation while the nest for validation of the schemes using station gauge observed rainfall data. The model top was set at 50 hPa with 40 vertical layers.

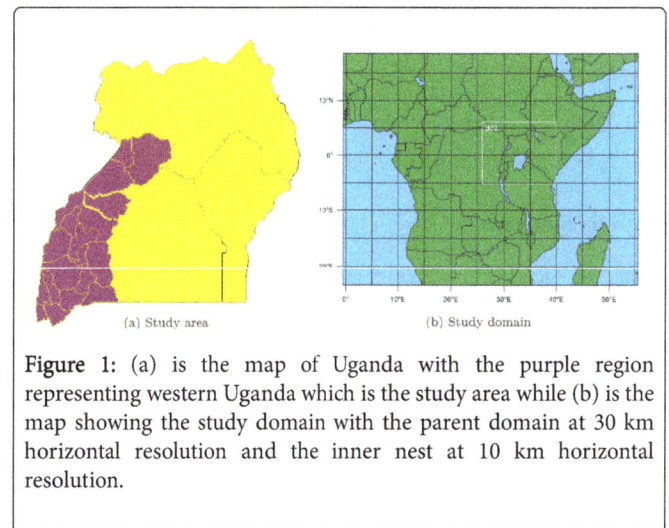

(a) Study area   (b) Study domain

Figure 1: (a) is the map of Uganda with the purple region representing western Uganda which is the study area while (b) is the map showing the study domain with the parent domain at 30 km horizontal resolution and the inner nest at 10 km horizontal resolution.

The study was carried out to assess the performance of cumulus parameterization schemes because they are vital to the model's ability to simulate precipitation and therefore needs accurate representation [29]. Thus the other physical parameterization schemes were set to default and only the cumulus schemes varied. A couple of cumulus schemes have been developed based on different closure assumptions [30-34].

For the study we used six cumulus schemes namely the Kain–Fritsch (KF), the Betts–Miller–Janji´c (BMJ), the Grell–Fretas (GF), the Grell 3D ensemble (G3), the New–Tiedke (NT) and the Grell–Devenyi (GD) (Table 2) and we allowed a spin–up period of 24 hours. For all the experiments, we used WRF Single Moment 3–class scheme as the microphysical scheme and the Yonsei University scheme as the planetary boundary layer scheme.

Evaluation of the cumulus schemes in different areas of the world has been carried out and results show varying performance of the different cumulus schemes. For example, Ratna et al. [35] evaluated KF, BMJ and GDE over South Africa and found all the schemes to produce positive biases and KF produced the largest biases and largest mean absolute error.

In a separate study of impacts of physical parameterization schemes on seasonal rainfall, Jankov et al. [13] noted that BMJ normally over estimated light rainfall events.

Scheme	Description
KF	The KF is a mass flux scheme that uses Langragian parcel method and is used at NOAA/NWS/SPC
BMJ	The BMJ is a moist column adjustment scheme relaxing towards a mixed profile but without explicit updrafts or downdrafts and no cloud detrainment [5]
GF	The GF is an improved GD scheme that smoothens the transition to cloud– resolving scales and has explicit updrafts and downdrafts including cloud ice and detrainment [5]
G3	Is an improved scheme of GD which is a one-dimensional mass flux scheme having a combination of multiple closures based on convective available potential energy, moisture convergence and allows subsidence to occur within the same grid column or spreading to neighboring grids
NT	The NT is a mass–flux scheme with CAPE removing time scale and also having a shallow component and momentum transport
GD	The GD is a multi–closure, multi–parameter ensemble with different updraft and downdraft entrainment and detrainment parameters

Table 2: Description of cumulus schemes.

Analysis methods

In the study, we used both parametric and non-parametric methods to comprehensively assess the performance of cumulus schemes. The parametric methods consider rainfall as a continuous random variable both in space and time and are based on difference between simulated and observed rainfall. For the study we used the Root Mean Square Error (RMSE); and mean error (ME or bias); as parametric methods which are popular statistical scores for assessing the skill of NWP models as observed by Mercader et al. [26] and the Sign Test Method (STM) as a non-parametric method as described by Mugume et al. [1] due to its insensitivity to outliers.

Model difference: For paired data-sets (i.e., simulated, M and observed, O), let the difference between simulated and observed data be d such that:

$$d_k^{(l)} = M_k^{(l)} - O_k^{(l)} \qquad (1)$$

Where l is the cumulus scheme and k is k^{th} data point ordered in time. A set of differences is thus obtained as:

$$d^{(l)} = \left\{ d_1^{(l)}, d_2^{(l)}, d_3^{(l)} ...., d_k^{(l)}, ..., d_{(n-1)}^{(l)}, d_n^{(l)} \right\}$$

Where n is the total number of data points ordered in time.

Root mean square error: The Root Mean Square Error (RMSE) is the square root of the mean square difference and is computed mathematically as:

$$RMSE_l = \sqrt{\frac{1}{n} \sum_{k=1}^{n} \left[d_k^{(l)} \right]^2} \qquad (2)$$

The model altitude was varied and the changes in RMSE (i.e., ΔRMSE) presented in a Table (8) below.

Mean error: The mean error (ME), also known as bias is the mean of the model differences $(d_k^{(l)})$ and is computed as:

$$ME_l = \frac{1}{n} \sum_{k=1}^{n} d_k^{(l)} \qquad (3)$$

The model altitude was varied and the changes in ME (i.e., ΔME) also presented in a Table (8) below.

Sign test method: The STM is described by considering a statistic, v:

$$v_k^{(l)} = \begin{cases} +1: & if\ d_k^{(l)} > 0 \\ 0: & if\ d_k^{(l)} = 0 \\ -1: & if\ d_k^{(l)} < 0 \end{cases} \qquad (4)$$

the values of $\vartheta_k^{(l)}$, i.e., $\left\{ \vartheta_1^{(l)}, \vartheta_2^{(l)}, \vartheta_3^{(l)}, ..., \vartheta_{n-1}^{(l)}, \vartheta_n^{(l)} \right\}$ are averaged to give STM for a given scheme (STMl), i.e.

$$STM_l = \frac{1}{n} \sum_{k=1}^{n} \vartheta_k^{(1)} \qquad (5)$$

The model altitude was varied and the changes in STM (i.e., ΔSTM) also presented in a Table (8) below.

The categorical method: We also used the categorical methods that discretize rainfall. The most common categorical tool is the 2 × 2 contingency table (Table 3) which considers rainy events in terms of occurrence (i.e., Yes/No) as explained by Done and Davis [36]. The problem of such a 2 × 2 contingency table when used for assessing the performance of an NWP model's ability to predict rainfall is that it treats all rainfall events the same and thus fails to capture the possible bias of the model. To attempt to address this limitation, we suggested and adopted a '5 × 5 contingency table' with different bins (i.e., rainfall=0, rainfall <10 mm, rainfall <20 mm, rainfall <30 mm and rainfall ≥30 mm). Although we used a '5 × 5 contingency table', this table can be adopted to have different ranges. We give a comprehensive account of how the contingency table can be adapted to extended ranges i.e., an 'n × n contingency Table' (Table 4).

Simulated		Yes	No
Observed	Yes	Hit	Miss
	No	False Alarm	Correct No

Table 3: The 2 × 2 contingency table.

	sim$_1$	sim$_2$	-	-	-	sim$_n$
obs$_1$	a$_{11}$	a$_{21}$	-	-	-	a$_{n1}$
obs$_2$	a$_{12}$	a$_{22}$	-	-	-	a$_{n2}$
-	-	-	-	-	-	-
-	-	-	-	-	-	-
-	-	-	-	-	-	-
obs$_n$	a$_{1n}$	a$_{2n}$	-	-	-	a$_{nn}$

Table 4: The n × n contingency table.

For n bins including a special case of 'no rain' (i.e., rainfall=0), we can construct an n × n square matrix obtained from an n × n contingency table. Suppose that n columns represent the simulated results and n rows represent the observed results and let i represent the ith column and j represent the jth row. The a_{ij}'s are the number of cases falling in the ith simulated and j^{th} observed category. We thus extract our performance matrix (M) as:

$$M = \begin{pmatrix} a_{11} & a_{21} & \cdots & a_{n1} \\ a_{12} & a_{22} & \cdots & a_{n2} \\ \cdot & \cdot & \cdots & \cdot \\ \cdot & \cdot & \cdots & \cdot \\ \cdot & \cdot & \cdots & \cdot \\ a_{1n} & a_{2n} & \cdots & a_{nn} \end{pmatrix}$$

Since the columns correspond to simulated values and the rows correspond to observed values, the performance matrix can be decomposed into '*Upper triangular matrix*' (UM), '*Diagonal matrix*' (DM) and '*Lower triangular matrix*' (LM).

$$UM = \begin{pmatrix} 0 & a_{21} & \cdots & a_{n1} \\ 0 & 0 & \cdots & a_{n2} \\ \cdot & \cdot & \cdots & \cdot \\ \cdot & \cdot & \cdots & \cdot \\ \cdot & \cdot & \cdots & \cdot \\ 0 & 0 & \cdots & 0 \end{pmatrix},$$

$$DM = \begin{pmatrix} a_{11} & 0 & \cdots & 0 \\ 0 & a_{22} & \cdots & 0 \\ \cdot & \cdot & \cdots & \cdot \\ \cdot & \cdot & \cdots & \cdot \\ \cdot & \cdot & \cdots & \cdot \\ 0 & 0 & \cdots & a_{nn} \end{pmatrix},$$

$$LM = \begin{pmatrix} 0 & 0 & \cdots & 0 \\ a_{12} & 0 & \cdots & 0 \\ \cdot & \cdot & \cdots & \cdot \\ \cdot & \cdot & \cdots & \cdot \\ \cdot & \cdot & \cdots & \cdot \\ a_{1n} & a_{2n} & \cdots & 0 \end{pmatrix}$$

We defined three performance measures thus, 'Probability of over Prediction' (POV), 'Probability of Prediction' (POP) and 'Probability of Under Prediction' (PUN). The POV is the average of values giving the Upper Triangular Matrix (UM) i.e., $< a_{ij} >$ for all and strictly i > j values, POP is the average of value making the Diagonal Matrix (DM) i.e., $< a_{ij} >$ for all i=j and PUN is the average of value making the Lower Triangular Matrix (LM) i.e., $< a_{ij} >$ for all i<j. Thus if N is the number of cases over which performance is being measured, we can calculate POV, POP and PUN using equations (6 to 8).

$$POV = \frac{1}{N}\sum_{all} a_{ij}, \ \forall i > j \qquad (6)$$

$$POP = \frac{1}{N}\sum_{all} a_{ij}, \ \forall i = j \qquad (7)$$

$$PUN = \frac{1}{N}\sum_{all} a_{ij}, \ \forall i < j \qquad (8)$$

An additional advantage of our proposed n × n contingency table is that it can help in determining the performance of a model in a given range of rainfall. As an illustration given a range e.g. obs_k we can investigate a_{ik} such that for all i>k values correspond to over prediction; for all i<k values correspond to under prediction and i=k is value simulated is in the range. In general we suggest that a good model (e.g. a cumulus scheme for our study) should have a higher POP compared to the rest and a smaller POV+PUN. A higher POV indicates a higher probability of over prediction in the same way a high PUN indicates a higher probability of under prediction.

Results and Discussion

The performance of cumulus schemes

The Tables 5 and 6 as well as (Figure 2) represent the performance of cumulus schemes. These results are obtained by comparing grid–point simulated rainfall, averaged over a point with corresponding observed rain gauge rainfall. Table 5 presents results obtained using a 5 × 5 contingency table described in analysis method while Table 6 presents statistical results (i.e., RMSE, ME and STM). Our results show that the GF scheme had the highest probability of giving rain falling in the same range as observed (i.e., POP) while NT had the smallest POP. The POP of the GF is marginally greater than that of BMJ (i.e., GF: 46% & BMJ: 45%). The drawback of BMJ is that it has a higher PUN compared to GF (i.e., BMJ: 40% & GF: 25%) which makes the GF a better cumulus scheme.

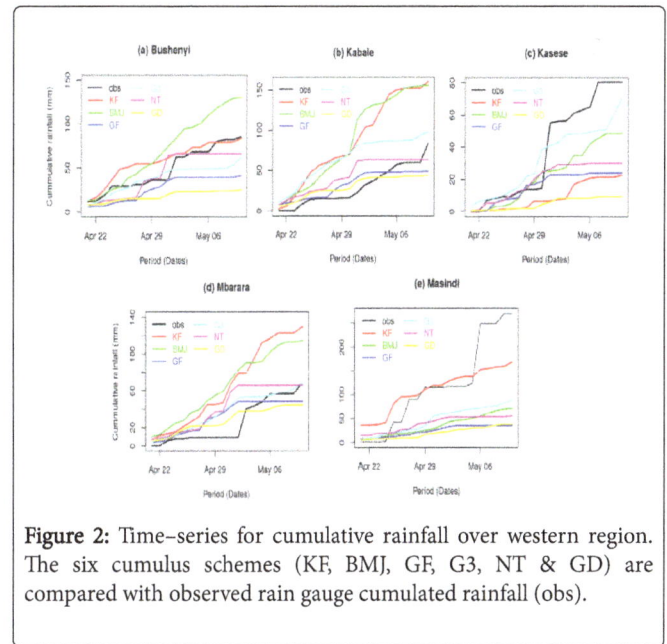

Figure 2: Time–series for cumulative rainfall over western region. The six cumulus schemes (KF, BMJ, GF, G3, NT & GD) are compared with observed rain gauge cumulated rainfall (obs).

	KF (%)	BMJ (%)	GF (%)	G3 (%)	NT (%)	GD (%)
POP	40	45	46	43	38	42
POV	20	15	29	16	29	29
PUN	40	40	25	41	33	29

Table 5: Results from a 5 × 5 contingency table for performance of cumulus schemes. The measures are in percentages.

Additional investigation using statistical indices (i.e., using the RMSE, ME and STM) as presented in Table 6 revealed that the NT cumulus scheme had the smallest RMSE (i.e., RMSE=12.71) and again the BMJ scheme had a marginally greater RMSE compared to GF (i.e., BMJ: 13.47 & GF: 13.31). Although the GD scheme had the smallest magnitude of ME (i.e., ME=-0.26), unfortunately it also had the highest magnitude of RMSE (i.e., RMSE=19.12). This GD scheme also had a high POP (42%) and a small STM (-0.01). Our problem with GD scheme is the high variability of the statistical indices (Table 6) over different stations.

RMSE						
	KF	BMJ	GF	G3	NT	GD
Bushenyi	6.64	7.14	13.14	13.64	9.97	18.17
Kabale	7.61	11.16	15.43	19.42	15.14	21.42
Kasese	9.88	9.91	9.83	14.35	10.78	15
Mbarara	10.49	8.17	14.35	18.74	15.11	23.5
Masindi	29.11	30.97	13.8	14.97	12.54	17.52
ME						
	KF	BMJ	GF	G3	NT	GD
Bushenyi	-0.07	2.29	2.7	5.35	-0.7	-1.38
Kabale	3.91	3.64	-0.3	3.5	-2.55	-5.05
Kasese	-2.9	-1.6	-1.35	4.4	0.6	-6.29
Mbarara	3.08	2.33	2.65	7.9	2.5	5.57
Masindi	-5.03	-9.9	2.15	10.45	1.75	5.86
STM						
	KF	BMJ	GF	G3	NT	GD
Bushenyi	0.35	0.4	-0.2	0.15	-0.25	-0.3
Kabale	0.6	0.4	-0.15	0.3	0	0.05
Kasese	-0.2	0.3	-0.1	0.2	-0.15	-0.3
Mbarara	0.45	0.55	0.2	0.4	0.15	0.2
Masindi	0.25	0.4	0.15	0.35	0.25	0.35

Table 6: Statistical Indices showing the RMSE, ME and STM of the cumulus schemes investigated.

We noted that the ME over individual stations are big but are compensated by averaging over the study region to appear to give a small ME. Overall, we find the GF scheme superior due to a fairly low RMSE (13.31), ME (1.17) and STM (-0.02).

Additional analysis of accumulated rainfall time series (Figure 2) generally revealed that the BMJ & KF schemes simulating excess accumulated rainfall with exception of Kasese station (Figure 2c). Over Masindi station (Figure 2e) all the cumulus schemes under–predicted but the KF scheme fairly simulated the rainfall over the period but like other schemes considered, it failed to simulate the heavy rainfall that occurred on 6th May 2016 (i.e., total rainfall of 124.9 mm).

Light and heavy rainfall events

The results for simulated heavy rainfall events and light rainfall events over the study period are presented in Table 7. A heavy rainfall day is one with total rainfall amount greater than 20 mm while a light rainfall day is one which total rainfall amount less than 1 mm [6]. We noted that all the schemes under–predicted the heavy rainfall events over the period of study and that the KF scheme had fairly the smallest magnitude of error compared to other schemes. This means that the KF scheme is better at simulating heavy rainfall events. For light rainfall events, we noted that KF scheme over–predicted rainfall amount (i.e., RMSE: 8.22; ME: 3.54 & STM: 0.21) followed by BMJ scheme (i.e., RMSE: 7.19; ME: 3.04 & STM: 0.38). These results confirm that KF and BMJ schemes over–predicted rainfall over the study period and are in line with a similar study by Ratna et al. [35] that showed that the KF scheme had a more unstable convective environment compared to the GD scheme thus over predicting convective rain compared to stratiform rain. Considering the GF scheme, we noted that it presented a higher mean error regarding heavy rainfall which probably indicates that the GF scheme is not good at simulating heavy rainfall events [36].

Schemes	Heavy rainfall events			Light rainfall events		
	RMSE	ME	STM	RMSE	ME	STM
KF	45.23	-33.28	-1	8.22	3.54	0.21
BMJ	49.38	-38.62	-1	7.19	3.04	0.38
GF	51.35	-41.17	-1	4.63	-0.53	-0.25
G3	50.11	-39.29	-1	5.56	0.81	0.08
NT	51.28	-41.27	-1	6.39	0.03	-0.29
GD	51.37	-40.86	-1	3.7	-1.33	-0.29

Table 7: Statistical results for simulation of heavy and light rainfall events over the study period.

Effect of altitude on simulated precipitation

The analysis of altitude (Table 1) and simulated precipitation showed that, Kabale which had the highest altitude (1,869 m) also had 4/6 schemes (i.e., BMJ, GF, G3 & NT) with the highest RMSE over it. Additional analysis of Figure 2b), for Kabale show that all the schemes over estimated rainfall. For Mbarara, which has altitude 1,420 m, all the schemes also over predicted (Figure 2d). Kasese and Masindi which were lower in altitude compared to the rest (i.e., 691 m and 1,147 m respectively) also had the schemes under predicting rainfall (Figure 2c and 2e respectively). Other studies e.g. Maussion et al. [37] found WRF model to over predict rainfall over high altitude and complex topography and called it orographic bias. Our results probably also confirm an orographic bias but the only problem is lack of validation data over high altitude areas like mountain Rwenzori.

To further investigate the effect of altitude on simulated precipitation, the altitude was decreased by 25%, 50% and also increased by 25% and 50%. The results for statistical indices (RMSE, ME & STM) are presented in Table 8; contingency results presented in Table 9 and time series for accumulated rainfall in (Figure 3). Our results do not necessary show improvement in simulation of precipitation but reveal changes in the amount of precipitation simulated (Table 8).

	RMSE for varied altitude					Change in RMSE (%)			
	GF-orig	GF-25	GF-50	GF+25	GF+50	GF-25	GF-50	GF+25	GF+50
Bushenyi	7.03	7.1	7.51	5.99	7.31	1	6.83	-14.79	3.98
Kabale	6.9	6.8	12.24	5.55	7.33	-1.45	77.39	-19.57	6.23
Kasese	9.3	10.23	9.33	10.22	10.08	10	0.32	9.89	8.39
Mbarara	8.59	8.55	8.5	8.46	8.01	-0.47	-1.05	-1.51	-6.75
Masindi	32.06	30.99	30.17	30.17	32.1	-3.34	-5.9	-5.9	0.12
	ME for varied Altitude					Change in ME (%)			
	GF-orig	GF-25	GF-50	GF+25	GF+50	GF-25	GF-50	GF+25	GF+50
Bushenyi	-2.17	-2.24	-1.07	-1.48	-2.5	-3.23	50.69	31.8	-15.21
Kabale	-1.71	-1.39	1.44	-0.78	-2.21	18.71	184.21	54.39	-29.24
Kasese	-2.82	-1.88	-1.54	-2.35	-3.27	33.33	45.39	16.67	15.96
Mbarara	-0.97	-0.47	0.99	0.87	-1.7	51.55	202.06	189.69	-75.26
Masindi	-11.71	-10.5	-8.79	-10.51	-12.25	10.33	24.94	10.25	-4.61
	STM for varied Altitude					Change in STM (%)			
	GF-orig	GF-25	GF-50	GF+25	GF+50	GF-25	GF-50	GF+25	GF+50
Bushenyi	-0.2	-0.15	0.05	-0.1	-0.25	25	125	50	-25
Kabale	-0.15	0	0	0.15	-0.1	100	100	200	33.33
Kasese	-0.1	0.05	0.2	-0.15	-0.35	150	300	-50	-250
Mbarara	0.2	0.25	0.6	0.4	0.2	25	200	100	0
Masindi	0.15	0.25	0.2	0.25	-0.1	66.67	33.33	66.67	-166.67

Table 8: The RMSE, ME and STM for various altitude considered of GF cumulus schemes. GF-orig is the original GF scheme with default altitude; GF-25 is GF scheme with altitude reduced by 25%; GF-50 is GF scheme with altitude reduced by 50%; GF+25 is GF scheme with altitude increased by 25% and GF+50 is the GF scheme with altitude increased by 50%.

We noted that when we varied altitude, there was a noticeable change in RMSE, ME and STM. Different stations experienced different changes and this could probably be attributed to changes in centers of formation of precipitating clouds since convective clouds are influenced by both diabetic heating, orography and moisture advection [38]. The largest positive changes in both ME and STM are observed when altitude is reduced by 50% whereas increase of altitude by 50% gave negative changes.

The Figure 3 also show that reduction of altitude by 50% resulted to over prediction of rainfall changing the bias from negative to positive especially for Mbarara and Kabale. These results are also confirmed by the changes in total rainfall over the study period due to variation in altitude Figure 4 but we noted that although reduction of altitude by 50% generally increased total rainfall, it was still under estimate for Bushenyi, Kasese and Masindi which could be due to the weaker ability of the GF scheme to simulate heavy rainfall events.

In terms of predictive ability, the results of extended contingency table (Table 9) shows that the predictive capacity of the GF scheme improved from 46% to 49% by increasing the altitude by 25%.

As already observed in Table 8 that the increase of altitude by 50% resulted to increased underestimate of rainfall, we also see in Table 9 that predictive ability declines to 40% and Probability of Underestimation (PUN) increases from 25% to 35%. Thus elevation can influence when and where precipitation happens [39].

Summary and Conclusion

The study assessed the performance of WRF model regarding rainfall simulation over Western Uganda by investigating six cumulus schemes, namely the Kain–Fritsch, the Betts–Miller–Janjíc, the Grell–Fretas, the Grell 3D ensemble, the New–Tiedke and the Grell–Devenyi using statistical indices (i.e., RMSE, ME and STM) as well as a '5 × 5 extended contingency table.

Additionally, we investigated the ability of the schemes to simulate heavy rainfall events (i.e., a heavy rainfall day is one with total rainfall amount greater than 20 mm) and light rainfall events (i.e., a light rainfall day is one with total rainfall amount less than 10 mm) over the study period.

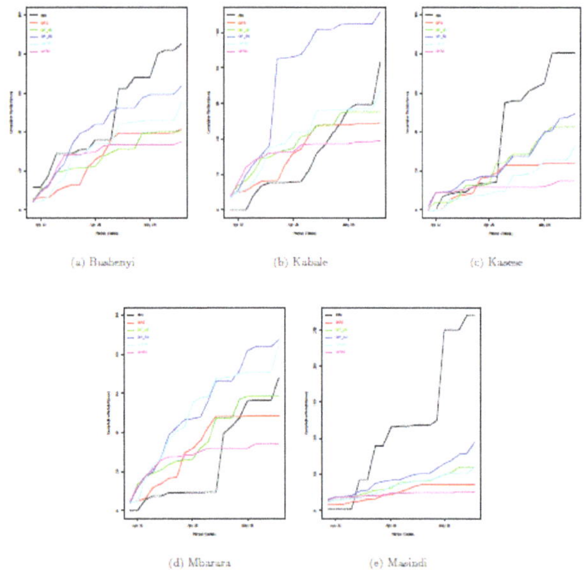

Figure 3: The Figures 3a-3e shows the time-series of cumulated rainfall for the stations used in the study. Each figure has respective graph of actual observations (obs), the simulated rainfall with model altitude (GF-0) and the simulated rainfall with varying altitudes i.e., reduced by 25% (GF 25), reduced by 50% (GF 50), increased by 25% (GF.25) and increased by 50% (GF.50)

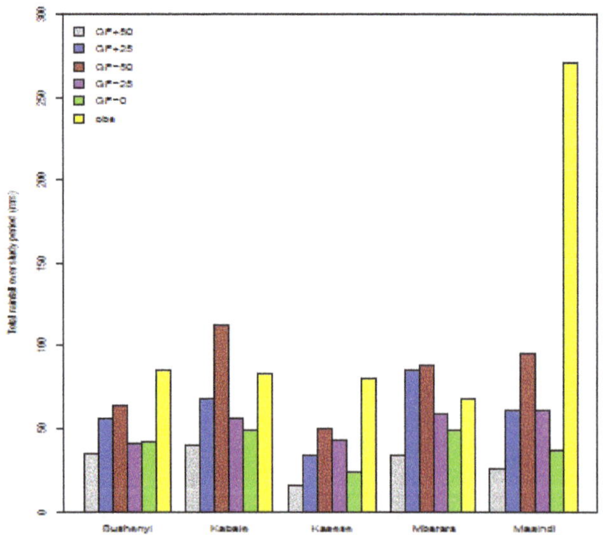

Figure 4: The bar chart showing changes in amount of total rainfall over the study period due to variation in altitude. GF+50 means that altitude was increased by 50%; GF+25 means that altitude was increased by 25%; GF−50 means that altitude was reduced by 50%; GF−25 means that altitude was reduced by 25%; GF−0 means that model altitude was maintained and obs is actual rainfall observation.

	GF-0 (%)	GF-25 (%)	GF-50 (%)	GF+25 (%)	GF+50(%)
POP	46	47	44	49	40
POV	29	29	34	31	25
PUN	25	24	22	20	35

Table 9: Contingency table for performance of cumulus schemes with varied altitude.

We found that the GF scheme had the highest Probability of Predicting (POP) rainfall while NT had the smallest POP. We also noted that although the GD scheme had the smallest magnitude of ME (i.e., ME=-0.26), it had the highest magnitude of RMSE (i.e., RMSE=19.12) and its other drawback was the high variability of the statistical indices over different areas which are compensated by averaging over the study region to appear to give a small ME. Generally we found the GF scheme a superior scheme due to a fairly low RMSE (13.31), ME (1.17) and STM (-0.02).

We also observed that all the schemes under-predicted heavy rainfall and that the KF scheme had fairly the smallest magnitude of error compared to other schemes which meant that the KF scheme is better at simulating heavy rainfall events. For the light rainfall events, we found the KF scheme to over-predict followed by BMJ scheme which confirmed the KF and BMJ schemes to over-predict rainfall over the study period and because the GF scheme presented a higher mean error regarding heavy rainfall, it probably indicated that it is not good at simulating heavy rainfall events.

Further, we found that the altitude had an influence on simulated precipitation and found that, Kabale which had the highest altitude also had 4/6 schemes (i.e., BMJ, GF, G3 & NT) with the highest RMSE and all the schemes over estimated rainfall over it. Kasese and Masindi which had a lower altitude compared to the rest also experienced under-prediction of rainfall which confirms an orographic bias. When the altitude was varied, the results did not necessary show improvement in precipitation simulation but revealed changes in the amount of precipitation simulated. We noted a change in RMSE, ME and STM which could probably be attributed to changes in centers of formation of precipitating clouds since convective clouds are influenced by many factors including orography. The largest positive changes in both ME and STM were observed when altitude was reduced by 50% whereas increase of altitude by 50% gave negative changes. The predictive ability of the GF scheme improved from 46% to 49% by increasing the altitude by 25% which emphasize the influence of elevation on convective rainfall

In the study we also adapted a '2 × 2 contingency table' to a 5 × 5 contingency table that enabled us to fairly assess whether the model over predicts, predicts rainfall in the a given range and or under-predicts. This adapted contingency table can also assist in assessing the ability of a model to predict rainfall in a given range (i.e., rainfall greater than a given threshold e.g. greater than 20 mm). Finally, we recommend the GF cumulus scheme as a basic deterministic scheme which can also be used as a convection scheme to give a background analysis during data assimilation but for heavy rainfall events, we recommend the BMJ scheme.

Acknowledgements

The authors are grateful to the project of "Improving Weather Information Management in East Africa for effective service provision through the application of suitable ICTs" (WIMEA-ICT project: UGA–13/0018) under the Norad's Programme for Capacity Development in Higher Education and Research for Development (NORHED) for technical support and the project of "Partnership for Building Resilient Ecosystems and Livelihoods to Climate Change and Disaster Risks" (BREAD project: SIDA/331) under the "Swedish International Development cooperation Agency" (SIDA) for the financial support and to the reviewers for the constructive feedback. We are also grateful to UNMA (https://www.unma.go.ug) for availing the rainfall data used in the study.

References

1. Mugume I, Basalirwa C, Waiswa D, Reuder J, Mesquita MDS, et al. (2016) Comparison of parametric and nonparametric methods for analyzing the bias of a numerical model. Mod Simulat Eng pp: 1-7.

2. Coiffier J (2011) Fundamentals of Numerical Weather Prediction. Cambridge University Press.

3. Flaounas E, Bastin S, Janicot S (2011) Regional climate modelling of the 2006 West African monsoon: sensitivity to convection and planetary boundary layer parameterisation using WRF. Clim Dyn 36: 1083-1105.

4. Sun X, Xie L, Semazzi HF, Liu B (2014) A numerical investigation of the precipitation over lake Victoria basin using a coupled atmosphere-lake limited-area model. Adv Meteorol pp: 1-15.

5. Mayor YG, Mesquita MD (2015) Numerical simulations of the 1 may 2012 deep convection event over cuba: sensitivity to cumulus and microphysical schemes in a high resolution model. Adv Meteorol pp: 1-16.

6. Mugume I, Mesquita MD, Basalirwa C, Bamutaze Y, Reuder J, et al. (2016) Patterns of dekadal rainfall variation over a selected region in lake victoria basin, Uganda. Atmosphere 7: 150.

7. Tao S, Shen S, Li Y, Wang Q, Gao P, et al. (2016) Projected crop production under regional climate change using scenario data and modeling: sensitivity to chosen sowing date and cultivar. Sustainability 8: 214.

8. Kizza M, Rodhe A, Xu CY, Ntale HK, Halldin S (2009) Temporal rainfall variability in the lake Victoria basin in East Africa during the twentieth century. Theor Appl Climatol 98: 119-135.

9. Bentzien S, Friederichs P (2012) Generating and calibrating probabilistic quantitative precipitation forecasts from the high-resolution NWP model COSMO-DE. Weather and Forecasting 27: 988-1002.

10. Beskow S, Caldeira TL, de Mello CR, Faria LC, Guedes HAS (2015) Multiparameter probability distributions for heavy rainfall modeling in extreme southern Braz J Hydrol Reg Stud 4: 123-133.

11. Ngailo T, Nyimvua S, Reuder J, Rutalebwa E, Mugume I (2016) Nonhomogeneous poisson process modelling of seasonal extreme rainfall events in Tanzania. Int J Sci Res 5: 1858-1868.

12. Grell GA, D´ev´enyi D (2002) A generalized approach to parameterizing convection combining ensemble and data assimilation techniques. Geophys Res Lett 29.

13. Jankov I, Gallus WA Jr, Segal M, Shaw B, Koch SE (2005) The impact of different WRF model physical parameterizations and their interactions on warm season MCS rainfall. Weather and Forecasting 20: 1048-1060.

14. Opijah FJ, Ogallo LA, Mutemi JN (2014) Application of the EMS-WRF model in dekadal rainfall prediction over the GHA Region. Afr J Phys Sci 1.

15. Ducrocq V, Ricard D, Lafore JP, Orain F (2002) Storm scale numerical rainfall prediction for five precipitating events over France: On the importance of the initial humidity field. Weather and Forecasting pp: 1236-1256.

16. Franzke CL, O'Kane TJ, Berner J, Williams PD, Lucarini V (2015) Stochastic climate theory and modeling. Wiley Interdisciplinary Reviews. Clim Change 6: 63-78.

17. Nimusiima A, Basalirwa C, Majaliwa J, Mbogga S, Mwavu E, et al. (2014) Analysis of future climate scenarios over central Uganda cattle corridor. J Earth Sci Clim Change 2014.

18. Hong SY, Kanamitsu M (2014) Dynamical downscaling: fundamental issues from an NWP point of view and recommendations. Asia Pac J Atmos Sci 50: 83-104.

19. Charabi Y, Al Yahyai S (2015) Evaluation of ensemble NWP models for dynamical down-scaling of air temperature over complex topography in a hot climate: A case study from the sultanate of Oman. Atmosfera 28: 261-269.

20. Awange J, Anyah R, Agola N, Forootan E, Omondi P (2013) Potential impacts of climate and environmental change on the stored water of lake Victoria basin and economic implications. Water Resour Res 49: 8160-8173.

21. Nishant N, Sherwood SC, Geoffroy O (2016) Radiative driving of shallow return flows from the ITCZ. J Adv Model Earth Syst 8: 831-842.

22. Williams K, Chamberlain J, Buontempo C, Bain C (2015) Regional climate model performance in the lake Victoria basin. Climate Dynamics 44: 1699-1713.

23. Lee AJ, Haupt S, Young G (2016) Down selecting numerical weather prediction multi physics ensembles with hierarchical cluster analysis. J Climatol Weather Forecasting 4: 1-16.

24. Kalnay E, Kanamitsu M, Kistler R, Collins W, Deaven D, et al. (1996) The NCEP/NCAR 40-year reanalysis project. Bull Am Meteorol Soc 77: 437-471.

25. Kerr RA (2013) Forecasting regional climate change flunks its first test. Science 339: 638-638.

26. Mercader J, Codina B, Sairouni A, Cunillera J (2010) Results of the meteorological model WRF-ARW over Catalonia, using different parameterizations of convection and cloud microphysics. J Weath Clim Western Medit 7: 75-86.

27. Krogsater O, Reuder J, Hauge G (2011) WRF and the marine planetary boundary layer. In 12th Annual WRF users workshop.

28. Rajasekhar M, Sreeshna T, Rajeevan M, Ramakrishna S (2016) Prediction of severe thunderstorms over Sriharikota island by using the WRF-ARW operational model. in SPIE Asia-Pacific Remote Sensing 988214–988214. International Society for Optics and Photonics.

29. Ma LM, Tan ZM (2009) Improving the behaviour of the cumulus parameterization for tropical cyclone prediction, Convection trigger. Atmos Res 92: 190-211.

30. Stensrud DJ (2009) Parameterization schemes: keys to understanding numerical weather prediction models. Cambridge University Press.

31. Kain JS (2003) The Kain-Fritsch convective parameterization, An update. J Appl Meteorol 43: 170-181.

32. Osuri KK, Mohanty UC, Routray A, Kulkarni AM, Mohapatra M (2011) Customization of WRF-ARW model with physical parameterization schemes for the simulation of tropical cyclones over north Indian Ocean. Nature Hazards 23: 1-20.

33. Yuan X, Liang XZ, Wood EF (2012) WRF ensemble downscaling seasonal forecasts of china winter precipitation during 1982–2008. Clim Dynam 39: 2041-2058.

34. Tiedtke M (1989) A comprehensive mass flux scheme for cumulus parameterization in large–scale models. Monthly Weather Review 117: 1779-1800.

35. Ratna SB, Ratnam J, Behera S, Ndarana T, Takahashi K, et al. (2014) Performance assessment of three convective parameterization schemes in WRF for downscaling summer rainfall over south Africa. Clim Dynam 42: 2931-2953.

36. Done ACWM, Davis J (2004) The next generation of NWP explicit forecasts of con-vection using the weather research and forecasting (WRF) model. Atmos Sci Lett 5: 110-117.

37. Maussion F, Scherer D, Finkelnburg R, Richters J, Yang W, et al. (2011) WRF simulation of a precipitation event over the tibetan plateau, China– an assessment using remote sensing and ground observations. Hydrol Earth Syst Sci 15: 1795-1817.

38. Fernandez-Gonzalez S, Wang P, Gascon E, Valero F, Sanchez J (2016) Latent cooling and microphysics effects in deep convection. Atmos Res 180: 189-199.

39. Houze RA (2012) Orographic effects on precipitating clouds. Rev Geophys 50.

21

Land Surface Heterogeneity and Tornado Formation: A Comparison of Tornado Alley and Dixie Alley

Frazier AE[1*]**, Hemingway B**[1] **and Brasher J**[2]

[1]*Department of Geography, Oklahoma State University, Stillwater, OK, 74074, USA*

[2]*Department of Geography, University of Tennessee, Knoxville, TN, 37996, USA*

[*]**Corresponding author:** Frazier AE, Oklahoma State University, Stillwater, OK, 74074, USA, E-mail: amy.e.frazier@okstate.edu

Abstract

Tornadoes are among the most destructive hazards to human life and property, and certain areas of the United States are more prone to these events. In particular, scientists use the terms "Tornado Alley" and "Dixie Alley" to refer to two general areas that experience higher incidences of tornadoes. While there is wide recognition that the two regions vary in the number, magnitude, and fatalities caused by tornadoes, more research is needed to better understand the reasons for these differences. The growing recognition that land surface heterogeneity may play a role in tornado formation provides motivation to compare the geographical characteristics of the two regions to determine whether there are significant differences in the landscape characteristics where severe storms form. To investigate the relationship between tornado formation and land surface heterogeneity in these two regions, we first delineate the spatial extent of Tornado Alley and Dixie Alley based on tornadic activity using a statistical test for the detection of significant clusters of spatial association. Next, using severe weather data for tornadoes and storms producing wind/hail (but no tornadoes), we investigate how land surface heterogeneity factors are related to tornado formation of weakly tornadic storms (EF0-EF1) and significantly tornadic storms (EF2-EF5) in each region using binary logistic regression. Lastly, we map probability surfaces for each region to show areas of greater risk. Results show that relationships between land surface heterogeneity and tornado formation vary from region to region. Elevation, slope, and distance to rivers are significant predictors of tornado formation, but the directionality of those relationships varies by region and storm severity. Urban land covers are associated with decreased tornado probability for all storm types in both regions. Spatial trends show an decreasing likelihood for EF0-EF1 from west to east in Tornado Alley but an increasing likelihood in that direction for EF2-EF5 storms.

Keywords: Severe weather; Environment; Terrain; Tornado touchdown; GIS

Introduction

Tornadoes are among the most destructive natural hazards to human life and property. In the United States, approximately 1,253 tornadoes occur every year (NCDC) causing an average of 50 fatalities annually and $400 million in yearly economic losses [1-3]. Most tornadoes occur in the central part of the United States in a region known as Tornado Alley [4]. While the exact geographic extent of Tornado Alley is debated in the literature [5], it is often loosely defined as encompassing northern Texas, Oklahoma, Kansas, and Nebraska. This region is not the only tornado prone area in the country though. A second area of high incidence, termed Dixie Alley [5,6], stretches across the Delta region of the United States and is generally considered to encompass Louisiana, Mississippi, and Alabama [7].

While the exact boundaries of Tornado Alley and Dixie Alley have been differentially defined in the literature [5], there are observable differences in the number of violent tornadoes affecting the two regions, both in terms of the seasonality of tornado occurrences as well as the time of day when tornadoes are most likely to occur [5]. For example, Gagan et al. [5] found that a greater number of stronger tornadoes form in Tornado Alley from April through October, while Dixie Alley experiences more stronger tornadoes during the remaining months. Tornado Alley also experiences the majority of its strong

tornadoes during the afternoon and early evening hours, whereas Dixie Alley experiences more tornadoes during the late night and early morning hours. These diurnal differences may also affect differences in the number of tornado fatalities, with hazard-related deaths concentrated in the southeastern United States [8,9]. While some research has focused on understanding the differences between the two regions in terms of their distinctive tornadic characteristics [5], more research is needed to investigate the relationships between environmental factors and meteorological processes [10] to understand why these geographic variations occur, with the ultimate goal of improving forecasting and warning.

A growing body of research recognizes the relationship between land surface heterogeneity and the evolution of mesoscale convective systems [11-18]. Land surface heterogeneity refers to the variation in terrain, vegetation, and land cover characteristics that combine to produce different environmental conditions (e.g., soil moisture, urban heat island, etc.), which in turn influence storm formation. For example, in a study comparing tornado frequencies in urban and rural counties, Aguirre et al. [19] suggest that terrain roughness (i.e., land surface heterogeneity measured through land cover) may explain higher frequencies of tornadoes in urban areas. More recently, Pielke et al. [15] document how land cover changes (e.g., conversion of forests to cropland) have altered surface climate fluxes, and Pielke et al. [18] describe how regional weather patterns are a function of land cover. In a study explicitly testing the role of land surface heterogeneity on tornado climatology, Kellner and Niyogi [17] found spatial relationships between tornado formation locations and surface roughness measured through the change in elevation and land cover in Indiana. These initial studies suggest that differences in land surface heterogeneity between Tornado Alley and Dixie Alley may help explain some of the geographical variation in tornado formation. While it remains extremely difficult to predict exactly where a tornado will form [20], a better understanding of the relationship between landscape surface heterogeneity and tornado formation from climate data remains an important factor in reducing risk [4,7].

The objective of this study is to investigate the relationship between tornado formation and land surface heterogeneity in Tornado Alley and Dixie Alley. We first delineate the spatial extent of both Tornado Alley and Dixie Alley using an established statistical test for the detection of significant clusters of spatial association. This extent delineation is not meant to supplant pre-existing conceptualizations but rather provide a spatially explicit area from which to undertake a systematic comparison of the two regions. Next, using severe weather data for tornadoes and storms producing wind/hail (but no tornadoes), we investigate how land surface heterogeneity factors are related to tornado formation in each region through binary logistic regression. Lastly, we map our results from both regions to visually compare the spatial trends of tornado formation probability in each region. Our ultimate goal is to derive useful information that can aid in identifying vulnerable geographic locations and ultimately inform risk assessment, warning dissemination, and infrastructure (e.g., tornado warning systems) development.

Data

Tornado data

The National Weather Service (NWS) Storm Prediction Center maintains a database of severe weather reports for tornado producing storms as well as non-tornado producing wind/hail events, which is available in GIS format. The tornado database contains path geometry for all tornadoes reported since 1950 along with attributes including date, time, and magnitude. Point locations of tornado formation were generated by isolating the starting vertex of each tornado path. The NWS also maintains a database of severe storms that produce wind/hail (but not tornadoes) dating back to 1955, which includes similar attributes for the date and time of the storm. To ensure there was no duplication between the two datasets, we queried the storm databases according to the date and time and removed any duplicate points in both datasets. Only storms reported during the 15-year period from 2001-2015 were included in the study to align with the land cover variables (discussed below). We divided the database of tornadoes by storm type into weakly tornadic supercells (EF0 and EF1) and significantly tornadic supercells (EF2 or greater) based on other studies using a similar typology [21-24]. For this study, we compare both the weakly tornadic and the significantly tornadic super cells to wind/hail storms that did not produce any tornadic activity.

Land surface heterogeneity data

Topographic variables: Three topographic/terrain variables are analyzed here: elevation, slope, and aspect. Elevation and slope have previously been included in tornado land surface heterogeneity studies [17], but the role of aspect (i.e., slope direction) is not typically considered. Since tornadoes typically have path directions oriented in a southwest to northeast direction [25], we hypothesize that slope direction may be related to tornado formation. Topography variables

were derived from a digital elevation model (DEM) obtained from the United States Geologic Survey (USGS) 3D Elevation Program (3DEP) at a spatial resolution of 1 arc second. The DEM was resampled to 90 m spatial resolution for improved processing time. Since there is some uncertainty in the exact location of tornado formation, this resampling is not expected to have a large impact on overall results. Slope and aspect were computed directly from the DEM using the suite of spatial analysis tools in ArcGIS. Slope is computed as the maximum rate of change (i.e., steepest downhill direction) and is reported in degrees. Aspect is the direction of maximum slope measured clockwise from 0 degrees (due north) to 360 degrees (also due north) with flat areas given a value of -1. Aspect data were reclassified into five categories: Flat, North (315 -45 degrees), East (45-135 degrees), South (135-225 degrees), and West (225-315 degrees).

Distance to major rivers: The channelling of wind through valleys may cause a jet of increased wind speed, thereby supporting tornadogenesis [26,27], and we hypothesize that tornado formation probabilities will increase closer to major rivers. Distance to river was computed using the National Hydrography Dataset (NHD), a linear vector shape file, to generate a raster surface using Euclidean distance computations where each cell in the contiguous United States represents the closest distance to a major river. The spatial resolution of the raster was generated at 90 m to match the topographic datasets.

Land cover: Early tornado studies [28] noted the importance of ground cover of tornado paths, and recent studies have continued to link land surface conditions with tornado formation [17,29-31]. Land cover investigations present a challenge for tornado-related research since detailed land cover maps have only become available in recent decades. Datasets produced by the Multi-Resolution Land Cover Consortium (MRLC) prior to 2001 are not compatible for direct comparison with more recent products, but by limiting our study to 2001-2015, we are able to utilize the three most recent, compatible NLCD datasets from 2001, 2006, and 2011 to assign land cover to storm locations. We assigned all storms occurring between 2001-2005 land cover values based on the 2001 version of the NLCD, storms occurring between 2006-2011 values from the 2006 version, and storms spanning 2011-2015 values from the 2011 version.

While it is possible that land covers changed within a single time window, using the multiple NLCD versions is preferred over assuming the current NLCD database represents historical land cover. NLCD land cover categories were reclassified into eight aggregated classes: Water, Developed, Barren, Forest, Shrub land, Grassland, Agriculture, and Wetlands.

Final data preparation

To prepare the data for logistic regression (discussed below) the EF0-EF1 tornadic storms were combined with the wind/hail storms into a single database where tornadic storms were coded 1 and wind/hail storms were coded 0.

The same process was repeated for the EF2-EF5 storms. All five of the land surface heterogeneity variables (elevation, slope, aspect, distance to river, and land cover) were then extracted to each of the storm points in each set using ArcGIS. The final datasets for analysis included each storm coded as a 1 or 0, depending on whether the storm produced a tornado, and several fields of data indicating the land surface characteristics at the geographical location of the storm.

Methods

Delineation of Tornado Alley and Dixie Alley

There are many different criteria and methods that can be used to define Tornado Alley versus Dixie Alley [4], and researchers have debated different boundary delineations following different statistical methods [4,6,32,33]. Our intent here is not to add to existing discourse but rather delineate spatially explicit boundaries for each region that permit systematic statistical comparisons between the two regions. Any definitive areas referenced here should be considered the authors' own interpretation based on the statistical analyses performed in this paper with recognition that there are several interpretations of the exact spatial extent of Tornado Alley and Dixie Alley.

We use an established statistical test for the detection of clusters of significant spatial association to delineate boundaries for Tornado Alley and Dixie Alley. Cluster detection tests are used to identify "hot spots" of activities [34,35] through local statistics that measure and test for spatial association of a variable (e.g., tornadoes) within a geographic neighbourhood (e.g., counties). Hot spots emerge in areas where neighboring values are unusually high [35], and cold spots occur in areas where values are unusually low compared to surrounding areas. Local spatial statistics are widely used to test hypotheses of clustering to determine whether there is a raised incidence of a phenomenon in an area or location of interest, but to our knowledge they have not yet been applied to tornado events in the United States.

The Getis-Ord generalized (local) G_i^* statistic [35,36] allows for analysis of local pockets of increased or decreased incidences and is computed as:

$$G_i^* = \frac{\sum_j^* - w_{ij}(d)x_j - W_i^* \underline{x}}{s\left\{\left[mS_{1i}^{*} - W_i^{*2}\right]/(m-1)\right\}^{1/2}}, \; all \; j \tag{1}$$

Where {wij(d)} is a symmetric one/zero spatial weights matrix with ones for all links defined as being within distance d of a given target region, i. All other links are set to zero. In the standardized version of the statistic used here, the target region i; included in the computation of the statistic. Therefore, Wij ≠ 0. The variables x and s are the sample mean and standard deviation of the observed set of xi, respectively. Gi* will produce high values and a high positive z-score when there is a dominant pattern of high values near other high values and will produce low values when there is clustering of low values [37]. One benefit of using the Gi* statistic over other commonly used measures, such as Moran's I [38], is that it can find both "hot" and "cold" spots of tornadic activity.

Binary logistic regression

Binary logistic regression is used here to analyze the impact of each land surface heterogeneity variable on tornado formation. Logistic regression uses a nonlinear function to explain the probability of tornado formation where the dependent variable is, whether or not a severe storm formed a tornado (1 or 0 for the database), and this outcome is influenced by a vector of five land surface heterogeneity variables. The dependent variable takes the value 1 if the storm produced a tornado (EF0-EF1 or EF2-EF5, depending on the analysis) and 0 if the storm was severe but produced only wind/hail and did not produce a tornado. In logistic regression, the statistical significance of the coefficients indicates whether the corresponding explanatory variable is significantly related to the dependent variable. However, it

must be noted that the coefficients cannot be directly compared to each other as a measure of relative importance to explain probability of demolition. Instead, the odds ratio can be interpreted in the context of the impact of each variable on the probability of tornado formation [39]. When the odds ratio for a particular variable is greater than 1, an increase in the variable of one unit will increase the odds of tornado formation by the amount of the odds ratio. When the odds ratio for a variable is less than one, an increase in the explanatory variable by one unit leads to a decrease in the odds of tornado formation. If the odds ratio is exactly one, the odds of tornado formation do not change as that particular variable changes. The exponent of a coefficient is the odds ratio; therefore the probability of tornado formation can be determined through Eqn. (2):

$$\Pr(Y = 1 \mid X1, X2...Xn) = \cfrac{1}{1 + \left(\cfrac{1}{e^{\left(\beta_0 + \beta_1 X1 + \beta_2 X2 + ...\beta_k Xk\right)}}\right)} \tag{2}$$

Logistic regression has been used extensively for predicting hazards [40,41] due to its ability to identify the degree of influence of all independent variables. It has also recently been used to investigate the influence of the El Nino/Southern Oscillation on tornado and hail frequency in the U.S. [42]. Here, we use the binary case of logistic regression to investigate the influence of land surface heterogeneity on tornado formation. Since logistic regression investigates the impact of one unit of change in the independent variable on the probability of encountering the dependent variable, we scaled elevation by 100 so that we can assess the impact of every 100 m change in elevation. We scaled the distance to river variable by 1,000 to create similarly meaningful units for interpretation, and we rounded slope to the nearest degree so every degree equals a one unit change. Additionally, the categorical variables (aspect and land cover) must have a reference category selected for comparison. In this case, we compared all land covers to cultivated land and all aspects to east-facing slopes.

Results and Discussion

The hot spot results from the Gi* cluster analysis show a large cluster of highly significant (z-score>1.96) counties stretching from Denver across eastern Colorado, north into Wyoming, through Nebraska, Kansas, Oklahoma, Arkansas, and down into northern Texas. This cluster, which we define as "Tornado Alley", is visually separate from the clusters of high values in Illinois and Iowa/South Dakota/Minnesota, but is less separated from the cluster of high values along the Gulf Coast, supporting research that has described the area of primary activity between the two regions as connected [6,21].

However, in order to compare "Tornado Alley" to "Dixie Alley", we sever the two regions at the Mississippi River using the state boundary of Arkansas. While our aim is only to compare Tornado Alley and Dixie Alley, it is interesting to note the locations of clusters of high tornadic activity in Illinois, Iowa, North Dakota/Northwestern Minnesota, Florida, and the Carolina coast. Our analysis did not indicate clusters of high activity in Indiana as other studies have found. It should be noted though that this statistical analysis highlights counties with high tornadic activity that are surrounded by other counties that also have high activity (i.e., clusters of tornadic events). Certain counties with large numbers of tornadoes may not be included in a cluster if they are not surrounded by other counties with high activity (Figures 1 and 2).

increase in elevation was accompanied by an increase in the likelihood of tornado formation (p<0.001), as evidenced by the odds ratio value of 1.018 (Table 2). Recall, odds ratio values above one indicate increases in likelihood of occurrence while values below one indicate decreases in likelihood. In Dixie Alley, a one unit increase in elevation decreased the odds ratio (0.742), and thus indicates a significant decrease in the likelihood of tornado formation (p<0.001). The finding that weakly tornadic storms are more likely to occur in higher elevations in Tornado Alley but lower elevations in Dixie Alley shows a difference in the effects of land surface heterogeneity on tornado formation across regions, supporting the need for local/regional analyses and comparisons.

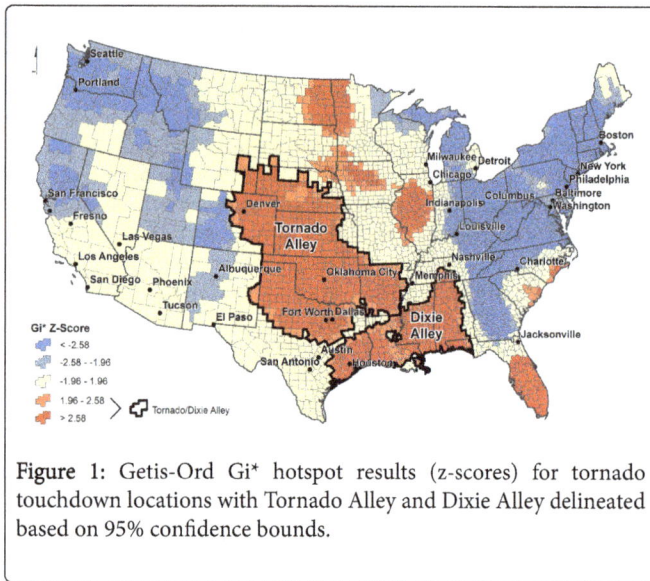

Figure 1: Getis-Ord Gi* hotspot results (z-scores) for tornado touchdown locations with Tornado Alley and Dixie Alley delineated based on 95% confidence bounds.

With explicit boundaries for Tornado Alley and Dixie Alley, we next analyzed the number and type of storm in each region. There were 100,421 total storms recorded in Tornado Alley during the study period (2001-2015), and 28,906 total storms recorded in Dixie Alley (Table 1). Of those storms, 94.8% were non-tornadic (only wind/hail) in Tornado Alley and 92.1% were non-tornadic in Dixie Alley. While the area of Tornado Alley is about 2.7 times larger than the size of Dixie Alley (based on our delineation procedure [1,058,126 km² vs. 400,149 km²]), Tornado Alley received approximately 3.5 times more severe storms than Dixie Alley over the study period, so based on area alone, Tornado Alley is receiving a larger number of severe storms. The remaining 5.2% of storms occurring in Tornado Alley and 7.9% in Dixie Alley were tornadic. Most of the tornadic storms were weakly tornadic (90.5% in Tornado Alley and 86.4% in Dixie Alley), while only 9.5% and 13.6% were significantly tornadic (EF2-5) in the two regions, respectively. When comparing the number of significantly tornadic events, the number in Tornado Alley (4,711) was about 2.4 times greater than the number in Dixie Alley (1,975), which roughly corresponds to their proportional sizes. So, while Tornado Alley receives proportionally more total storms that Dixie Alley, the two regions receive roughly equivalent EF2-EF5 storms according to their respective sizes.

	Total No. Storms	Wind/Hail (% Total Storms)	All Tornadoes (% Total Storms)	EF0-EF1 (% of Tornadoes)	EF2-EF5 (% of Tornadoes)
Tornado Alley	100,421	95,213 (94.8%)	5,208 (5.2%)	4,711 (90.5%)	497 (9.5%)
Dixie Alley	28,906	26,620 (92.1%)	2,286 (7.9%)	1,975 (86.4%)	311 (13.6%)

Table 1: Observation totals for Tornado Alley and Dixie Alley (2001-2015).

Logistic regression results: weakly tornadic supercells

For the weakly tornadic supercells (EF0-EF1), elevation, distance to rivers, and urban, forest, and wetland land covers were all significant predictors in both regions. In Tornado Alley, a one unit (100 m)

Variable	Tornado Alley			Dixie Alley		
	β Coef.	Odds Ratio	Std. Error	β Coef.	Odds Ratio	Std. Error
Intercept	-2.787	0.062***	0.044	-1.958	0.141***	0.073
Elevation	0.018	1.018***	0.004	-0.299	0.742***	0.038
Slope	0.016	1.016*	0.008	-0.006	0.994	0.015
Distance to River	-0.009	0.991.	0.005	0.035	1.035**	0.013
Aspect						
Flat	0.194	1.214	0.279	0.532	1.703.	0.306
East	-0.048	0.953	0.040	-0.075	0.928	0.065
North	0.005	1.005	0.046	-0.006	0.994	0.070
South	-0.107	0.898*	0.045	-0.015	0.985	0.065
Land Cover						
Water	-0.146	0.864	0.162	-0.029	0.971	0.179
Urban	-1.491	0.225***	0.056	-0.790	0.454***	0.068
Barren	0.295	1.344	0.303	0.238	1.268	0.408
Forest	-0.141	0.868*	0.059	-0.222	0.801***	0.064
Shrubland	-0.003	0.994	0.070	-0.197	0.821*	0.098
Grassland	-0.045	0.965	0.038	-0.075	0.927	0.164
Wetland	-0.121	0.886*	0.058	-0.264	0.768*	0.134

Table 2: Logistic regression results for weakly tornadic supercells (EF0-EF1) in Tornado Alley versus Dixie Alley.

Distance to river was also a significant variable in both regions with the directionality of the relationship again reversed. In Tornado Alley, as the distance from a major river increased (one unit increase equals 1000 m), the probability of tornado formation decreased (p<0.1), which supports our hypothesis that tornadoes are more likely to form in riverbeds where winds can be channelled. However, in Dixie Alley, moving away from a major river increased the likelihood of tornado occurrence (p<0.01).

This disparity can possibly be explained by the topographical differences between the two regions. The rivers in Tornado Alley are located in relatively flatter areas compared to Dixie Alley, and so it is possible the significance of this relationship may indirectly be related to terrain with flatter areas being the primary contributing factor rather than a channelling of winds through river valleys. However, this is an area in need of further investigation.

For land cover, urban land, forests, and wetlands all significantly decreased the likelihood of tornado formation in both regions. The extremely low odds ratios for urban land cover in both regions (0.225 and 0.454, respectively, p<0.001) are significant because many prior studies have suggested that tornado frequencies may be greater around urban areas [17,19]. Urban and developed land covers are known to increase the direct heating of the lower atmosphere [15] and have been linked to enhanced thunderstorm activity [43].

However, our analysis indicated no increase in tornado formation for weakly tornadic supercells in urban areas compared to other land covers. These results may partially be explained by Niyogi et al. [44] who found that storms may diverge, or split, when passing over urban areas, which may lead to fewer relative tornado touchdown events in the actual urban area. Early research on the impact of surface heterogeneity also found that hotter areas over cities suppressed development of tornadoes in certain areas of London [45,46], which may help explain the significantly decreased probabilities of tornado formation in urban areas.

Forested land covers were negatively related to tornado formation in both regions (Tornado Alley: p<0.05; Dixie Alley, p<0.001). Forests have a darker albedo and increased surface roughness compared to other land covers and therefore reflects less electromagnetic radiation. This reduction in reflectance results in a higher atmospheric boundary layer and higher sensible heat flux (lower latent heat flux) over wooded areas compared to cultivated lands [47]. The higher these conditions, the less likely tornadoes are to form, which would result in decreased likelihood of tornado formation over forested areas. Our results corroborate these findings with forested areas decreasing the likelihood of tornado formation in both regions.

Lastly, wetlands were associated with a significant decrease in storms for both regions (p<0.05), and south-facing slopes were associated with a significant decrease in Tornado Alley only.

Overall, the results of the analysis for weakly tornadic supercells (EF0-EF1) highlight several key findings. First, the relationships between land surface heterogeneity and tornado formation varied across the two regions. Specifically, elevation and distance to rivers were significant predictors in both regions, but the directionality of the relationships was reversed in Tornado Alley versus Dixie Alley. Knowledge that such differences exist may help target future forecasting and warning tools according to specific geographic criteria. Second, the low odds ratios for both urban and forest land covers suggest that these areas are not at greater risk for tornado formation compared to cultivated areas.

Logistic regression results: significantly tornadic storms

For the EF2-EF5 storms, elevation remained highly significant in Tornado Alley (p<0.001), but the direction of the relationship changed from that of the EF0-EF1 tornadoes. An increase in elevation decreased the probability of EF2-EF5 tornado formation in Tornado Alley (Table 3). Highest elevations in Tornado Alley are located along the western edge of the region in the Colorado plateau.

This area experiences many weakly tornadic super cells and not as many significantly tornadic super cells, which is likely driving this

relationship. Elevation was not a significant predictor of EF2-EF5 storms in Dixie Alley.

Variable	Tornado Alley			Dixie Alley		
	β Coef.	Odds Ratio	Std. Error	β Coef.	Odds Ratio	Std. Error
Intercept	-4.548	0.011***	0.135	-4.070	0.017***	0.184
Elevation	-0.105	0.900***	0.015	-0.088	0.916	0.085
Slope	0.047	1.048*	0.020	0.069	1.071**	0.027
Distance to River	-0.012	0.988	0.025	-0.221	0.802*	0.096
Aspect						
Flat	0.911	2.488	0.717	-12.238	0.000	253.0
East	-0.080	0.923	0.122	-0.177	0.838	0.158
North	-0.079	0.924	0.139	0.152	1.164	0.160
South	-0.017	0.984	0.129	-0.140	0.869	0.161
Land Cover						
Water	-0.607	0.545	0.556	0.498	1.645	0.405
Urban	-1.147	0.318***	0.157	-0.834	0.434***	0.178
Barren	0.185	1.203	1.009	-12.408	0.000	345.0
Forest	0.231	1.259	0.149	-0.128	0.880	0.157
Shrubland	0.120	1.127	0.228	0.021	1.021	0.222
Grassland	0.053	1.054	0.125	0.196	1.216	0.359
Wetland	0.395	1.485*	0.168	0.155	1.168	0.308

Table 3: Logistic regression results for significantly tornadic super cells (EF2-EF5).

Increases in slope slightly increased the probability of strong tornadoes (EF2-EF5) in both regions. Logic suggests that flatter areas would be more prone to tornado formation and in their study of tornadoes in Indiana, Kellner and Niyogi [17] found that topographic changes over short distances did not strongly correlate with tornado formation. However, in this study, we found a significant relationship between tornado formation and slope for weaker storms in Tornado Alley (Table 2) and for stronger storms in both Tornado and Dixie Alleys (Table 3). In all cases, an increase in slope signalled an increase in the odds of formation. These regional differences further support the need for targeted investigations in specific geographic areas as relationships between land surface heterogeneity and tornado formation do not appear to be constant across space.

Distance to river remained a significant predictor in Dixie Alley (p<0.05) with the probability of tornado formation decreasing as distance from river increased. In Dixie Alley, rougher terrain supports the formation of valleys through which wind may be channelled, possibly leading to increased probabilities of significant tornado formation closer to rivers. In Tornado Alley, where the terrain is much less variable, distance to river was insignificant. Lastly, urban land cover was again a highly significant predictor of tornado formation in both regions with urban lands signaling a decrease in the probability of tornado formation.

Several relationships that were not significant in the model are nonetheless interesting for understanding differences between Tornado Alley and Dixie Alley. First, our hypothesis that flat areas would experience an increased probability of tornadoes was true in Tornado Alley (odds ratio=2.488) but not in Dixie Alley. The high standard error in Dixie Alley (253.0) suggests that the mean is not reliable, which may be due to a lack of truly flat areas in this region as there were only 55 total instances of storms in flat areas (out of 28,906 total storms). The low numbers issue is also likely causing the high standard error for the Barren land cover in Dixie Alley where there were only 102 instances (Table 3). Water also showed contrasting (but not significant) relationships between the two regions with an odds ratio below one in Tornado Alley (decrease in probability) and above one in Dixie Alley (increase in probability). This dichotomy may be due to the presence of more water in Dixie Alley compared to Tornado Alley.

Probability maps

Tornado probabilities were computed for each pixel using the β coefficients computed from Eqn. 2 and presented in Tables 2 and 3. We mapped the resulting probability surfaces for the two categories of storms (EF0-EF1 and EF2-EF5) across the two study areas (Figure 2). Several interesting patterns are apparent. In Tornado Alley, weakly tornadic storms (Figure 2a) have the highest probabilities of forming in the western part of the region, with probabilities decreasing from west to east.

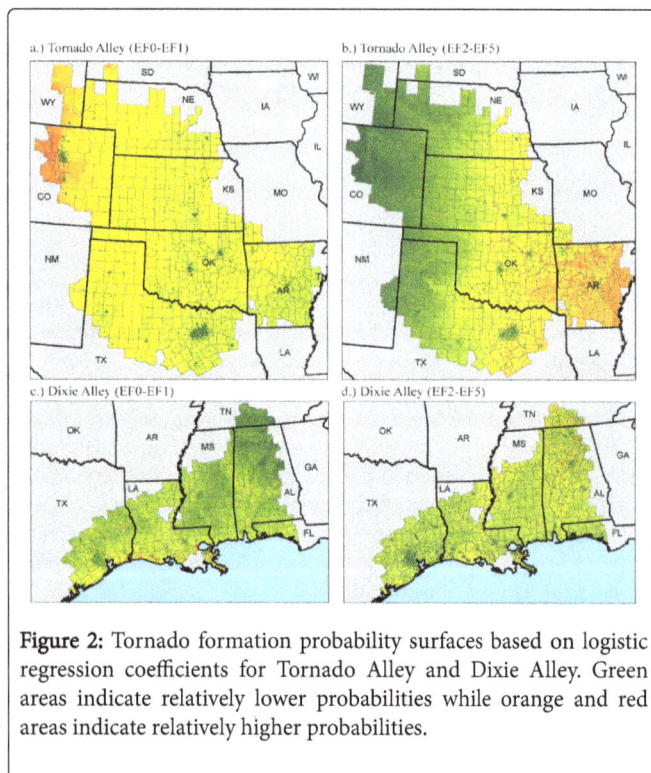

Figure 2: Tornado formation probability surfaces based on logistic regression coefficients for Tornado Alley and Dixie Alley. Green areas indicate relatively lower probabilities while orange and red areas indicate relatively higher probabilities.

However, for the stronger storms (Figure 2b), probabilities of formation are highest in the eastern part of the region, notably in Arkansas and southeastern Oklahoma. The reverse patterns for the weak versus strong storms in Tornado Alley appear to be strongly influenced by the elevation gradient. For Dixie Alley, the patterns are

not as stark as in Tornado Alley, but there are clear differences in the spatial distribution of probabilities for weaker storms versus stronger storms. The weaker storms (Figure 2c) show a slightly higher probability of forming in the western portion of the region, with several small pockets of very high probabilities along the Gulf Coast (areas in red). The stronger storms (Figure 2d) show a slight gradient of increasing probability from west to east but no strong trends.

Limitations

This study has several limitations that warrant consideration. First, some uncertainty and error exist in the spatial locations of tornado data provided by the NWS. The 90 m spatial resolution of the four terrain variables (elevation, slope, aspect, and distance to river) minimizes some of this uncertainty, but the land cover datasets from MRLC are generated at a nominal 30 m spatial resolution, and since land cover is categorical, it cannot easily be aggregated to coarser resolutions. However, moderately high spatial and temporal resolution satellite imagery have been available since well before the start date of this study, and these ancillary datasets are frequently used to verify touchdown locations, which can increase positional accuracy.

Second, the Getis-Ord statistical test for the detection of clusters identifies counties in the United States with high tornado activity surrounded by other counties with high activity. Sampling biases can occur when the probability of observing phenomena depends on the shape and size of the geographic area (counties). These biases were reduced by ensuring that every county was analyzed with at least one neighbour, and larger counties primarily occur in western states where tornado activity is reduced. Edge effects can also occur for areas that do not have physical neighbours (e.g., counties along the coast), but since most of the tornado activity in the United States occurs inland [48], edge effects are limited.

Lastly, the Fujita and Enhanced Fujita scales used to assign magnitude values to tornadoes are damage-rating systems, not intensity-rating systems [49]. While damage and intensity are highly correlated, there can be differences in magnitude depending on the environment through which the tornado passed. For example, tornadoes in urban areas are more likely to obtain higher F-scale (or EF-scale) ratings than rural areas because the potential for damage is greater [50]. Aside from the lowest magnitude tornadoes, we did not find significant differences between different magnitudes for the environmental and land cover variables, so this discrepancy is not likely impacting our analysis disproportionately.

Conclusion

In this study, we investigated the relationship between tornado formation and land surface heterogeneity in Tornado Alley and Dixie Alley using binary logistic regression and mapped the results of that analysis to probability surfaces of tornado formation. Our key findings are highlighted below:

Relationships between land surface heterogeneity and tornado formation vary from region to region.

Increases in elevation increase the likelihood of EF0-EF1 storms but decrease the likelihood of EF2-EF5 storms in Tornado Alley. Increases in elevation increase the likelihood of EF0-EF1 storms in Dixie Alley but have no significant relationship with EF2-EF5 storms in that region.

Increases in distance to rivers (i.e., being further from a river) increase the likelihood of EF0-EF1 storms decreases in Tornado Alley and Dixie Alley but decrease the likelihood of EF2-EF5 storms in Dixie Alley. There was no significant relationship between distance to major rivers and EF2-EF5 storms in Tornado Alley.

Increases in slope (i.e., steeper slopes) slightly increased the probability of EF0-EF1 storms in Tornado Alley and EF2-EF5 storms in both regions but did not have a significant impact on EF0-EF1 storms in Dixie Alley.

Urban land was associated with a significant decrease in the likelihood of tornado formation for all tornadic storms (EF0-EF1 and EF2-EF5) in both regions.

Forest land cover was associated with a significant decrease in the likelihood of tornado formation for EF0-EF1 storms in both regions but was not a significant predictor of significantly EF2-EF5 storms in either region.

Wetlands were associated with a decrease in the likelihood of EF0-EF1 storms in both regions but an increase in EF2-EF5 storms in Tornado Alley.

Aspect does not appear to have a strong relationship with tornado formation.

The results of this study are not intended to pinpoint locations where tornadoes will form but are meant to aid scientists in developing more accurate warning systems with increased information to reduce false alarms and provide adequate lead times for sheltering. In the future, more accurate information on the relationships between tornado formation and land surface heterogeneity may help researchers transition from the current 'warn on detection' model to an improved 'warn to forecast' system, ultimately leading to reduced fatalities.

Acknowledgements

This research was supported by a National Science Foundation (NSF) EPSCoR Track 2 Research Improvement Infrastructure grant 1539070 for CLOUD-MAP: Collaborative Leading Operational UAS Development for Meteorology and Atmospheric Physics. Some of the computing for this project was performed at the OSU High Performance Computing Center at Oklahoma State University supported in part through NSF grant OCI–1126330.

References

1. Brooks HE, Doswell CA (2000) Normalized damage from major tornadoes in the United States: 1890-1999. Weath Forecast 16: 168-176.

2. Boruff BJ, Easoz JA, Jones SD, Landry HR, Mitchem JD, et al. (2003) Tornado hazards in the United States. Clim Res 24: 103-117.

3. Brooks HE, Dotzek N (2008) The spatial distribution of severe convective storms and an analysis of their secular changes. Clim Extremes Soc 35: 53.

4. Dixon GP, Mercer AE, Choi J, Allen JS (2010) Tornado risk analysis: Is Dixie Alley an extension of Tornado Alley. Bull Amer Meteor Soc 92: 433-441.

5. Gagan JP, Gerard A, Gordon J (2010) A historical and statistical comparison of "Tornado Alley" to "Dixie Alley". Natl Wea Dig 34: 145-155.

6. Dixon PG, Mercer AE (2012) Reply to "Comments on 'Tornado Risk Analysis: Is Dixie Alley and Extension of Tornado Alley?'". B Am Meteorol Soc 3: 408-410.

7. Brooks HE, Lee JW, Craven JP (2003) The spatial distribution of severe thunderstorm and tornado environments from global reanalysis data. Atmos Res 67: 73-94.

8. Ashley WS (2007) Spatial and temporal analysis of tornado facilities in the United States: 1880-2005. Weather Forecast 22: 1214-1228.

9. Sutter D, Simmons KM (2010) Tornado fatalities and mobile homes in the United States. Nat Hazards 53: 125-137.

10. Yuan M (2005) Beyond mapping in GIS applications to environmental analysis. Amer Meteor Soc 86: 169-170.

11. Cheresnick DR, Basara JB (2005) The impact of land-atmosphere interactions on the Benson. Bul Amer Meteorol Soc 86: 637-642.

12. Holt T, Niyogi D, Chen F, LeMone MA, Manning K, et al. (2006) Effect of land-atmosphere interactions on the IHOP 24-25 May 2002 convection case. Mon Wea Rev 134: 113-133.

13. Niyogi D, Holt T, Zhong S, Pyle PC, Basara J (2006) Urban and land surface effects on the 30 July 2003 mesoscale convective system event observed in the southern Great Plains. J Geophys Res 111: D19107.

14. Wu Y, Nair UA, Pielke Sr RA, McNider RT, Christopher SA, et al. (2009) Impact of land surface heterogeneity on mesoscale atmospheric dispersion. Bound-Layer Meteorol 133: 367-389.

15. Pielke Sr RA, Pitman A, Niyogi D, Mahmood R, McAlpine C, et al. (2011) Land use/land cover changes and climate: Modeling analysis and observational evidence. Wiley Interdis Rev Clim Change 2: 828-850.

16. Bozeman ML, Niyogi D, Gopalakrishnan S, Marks FD, Zhang X, et al. (2012) An HWRF-based ensemble assessment of the land surface feedback on the post-landfall intensification of Tropical Storm Fay. Nat Hazards 63: 1543-1571.

17. Kellner O, Niyogi D (2014) Land surface heterogeneity signature in tornado climatology? An illustrative analysis over Indiana, 1950-2012. Earth Interact 18: 1-32.

18. Pielke Sr RA, Mahmood R, McAlpine C (2016) Land's complex role in climate change. Phys Today 69: 40-46.

19. Aguirre BE, Saenz R, Edmiston J, Yang N, Agramonte E, et al. (1993) The human ecology of tornadoes. Demography 30: 623-633.

20. Brotzge J, Erickson S, Brooks H (2011) A 5-yr climatology of tornado false alarms. Weather Forecast 26: 534-544.

21. Concannon PR, Brooks HE, Doswell III CA (2000) Climatological risk of strong and violent tornadoes in the United States. Second Symp Env App 9: 4.

22. Thompson RL, Edwards R, Hart JA, Elmore KL, Markowski P (2003) Close proximity soundings within supercell environments obtained from the rapid update cycle. Weather Forecast 18: 1243-1261.

23. Thompson RL, Mead CM, Edwards R (2007) Effective storm-relative helicity and bulk shear in super cell thunderstorm environments. Weather Forecast 22: 102-115.

24. Coleman TA, Dixon PG (2014) An objective analysis of tornado risk in the United States. Weather Forecast 29: 366-376.

25. Thom HCS (1963) Tornado probabilities. Mon Wea Rev 91: 730-736.

26. Bosart LF, LaPenta K, Seimon A, Dickinson M, Galarneau TG (2004) Terrain-influenced tornadogenesis in the northeastern United States.

27. Coleman TA (2010) The effects of topography and friction on mesocyclones and tornadoes. Preprints, 25th Conf. on Severe Local Storms, Denver, Colorado.

28. Finley JP (1884) Report on the character of six hundred tornadoes. Prof Papers Sig Serv No VII, US Signal Service, Washington, DC, USA.

29. Dessens J (1972) Influence of ground roughness on tornadoes: A laboratory simulation. J Appl Meteorl 11: 72-75.

30. Diamond CJ, Wilkins EM (1984) Translation effects on simulation tornadoes. J Atmos Sci 41: 2574-2580.

31. Clark C, Arritt R (1995) Numerical simulations of the effect of soil moisture and vegetation cover on the development of deep convection. J Appl Meteorol 34: 2029-2045.

32. Marsh PT, Brooks HE (2012) Comments on "Tornado Risk Analysis: Is Dixie Alley and Extension of Tornado Alley?" B Am Meterol Soc 3: 405-407.

33. Deng Y, Wallace B, Maassen D, Werner J (2016) A Few GIS Clarifications on Tornado Density Mapping. J App Meteorol Climatol 55: 283-296.

34. Besag J, Newell J (1991) The detection of clusters in rare diseases. J Roy Stat Soc A 154: 143-155.

35. Ord J, Getis A (1995) Local spatial autocorrelation statistics: Distributional issues and an application. Geogr Anal 27: 286-306.

36. Getis A, Ord J (1992) The analysis of spatial association by use of distance statistics. Geogr Anal 24: 189-206.

37. Rogerson P, Yamada I (2008) Statistical detection and surveillance of geographic clusters. Int stat rev 23.

38. Moran, PAP (1948) The interpretation of statistical maps. J Roy Stat Soc A 10: 243-251.

39. Hailpern SM, Visintainer PF (2003) Odds ratios and logistic regression: further examples of their use and interpretation. Stat J 318: 356.

40. Ohlmacher GC, Davis JC (2003) Using multiple logistic regression and GIS technology to predict landslide hazard in northeast Kansas, USA. Eng Geol 69: 331-343.

41. Gascón E, Merino A, Sánchez JL, Fernández-González S, García-Ortega E, et al. (2015) Spatial distribution of thermodynamic conditions of severe storms in southwestern Europe. Atmos Res 164: 194-209.

42. Allen JT, Tippett MK, Sobel AH (2015) Influence of the El Nino/Southern Oscillation on tornado and hail frequency in the United States. Nat Geosci 8: 278-283.

43. Chen TC, Wang SY, Yen MC (2007) Enhancement of afternoon thunderstorm activity by urbanization in a valley: Taipei. J Appl Meteorol Clim 46:1324-1340.

44. Niyogi D, Pyle P, Lei M, Arya SP, Kishtawal CM, et al. (2011) Urban modification of thunderstorms: An observational storm climatology and model case study for the Indianapolis urban region. J Appl Meteorol Climatol 50: 1129-1144.

45. Changnon SA (1978) Urban effects on severe local storms. J Appl Meteorol 17: 578-586.

46. Elsom DM, Meaden GT (1982) Suppression and dissipation of weak tornadoes in metropolitan areas: A case study of greater London. Mon Weather Rev 110: 745-756.

47. Lyons TJ, Schwerdtfeger P, Hacker JM, Foster IJ, Smith RCG, et al. (1993) Land-Atmosphere Interaction in a Semiarid Region: The Bunny Fence Experiment. B Am Meterol Soc 74: 1327-1334

48. Goliger AM, Milford RV (1998) A review of worldwide occurrence of tornadoes. J Wind Eng Ind Aerodyn 76: 111-121.

49. Farney TJ, Dixon PG (2015) Variability of tornado climatology across the continental United States. Int J Climatol 35: 2993-3006.

50. Doswell III CA, Burgess DW (1988) On some issues of United States tornado climatology. Mon Weather Rev 116: 495-501.

22

Rainfall Variability and Its Impact on Crop Agriculture in Southwest Region of Bangladesh

Kabir H[1]* and Golder J[2]

[1]*International Islamic University Malaysia, Kulliyyah of Architecture and Environment Design (KAED), Malaysia*

[2]*Department of Geography and Environment, Faculty of Earth and Environmental Sciences, University of Dhaka, Dhaka*

***Corresponding author:** Kabir H, International Islamic University Malaysia, Kulliyyah of Architecture and Environment Design (KAED), Malaysia
E-mail: hksmmm@gmail.com

Abstract

The amounts of annual rainfall in the southwestern part of Bangladesh are almost equal having very little spatial variation. Significant decreasing trend (Mann-Kendal) of annual rainfall over the region is found (-4.5 mm/year at Mongla, -9.14 mm/year at Jessore, -15.71 mm/year at Madaripur) except Khulna, Satkhira and Khepupara, where the trend is positive over the long period (1948-2014) but it exhibits a decreasing trend during the recent period i.e., after 1980s. It clearly indicates a gradual decrease of the amount of rainfall over southwestern part, which has become most prominent in the recent climatic period of 1981-2014. The trend is more intense at the upper southwestern part, i.e., places like Jessore and Madaripur. Annual rainfall deviates within the range -42.6% to +48.6% over the region. Most wide annual deviation is observed in Khulna from 48.6% to -35.5%. Like annual rainfall, seasonal rainfall also has anomalous behaviour over the region. Monsoon rainfall at maximum locations are syncline with annual trend, but at Khulna, Satkhira and Mongla though the annual trend is negative but the monsoon trend is positive, it clearly indicates, intensification of rainfall in monsoon period over the fore southwestern part. In pre-monsoon season, overall rainfall trend is significantly negative at maximum places (-8.49 mm/year at Jessore, -2.02 mm/year at Barisal, -7.80 mm/year at Madaripur) over the region except Khulna, satkhira and Khepupara. But in Khulna and Satkhira it is also significantly negative over recent climatic period.

Rainfall deviation is higher in seasonal scale than the annual scale. Among the seasons wider deviation is observed in winter (from -100.0% to +586%) and narrower in monsoon (from -43.0% to +62.1%). The coastal region of southwest Bangladesh has been facing various natural extremes like salinity intensification, drainage congestion, inundation, water logging etc. Anomalous behavior of rainfall in combination with these problems, affecting agricultural crop production in the upazilas under study. The problem is serious in Rabi season, where groundwater irrigation facility is limited for salinity problem. In these areas agriculture is totally dependent on surface water (irrigation canal), which becomes scarce during the month of January, February, March and rainfall is also becoming low in this time (supports by public opinion and rainfall trend analysis) imposing water shortage in crop field and as a result farmers have to incur yield reduction. In monsoon season problem is different. Drainage congestion is a major problem in this time. A little heavy rainfall causes inundation and water logged condition. Besides this problem, increasing trend of rainfall in monsoon is very likely to intensify the risk of inundation. And the farmers of the study area have already faced total damage of Aus crop due to flooding in some years with comparatively high rainfall or fallow due to water logged condition. Moreover, overall decreasing trend of rainfall is more likely to intensify the risks of salinization due to decreasing upstream flow as well as sedimentation on river bed and consequent poor drainage and water logging. This intensified environmental problem is further likely to intensify the detrimental effects on crop production. So, changing pattern of rainfall in combination with the local environmental stress is being imposed on risk of agricultural crop production over the study area. It is also occasionally responsible for crop failure over the study area. It is also likely to further intensify the risk for future time.

Keywords: Southwestern; Khulna; Sathkhira; Jessore; Khepupara

Introduction

Rainfall is one of the major climatic parameters and is also a major influencing factor for crop production [1]. Crop agriculture practices of a particular region are normally dependent on the precipitation pattern of that area. But now-a-days warming of the climate system is unequivocal, and since the 1950s, many of the observed changes are unprecedented over decades to millennia [2].

IPCC estimated that changes in precipitation in a warming world will not be uniform. In many mid latitude and subtropical dry regions, mean precipitation will likely to decrease, while in many mid-latitude and tropical wet regions, mean precipitation will likely to increase. Besides, extreme precipitation events over wet tropical regions will very likely become more intense and more frequent as global mean surface temperature increases [2].

This change in precipitation, combined with increasing demand for food, would pose large risks to food security globally (high confidence). Moreover, it is projected to reduce renewable surface water and groundwater resources in most subtropical regions (robust evidence, high agreement), intensifying competition for water among ssectors.

Variability of Rainfall

Rainfall variability is the fluctuations of rainfall occurrence annually or seasonally above or below a long term normal value. Every year, in a specific time period the rainfall of a location can be different, either above or below normal, it is the variability [2], but round the year, the mean is not different.

The severity of impact of a climatic change depends on the specific physiographic setting of a particular area, i.e., the way, how the physiography interplay with the change and also the socio-economic and cultural strength of the people living there to act with it. The physiography of the southwest coastal region of Bangladesh is almost different from rest of the landmass of the country. It is a floodplain delta, having numerous beels, peat basins, swamps, rivers and canals. Elevation of the total landmass is very low and it is within 1 m above mean sea level [3].

The area covers 47,201 square kilometer land area, which is 32% of total landmass of the country [4]. Water area covers 370.4 square km (200 nautical miles) from the coastline including estuaries and the internal river water [5]. The landmass is very young and is also growing through erosion and accession activities of major rivers. A part of the coastal area, the Sundarbans, is a mangrove forest covering about 4,500 km², and the remaining part of the area is used for agriculture [1].

Bangladesh is considered as a one of the most disaster prone countries in the world [6]. Climate change is a major concern here. Climate change assumed to amplify existing risks and create new risks for natural and human systems.

But risks are unevenly distributed and are generally greater for disadvantaged people and communities in the country. The southwest region of Bangladesh is more vulnerable to climate change than other parts of the country due to high spatial and temporal climatic variability, extreme weather events, environmental stress, high population density, high incidence of poverty and social inequity, inadequate financial resources, and poor infrastructure [7]. So any adverse change is likely to influence tremendously the life and livelihood of people living here.

The impact of climate change on food production is of global concern, and also very important for Bangladesh. Agriculture in Bangladesh is already under pressure both from huge and increasing demands for food, and from problems of agricultural land and water resources depletion [8].

Bangladesh needs to increase the rice yield in order to meet the growing demand for food emanating from population growth. Irrigated rice or Boro is a potential area for increasing rice yield, which currently accounts for about 50% of the total rice production in the country [9]. However, change in hydrological cycle is a potential threat towards attaining this objective. It is therefore very important to understand the effect of rainfall variability on crop production NIDOS [10], in southwest coastal area to get a comprehensive knowledge on how rainfall variability is interplaying with coastal crop agriculture and its underlying risks.

Various studies have already been conducted to examine rainfall variability [11-13]. Some have also run GCM model to identify rainfall trend and to make future projections [8,14]. But most of these researches are either conducted over the whole country and generalizes overall feature not focusing on physiographic and environmental issues of coastal region or over a specific season or not considered the impact perceived at local level.

Some studies also conducted over southwest region focusing agriculture, climatic and hydrological problems like drought, salinity intrusion etc., [1,6,15,16] but not considered the rainfall variability interplay with crop agriculture. The present study aimed to analyze rainfall variability trend and its consequent impact on crop agriculture in the environmental condition of the southwest coastal region.

A deeper understanding of the characteristics and distribution pattern of rainfall will support water management, agricultural development and disaster management planning in Bangladesh in the context of global climate change in southwest region of the country (Tables 1-4).

Name of AEZS	Areal Coverage	Organic Matter Content	Fertility Level	Suitable Crops
High Ganges River Floodplain	It covers study area districts of Jessore, Satkhira and Khulna together with minor areas of Narail district	Low	Low	Kharif: B. Aus, T. Aman, T. Aus, Mungbean, Jute, Cotton. Rabi: Wheat, Mustard, Chickpea, Lentil, Boro rice
Lower Ganges River Floodplain	It covers Narail, North eastern part of Khulna, and Bagerhat and northern Barisal	Medium to high	Medium	Kharif: B. Aus, B. Aman, T. Aman, Jute, Kaon, GM. Rabi: Pulses, Wheat, Mustard, Linseed, Boro rice
Ganges Tidal Floodplain	All or most of the Barisal, Jhalakathi, Pirojpur, Patuakhali, Bagerhat, Khulna and Satkhira districts. Ti includes the Khulna and Bagerhat Sundarbans reserve forest	Medium to high	High	Kharif: B. Aus, T. Aman, Green manures. Rabi: Boro rice, Wheat, Mungbean, Grasspea, Cowpea, Chili
Gopalganj Khulna Beels	A number of separate basin areas in Madaripur, Gopalganj, Narail, Jessore, Bagerhat and Khulna districts	Medium to high	Medium	Kharif: T. Aman, T. Aus, Jute, Sesame, B. Aman. Rabi: Boro rice, Bean, Wheat, Grasspea

Table 1: Agro-ecological zones lied in the study area.

Station	Lat./Long.	Area Category	Period of Record	Record Length
Jessore	23.17°N 89.22°E	ICZ	1948-2014	67
Satkhira	22.68°N 89.07°E	ICZ	1948-2014	67
Khulna	22.80°N 89.58°E	ICZ	1948-2014	67
Madaripur	23.16°N 90.18°E	NCZ	1977-2014	38
Mongla	22.46°N 89.6°E	ECZ	1991-2014	24
Barisal	22.70°N 90.36°E	ICZ	1949-2014	66
Patuakhali	22.36°N 90.34°E	ECZ	1973-2014	42
Khepupara	21.98°N 90.22°E	ECZ	1974-2014	41
ICZ: Interior Coastal Zone; ECZ: Exposed Coastal Zone; NCZ: Non Coastal Zone				

Table 2: Location of rainfall records of BMD weather stations under study.

Station	Lat./Long.	Area Category	Period of Record	Record Length
Jessore	23.17°N 89.20°E	ICZ	1998-2014	17
Narail	23.13°N 89.50°E	ICZ	1998-2014	17
Gopalganj	23.20°N 89.80°E	ICZ	1998-2014	17
Madaripur	23.17°N 90.10°E	NCZ	1998-2014	17
Satkhira	22.35°N 89.08°E	ICZ	1998-2014	17
Khulna	22.35°N 89.30°E	ICZ	1998-2014	17
Bagerhat	22.06°N 89.08°E	ICZ	1998-2014	17
Barisal	22.80°N 90.37°E	ICZ	1998-2014	17
Jhalakathi	22.6431°N 90.20°E	ICZ	1998-2014	17
Pirojpur	22.58°N 89.97°E	ICZ	1998-2014	17
Patuakhali	22.354°N 90.3181°E	ECZ	1998-2014	17
Barguna	22.1508°N 90.1264°E	ECZ	1998-2014	17
Koyra	22.20°N 89.40°E	ECZ	1998-2014	17
Karamjol	22.43°N 89.6°E	ECZ	1998-2014	17
Pashurtala	22.244°N 89.443°E	ECZ	1998-2014	17
Arpangachia	22.313°N 89.461°E	ECZ	1998-2014	17
Sibsa Point	21.983°N 89.533°E	ECZ	1998-2014	17
Supati	22.057°N 89.818°E	ECZ	1998-2014	17

Table 3: Selected locations of TRMM rainfall records under study.

Rainfall Variability over Southwestern Part of Bangladesh

The observed annual rainfall over southwestern part of Bangladesh varies from location to location. It is the lowest of 1607.8 mm (at Satkhira) and the highest of 2608.7 mm (at Patuakhali). The standard deviation of annual rainfall over this region varies from 334.0 to 586.3 mm. The mean annual rainfall of each of the location under study and their standard deviations are given in Figure 1. TRMM rainfall at each of district headquarters as well as some selected locations of southwestern part of Bangladesh exhibits almost similar scenario. From TRMM data it is found that the mean annual rainfall varies between from 1680.3 to 2467.3 mm during the study period of 1998-2014.

The highest amount of rainfall is observed at Patuakhali but the lowest is at Jessore. The mean annual rainfall was higher over the eastern part and lower over western part of the region. The standard deviation of rainfall was maximum at Supati but it is lowest at Jessore. Long periodic (1948-2014) annual rainfall of the Jessore, Satkhira, Khulna and Barisal indicates that the rainfall variability is high at these locations as it is found for the recent climate period (1981-2014). Accordingly, the mean rainfall for long period over Jessore, Satkhira, Khulna and Barisal are 2157.2, 1705.3, 1751.9 and 2115.6 mm respectively and their STDs are 457.4, 308.8, 368.1 and 426.0 mm respectively.

The mean and STDs of the rainfall of long and short period of these stations fluctuates slightly that is the indication of the similar annual rainfall variability over this region (Figure 2).

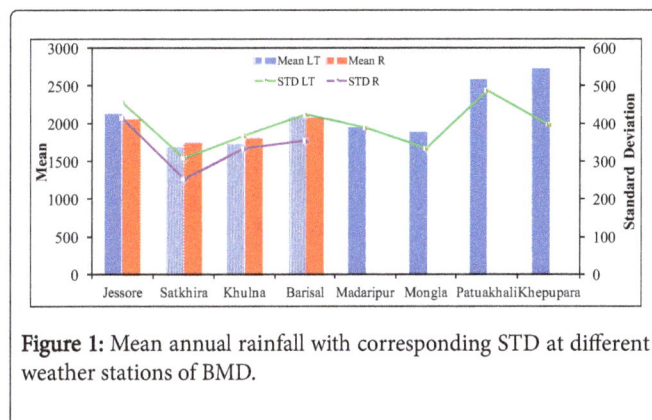

Figure 1: Mean annual rainfall with corresponding STD at different weather stations of BMD.

Description	Total cultivable area (ha)	Saline Area	Area of each salinity class (dS/m)				
			S1 (2.0-4.0)	S2 (4.1-8.0)	S3 (8.1-12.0)	S4 (12.1-16.0)	S5 (>16.0)
Non-saline with very slightly saline	4,25,490	1,15,370 (27%)	82,260 (72%)	31,590 (27%)	1,520 (1%)	0	0
Very slightly saline with slightly saline	4,20,420	3,09,190 (73%)	1,70,380 (55%)	1,10,390 (35%)	29,420 (10%)	0	0
Slightly saline with moderately saline	2,57,270	2,40,220 (93%)	35,490 (15%)	1,13,890 (47%)	61,240 (26%)	25,870 (11%)	2,650 (1%)
Moderately saline with strongly saline	1,98,890	1,98,890 (100%)	1,630 (1%)	36,060 (18%)	73,400 (37%)	55,130 (28%)	32,750 (16%)

Table 4: Salinity affected areas in the coastal regions of Bangladesh.

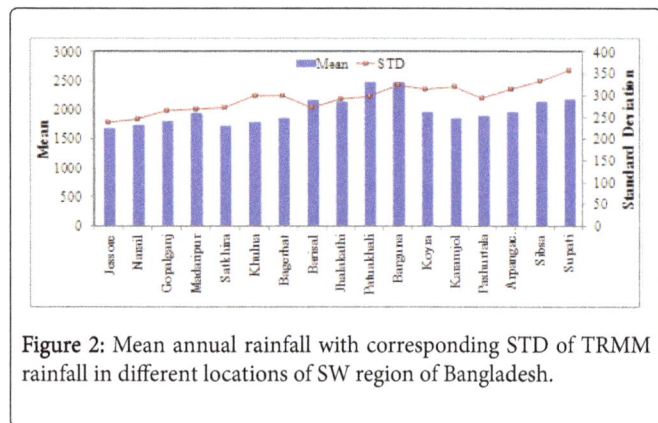

Figure 2: Mean annual rainfall with corresponding STD of TRMM rainfall in different locations of SW region of Bangladesh.

Analysis also reveals that the higher amounts of rainfall are recorded in the monsoon months with the maximum in July followed by June or August (Figure 3). Accordingly, about 69.1, 71.3, 70.9, 69.8, 67.8, 71.9, 74.0 and 72.9% rainfall are recorded in monsoon, about 23.0, 15.3, 17.1, 16.9, 21.3, 15.0, 15.4 and 14.5% in pre-monsoon, about 9.6, 9.7, 9.5, 10.6, 9.6, 11.1, 9.6 and 11.5% in post-monsoon and about 1.7, 3.8, 2.2, 1.7, 1.8, 2.0, 1.3 and 1.3% in Patuakhali and Khepupara. The mean monthly, seasonal and annual rainfall and their STDs of the stations are under study is given in Figures 1, 3 and 4 and in Tables 5 and 6. Winter season respectively in Jessore, Satkhira, Khulna, Barisal, Madaripur, Mongla, Mean seasonal rainfall is highest in Monsoon and lowest in winter. Long term mean monsoon rainfall varies between 1228.6 mm to 2001.4 mm (Figure 4). Highest value was observed in Khepupara and lowest at Satkhira. It deviates within the range 135.0 mm to 388.3 mm (Table 1). Mean Pre-monsoon rainfall was found within 271.0 mm (Satkhira) to 519.5 mm (Jessore) range. Mean Post-monsoon rainfall was within 165.8 mm (Khulna) to 314.9 mm (Khepupara). Mean winter rainfall was observed within a very short range of 34.0 mm (Madaripur) to 50.4 mm (Khulna). In most of the cases STD value becomes lower in recent climatic period reveals decline of in fluctuation in rainfall amount (Figure 4).

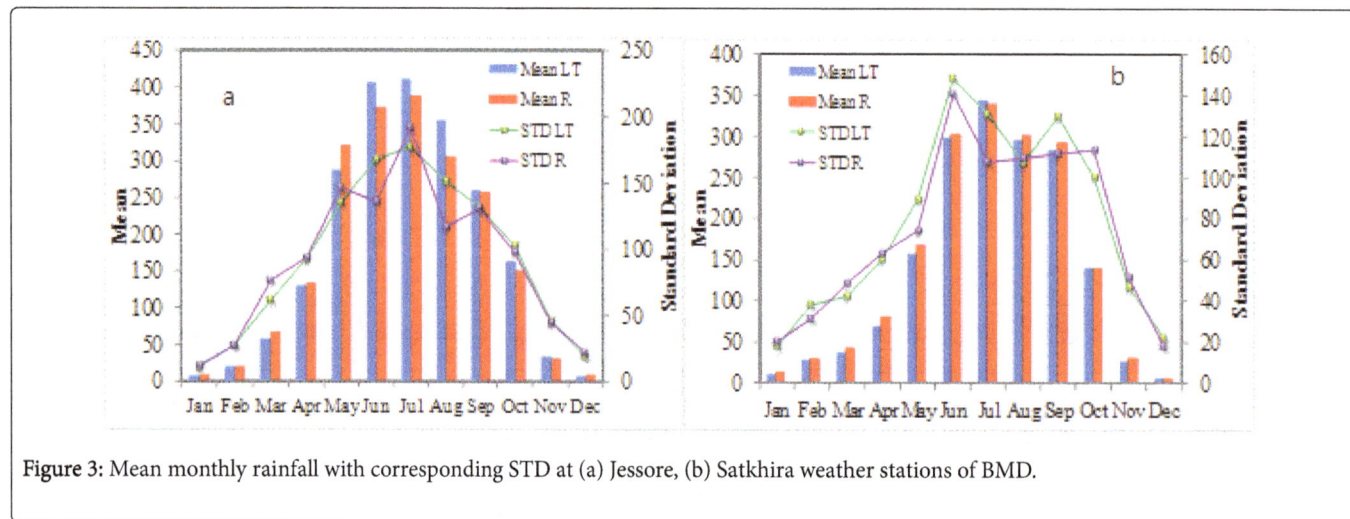

Figure 3: Mean monthly rainfall with corresponding STD at (a) Jessore, (b) Satkhira weather stations of BMD.

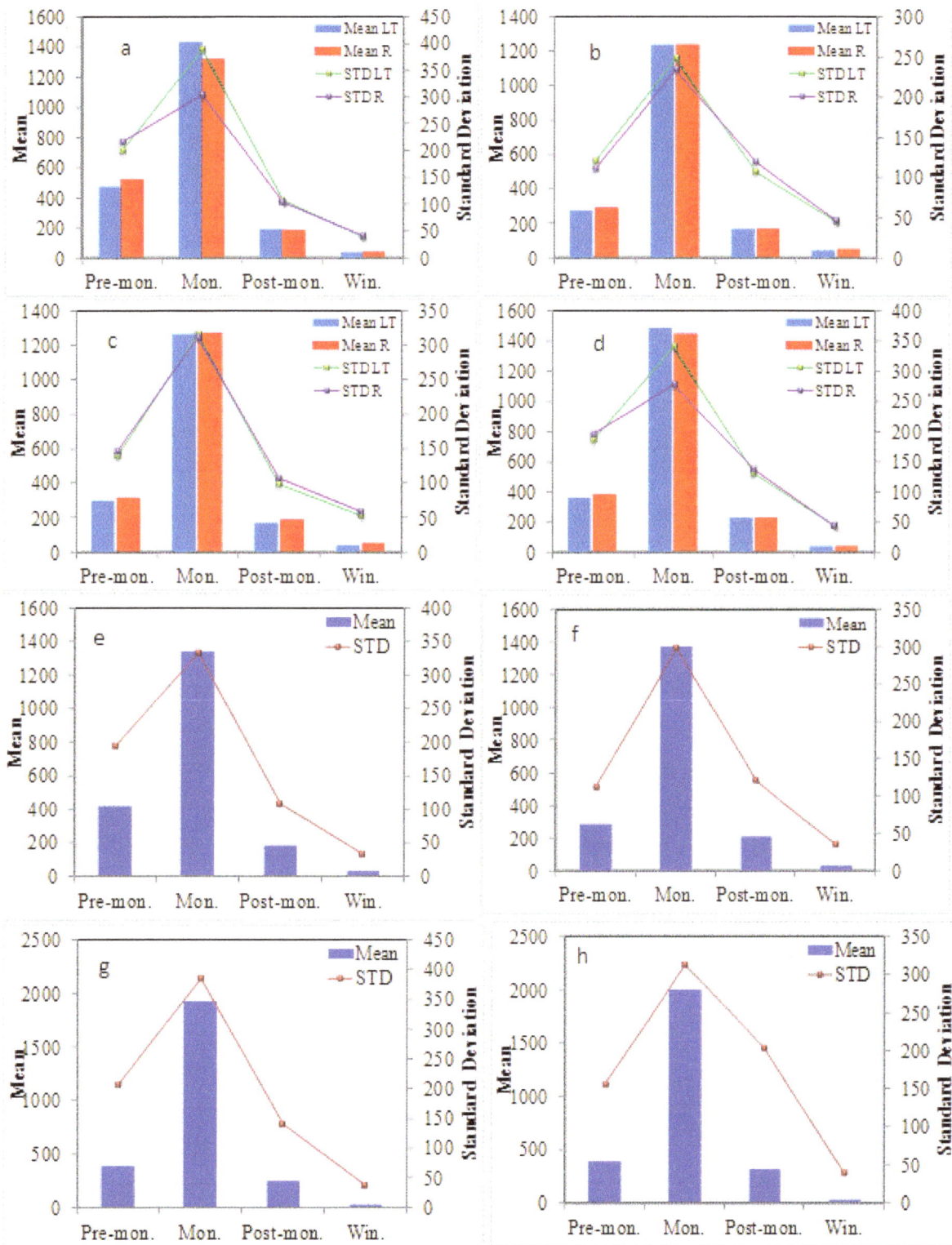

Figure 4: Mean seasonal rainfall with corresponding STD at (a) Jessore, (b) Satkhira, (c) Khulna, (d) Barisal, (e) Madaripur, (f) Mongla, (g) Patuakhali and (h) Khepupara weather stations of BMD.

Time Period		Jessore	Satkhira	Khulna	Barisal	Madaripur (1977-2014)	Mongla (1991-2014)	Patuakhali (1973-2014)	Khepupara (1974-2014)
Jan	1948-14	7.7	10.6	11	9.5	8	12.8	7.7	8.2
	1981-14	8.4	14	14.9	10.8				
Feb	1948-14	20	27.6	23.5	19.3	23.1	22.8	21.2	20.4
	1981-14	19.6	29	30.2	20.9				
Mar	1948-14	57.2	36.2	42.9	49.9	49.7	41.5	37.5	43.1
	1981-14	66.6	42.4	49	55.1				
Apr	1948-14	130.4	68.2	74.8	94.2	120.9	54.7	100.9	88.3
	1981-14	133.5	79.6	75.4	103.2				
May	1948-14	286.3	156.2	181.4	213.5	148.2	190.9	262.2	265.8
	1981-14	319.3	167.3	188.1	217				
Jun	1948-14	404.4	297.3	335.5	409.5	374.2	340.7	535.3	509.9
	1981-14	371.3	302.1	330.5	397.9				
Jul	1948-14	409	342.6	351	412.8	381.7	373.8	560.3	637.3
	1981-14	386.4	337.4	336.8	408				
Aug	1948-14	353	293.6	291.3	353.5	319.6	331.7	450.8	467.2
	1981-14	303.3	299.9	315.1	350.3				
Sep	1948-14	258.4	282	264.4	300.6	259.7	326	382.2	385.8
	1981-14	255.9	291	287.5	281.5				
Oct	1948-14	162.6	139.7	138.4	187.4	158.7	175.1	200.2	263.6
	1981-14	150.6	139	147.2	186				
Nov	1948-14	34.1	26.4	27.2	37.5	30.3	36	51.4	51.3
	1981-14	31.1	30.5	35	43.5				
Dec	1948-14	6.9	6.6	3.7	8	4.1	2.7	5	6.6
	1981-14	8.8	6.6	5.8	6.4				
Pre.	1948-14	474.1	271	297.8	359.6	418.7	287.1	394	397.2
	1981-14	519.5	289.2	312.4	375.3				
Mon.	1948-14	1428.5	1228.6	1262.5	1479.2	1339.3	1372.2	1922.8	2001.4
	1981-14	1311.2	1230.3	1268	1437.7				
Post.	1948-14	192.6	166.4	165.8	224.8	189	211.2	251.5	314.9
	1981-14	181.7	169.6	182.2	221.4				
Win.	1948-14	35.4	42.6	38.3	37.9	34	37.3	34.4	36.2
	1981-14	36.9	49	50.4	37.2				
Ann.	1948-14	2157.2	1705.3	1751.9	2115.6	1969.8	1908.6	2607.8	2745.2
	1981-14	2053.2	1738.7	1805.4	2072.5				

Table 5: Mean monthly, seasonal and annual rainfall at different BMD weather station.

Time Period		Jessore	Satkhira	Khulna	Barisal	Madaripur (1977-2014)	Mongla (1991-2014)	Patuakhali (1973-2014)	Khepupara (1974-2014)
Jan	1948-14	11.5	17.5	20.9	21.4	14.9	16.5	13.6	17.1
	1981-14	11.3	20.2	22.6	22.4				
Feb	1948-14	26.6	37.8	40.4	25	24	27.8	30.4	31
	1981-14	26.6	31.3	42	26.4				
Mar	1948-14	61.7	41.8	62	66.5	57.7	60.9	57	71.2
	1981-14	76.2	48.5	70.1	73.3				
Apr	1948-14	92.8	60.3	70.9	79	95.3	44.3	91.6	71.1
	1981-14	93.1	63.1	66.1	83.3				
May	1948-14	134.5	88.9	87.9	124.6	141.1	84.2	175.3	142.7
	1981-14	145.9	74.1	83.8	125.2				
Jun	1948-14	167.5	148.4	149.4	180	170.9	159.3	266.6	201.4
	1981-14	135.6	140.6	165.5	188.7				
Jul	1948-14	177.2	130.6	138	136	127.8	99.7	155.9	189.8
	1981-14	192.1	107.4	108.6	122.8				
Aug	1948-14	151.2	107	130	143.1	132	111.3	171.1	182.9
	1981-14	117.1	109	124.7	133.6				
Sep	1948-14	131.1	129.6	157.4	167.9	149.9	134	179.4	161.7
	1981-14	130.6	111.4	161.8	169.2				
Oct	1948-14	130	99.8	92	117.7	103.8	105.2	129.5	175.2
	1981-14	98	113.3	104.1	127.4				
Nov	1948-14	46.4	46.6	47.8	58.4	42.2	63.8	71.4	96.3
	1981-14	44.4	51.5	54.7	69.5				
Dec	1948-14	18.5	21.6	10.8	26	9	6.6	12.2	13.3
	1981-14	21.9	18.1	14	12.4				
Pre	1948-14	199.9	120.8	138.6	185.3	194.4	111.2	205.9	155.5
	1981-14	216.4	110.1	146.7	196.3				
Mon	1948-14	388.3	248.4	135	340	333.6	298.2	385	312.9
	1981-14	304	233.7	311	276.8				
Post	1948-14	107.5	106.8	97.8	129.7	108.4	120.8	140.5	203.3
	1981-14	104.3	119.5	106.7	137.6				
Win	1948-14	39.1	45.3	52.4	43.5	32	35	35.9	38.3
	1981-14	41.6	47	58.3	43				
Ann	1948-14	457.4	308.8	368.1	426	388.9	334	488.2	397.9
	1981-14	416.1	254.6	334.4	355.9				

Table 6: Standard deviation of monthly, seasonal and annual rainfall at different BMD weather stations.

Rainfall Occurrence Trend over Southwest in Bangladesh

Annual rainfall trend over the southwestern part of Bangladesh and their magnitudes are calculated using Mann-Kendal Tests, Sen's slope estimator and simple linear regression analysis (Least squares).

The annual trend from these three different tests indicates similar results. Predominantly negative trends of annual rainfall are found over the region except Khulna, Satkhira and Khepupara where, trends are going positive with a magnitude of 5.36 mm/year (at 95% confidence level, 1948-2014), 1.15 mm/year (1948-2014) and 12.67 mm/year (at 90% confidence level, 1974-2014) respectively (Figures 5i and 5ii).

The highest significant negative trend is observed at Madaripure (at 99% confidence level) with a magnitude of 15.71 mm/year over the period 1977-2014. In case of Mongla Mann-Kendal test indicates no significant trend in annual rainfall but a negative Sen's slope (4.74 mm/year) (Table 6).

Although in Khulna and Satkhira trends are positive over long period (1948-2014), but it turned to negative over recent climatic period (1981-2014) with a magnitude of 1.43 mm/year to 4.08 mm/year respectively (Figure 6).

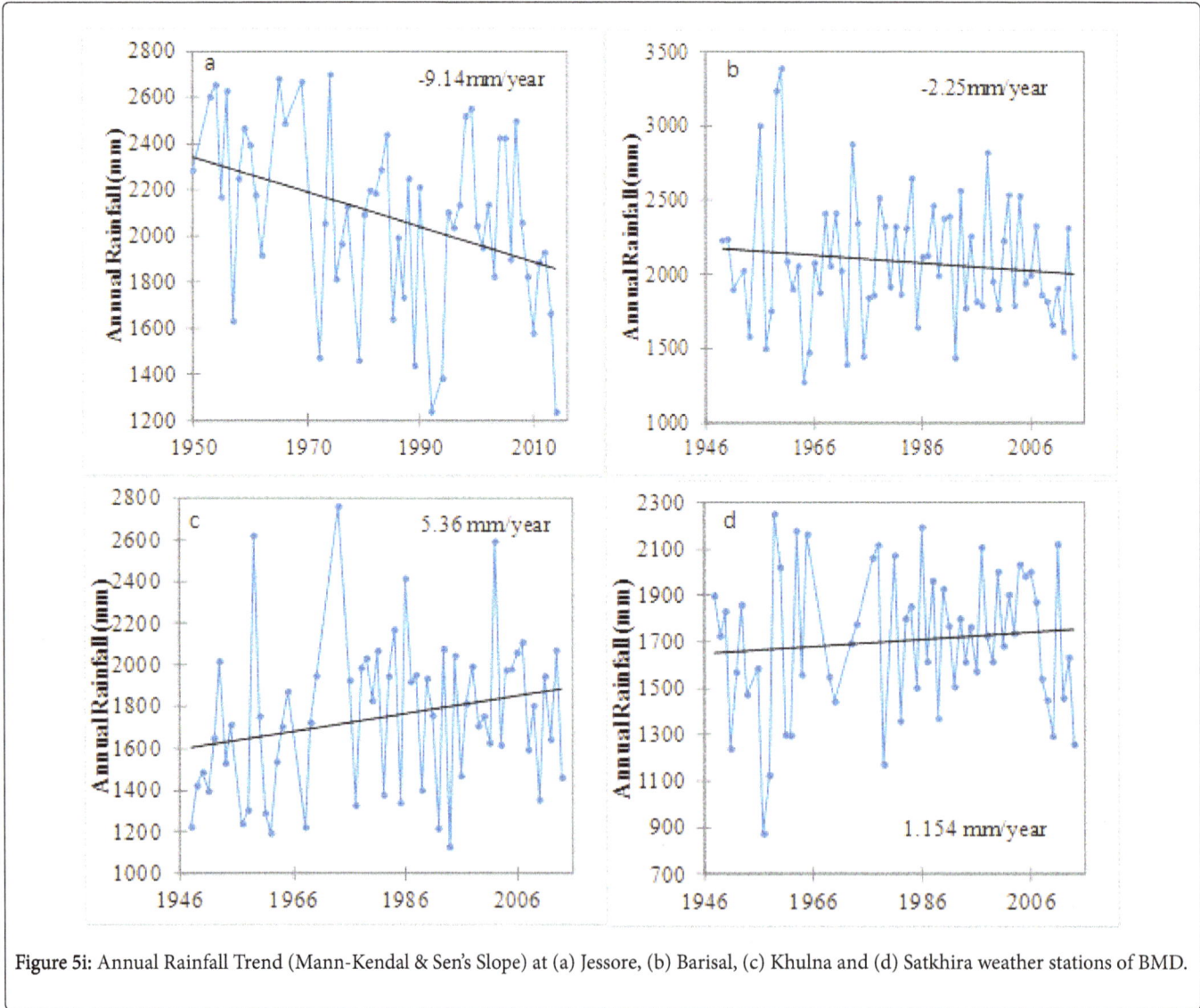

Figure 5i: Annual Rainfall Trend (Mann-Kendal & Sen's Slope) at (a) Jessore, (b) Barisal, (c) Khulna and (d) Satkhira weather stations of BMD.

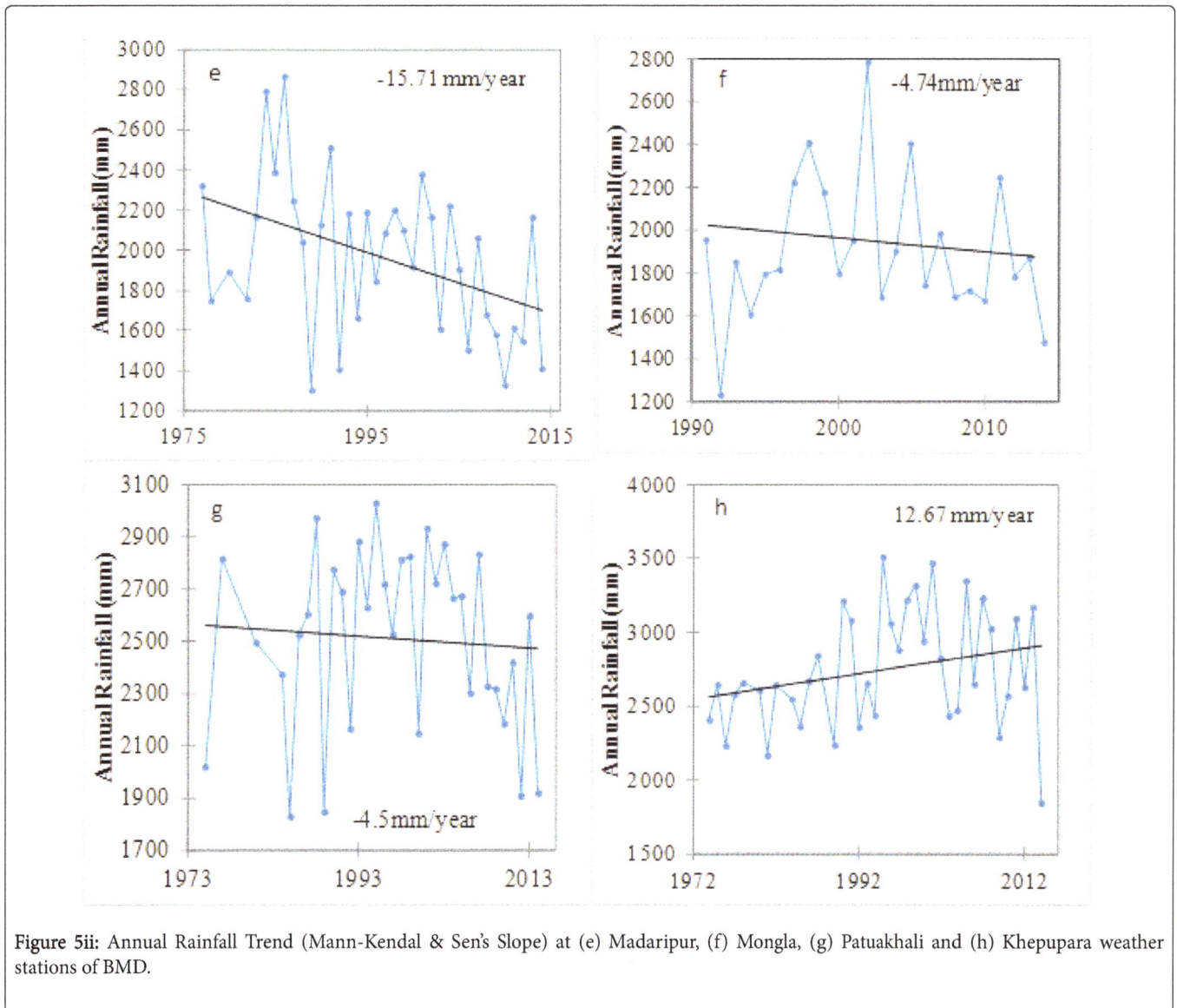

Figure 5ii: Annual Rainfall Trend (Mann-Kendal & Sen's Slope) at (e) Madaripur, (f) Mongla, (g) Patuakhali and (h) Khepupara weather stations of BMD.

Mann-Kendal and Sen's slope tests are also applied over the annual rainfall time series of TRMM records of different study points. Significant negative trends are observed at Gopalganj, Madaripur, Barisal, Jhalakathi, Patuakhali, Barguna and Sibsa point. While other places negative slopping with no significance.

For the month of January, February, March, November and December no significant trend (Mann-Kendal) is observed for maximum cases. While for rest of the months significant trend is found for maximum cases. In these months at Khulna, Satkhira and Khepupara trends are observed positive (i.e., increase) for maximum cases but in other stations significant negative trends are found for maximum cases.

Trends in seasonal rainfall also varied over the study area. In Pre-monsoon season highest negative trend is observed at Madaripure

(6.91 mm/year, at 99% confidence level), and positive at Satkhira (1.46 mm/year, at 99% confidence level). But in Monsoon highest negative trend is observed at Jessore (10.49 mm/year, at 99% confidence level) and positive at Khepupara (4.85 mm/year, at 99% confidence level). In Post-monsoon trends turned positive for maximum cases, but only significantly negative at Jessore. And in winter significant negative trends are present at Madaripur, Mongla and Patuakhali but positive over Khulna and Satkhira (Table 7).

Station wise linear regression analysis of monthly, seasonal and annual rainfall is under study and negative trends are observed for maximum cases. In Khulna annual trend is positive over long term period but becomes negative over the recent climatic period. Monthly and seasonal trends are anomalous but predominantly negative for maximum cases (Table 8).

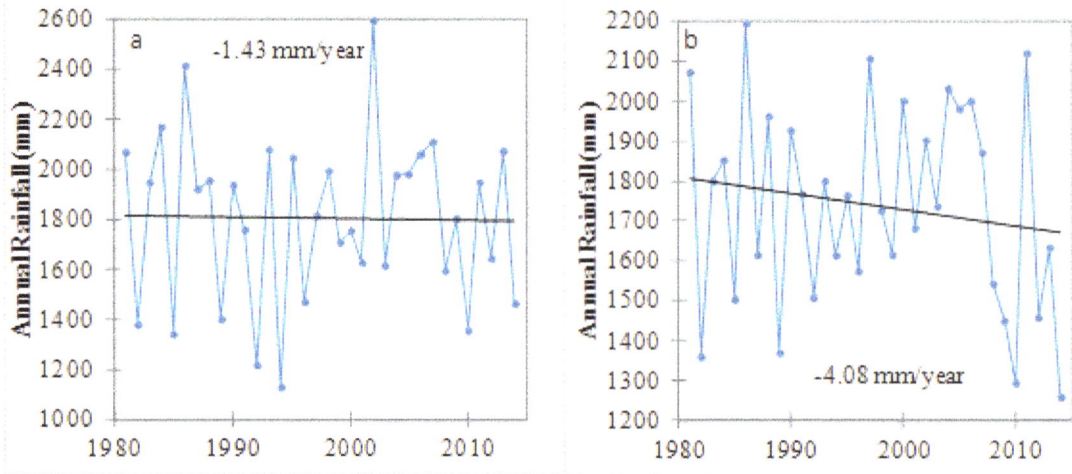

Figure 6: Annual rainfall trend over recent climatic period (1981-2014) at (a) Khulna and (b) Satkhira.

Time Period	Jessore	Satkhira	Khulna	Barisal	Madaripur	Mongla	Patuakhali	Khepupara
Jan	-	-	-	-	-	-0.24***	-	-
Feb	-	-	0.05**	-	-0.55***	-1.18***	-	-
Mar	-0.178	0.06	-	0.02	-0.52**	-1.06***	-	-
Apr	-0.86***	0.15	-0.39**	0.06	-3.73***	-1.33***	-2.42***	-1.06**
May	0.63**	1.25***	0.68***	-0.06	-3.29***	2.26**	-1.48	0.673
Jun	-3.02***	0.02	-1***	-1.01***	-2.64**	-4.38***	-4.89***	-0.495
Jul	-3.91***	0.14	0.273	-0.22	-3.16***	2.81**	-0.18	0.919
Aug	-2.95***	-0.17	1.8***	-0.26	-2.52***	-0.65	-1.6*	-0.722
Sep	-1.11**	1.14***	2.33***	-0.5*	-0.02	2.44	-0.31	3.72***
Oct	-0.86**	-0.25	0.71***	-0.84**	1.53***	1.51*	4.5***	4.86***
Nov	-	-	-	-	-	-0.09*	-1.13***	-
Dec	-	-	-	-	-	-	-	-
Pre	-0.34	1.46***	0.143	0.293	-6.91***	-3.53	-4.66***	-0.58
Mon	-8.49***	0.77	3.69***	-2.02***	-7.80**	0.06	-3.6	4.85***
Post	-0.99**	-	1.15***	-0.469	1.95***	0.2	2.08***	4.44***
Win	0.05	0.39***	0.33***	-	-0.81***	-1.27***	-0.37**	0.11
Ann.	-9.14***	1.15	5.36**	-3.67	-15.71***	-4.734	-4.5*	12.67*

***Significant at 99% confidence level; **Significant at 95% confidence level; *Significant at 90% confidence level.

Table 7: Mann-Kendal trend test results over the study area.

Time Period		Jessore	Satkhira	Khulna	Barisal	Madaripur (1977-2014)	Mongla (1991-2014)	Patuakhali (1973-2014)	Khepupara (1974-2014)
Jan	1948-14	0.01	0.13	0.14	-0.01	-0.19	-0.32	-0.08	0.16
	1981-14	-0.05	-0.18	-0.23	-0.12				
Feb	1948-14	-0.26	0.11	0.28	0.01	-0.69	-1.56	-0.02	-0.01
	1981-14	-0.27	-0.569	-0.803	-0.61				
Mar	1948-14	0.14	0.27	0.13	-0.02	-1.21	-2.82	0.01	-0.55
	1981-14	-1.40	-0.31	-1.21	-0.95				
Apr	1948-14	-1.39	0.11	-0.75	0.04	-4.32	-1.59	-3.02	-1.40
	1981-14	-4.36	-2.51	-3.89	-3.64				
May	1948-14	0.60	0.68	0.73	0.06	-1.50	2.05	-2.92	2.34
	1981-14	-2.49	1.45	-0.06	-0.32				
Jun	1948-14	-4.02	-0.43	-0.45	-1.69	-3.76	-3.89	-4.96	-0.30
	1981-14	1.84	-2.48	-2.73	-3.22				
Jul	1948-14	-3.84	0.15	-0.16	-0.71	-2.91	3.02	-0.62	1.29
	1981-14	-1.74	0.02	2.06	-0.86				
Aug	1948-14	-3.15	0.19	2.03	-0.77	-3.39	1.05	-2.42	-0.35
	1981-14	-1.28	-0.8	0.39	-3.54				
Sep	1948-14	-1.04	0.81	2.24	-0.31	0.77	2.67	-1.26	3.26
	1981-14	0.41	0.57	3.94	2.77				
Oct	1948-14	-1.26	0.20	0.78	-0.54	1.57	1.75	4.23	4.17
	1981-14	1.03	1.55	1.54	1.18				
Nov	1948-14	-0.44	0.04	0.22	0.22	-0.11	-1.35	-2.39	-0.50
	1981-14	-0.49	-0.37	-0.57	-1.22				
Dec	1948-14	0.04	-0.03	0.05	-0.04	-0.16	0.07	-0.38	-0.08
	1981-14	-0.40	-0.47	-0.43	-0.26				
Pre	1948-14	-0.79	1.21	0.25	0.83	-7.03	-2.36	-6.76	0.38
	1981-14	-8.24	-1.36	-5.17	-4.71				
Mon	1948-14	-12.05	1.32	4.10	-3.83	-10.41	2.847	-3.24	4.32
	1981-14	-0.77	-2.69	5.02	-5.76				
Post	1948-14	-1.70	0.29	1.02	-0.43	1.49	0.40	1.84	3.67
	1981-14	0.54	1.18	0.97	-0.09				
Win	1948-14	0.05	0.53	0.65	-0.25	-0.90	-1.75	-0.42	0.05
	1981-14	-0.73	-1.14	-1.46	-0.92				
Ann	1948-14	-7.31	-0.06	6.41	-2.70	-12.94	-0.92	-3.00	8.86
	1981-14	-9.18	0.03	-0.64	-10.48				

Table 8: Linear regression trend values of monthly, seasonal, and annual rainfall.

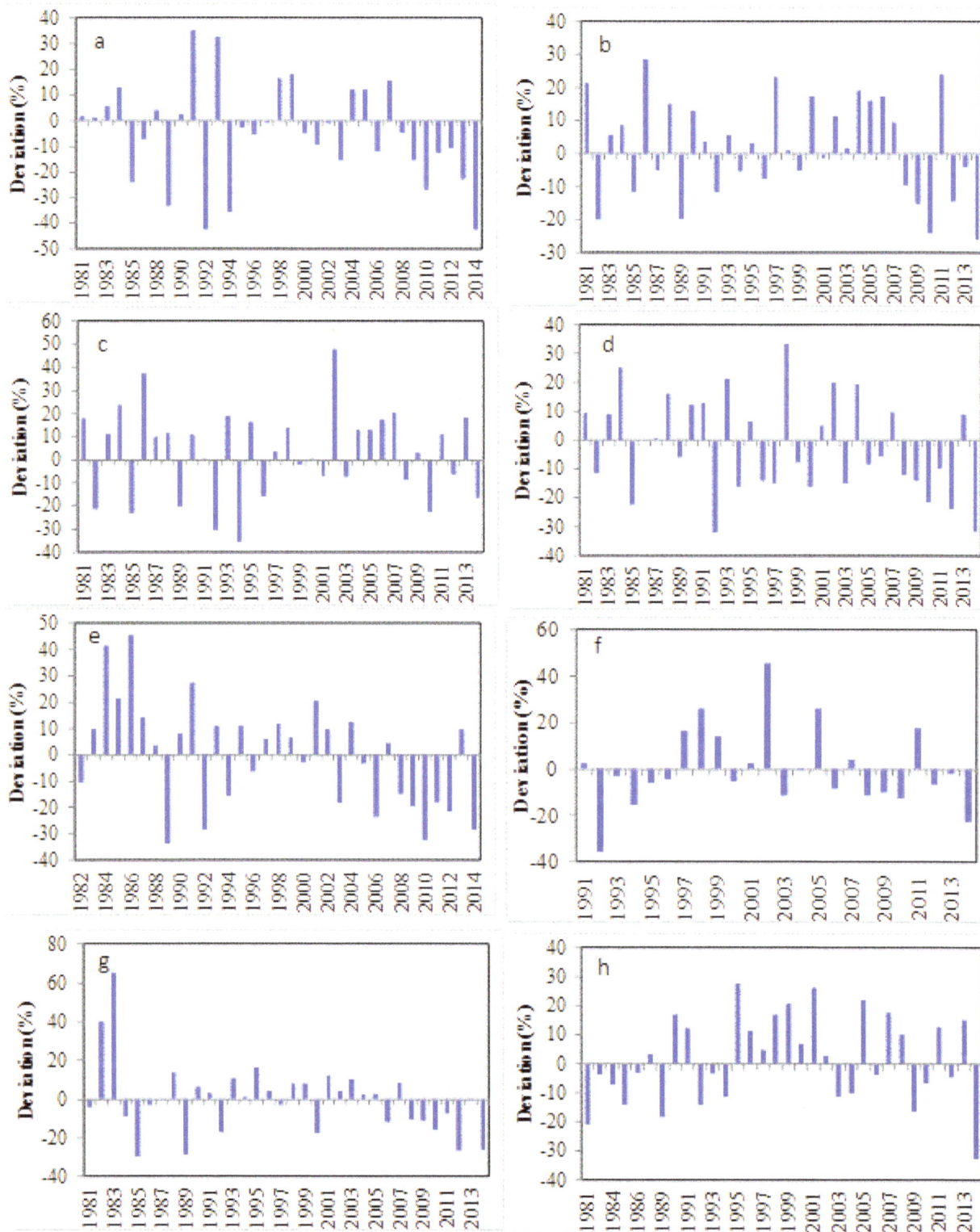

Figure 7: Annual rainfall deviation (%) in recent period at (a) Jessore, (b) Satkhira, (c) Khulna, (d) Barisal, (e) Madaripur, (f) Mongla (1991-2014), (g) Patuakhali and (h) Khepupara weather stations of BMD.

Deviation of Rainfall in Recent Period

Rainfall in the southwestern part of Bangladesh is very much variable in nature. In recent time annual rainfall varies within a range of 48.1% to -42.6%. Most wide annual deviation is observed in Khulna from 48.6% to -35.5%, and most narrow deviation is found in Satkhira, from 28.7% to -26.1% (Figures 6 and 7).

Rainfall deviation is wider in seasonal scale than annual scale. It is observed lowest in monsoon and highest in winter, i.e., rainfall is most consistent in monsoon and most variable in winter. In monsoon positive deviation is found as high as 72.0% in Madaripur and negative deviation as low as -43.0% in Khulna.

In pre-monsoon it is found within 216.5% (Patuakhali) to -71.6% (Barisal), and in post-monsoon within 181.9% (Satkhira) to -100.0% (Jessore & Mongla), and in winter within 586.0% (Khulna) to -100.0% (all station) (Table 9).

Station	Deviation Category	Pre-monsoon	Monsoon	Post-monsoon	Winter	Annual
Jessore	Highest (%)	114	40.3	88	399.6	35.1
	Lowest (%)	-66.2	-42.9	-100	-100	-42.6
Satkhira	Highest (%)	117.3	40.7	181.9	278.1	28.7
	Lowest (%)	-51.7	-33.5	-95.2	-100	-26.1
Khulna	Highest (%)	173.7	61.4	152.7	586	48.1
	Lowest (%)	-61	-43	-94.6	-100	-35.5
Barisal	Highest (%)	142.5	43.3	145.1	465.1	33.4
	Lowest (%)	-71.6	-30.7	-95.1	-100	-32
Madaripur	Highest (%)	94.4	72	121.2	285.6	45.4
	Lowest (%)	-54.9	-40	-91	-100	-33.8
Mongla	Highest (%)	67.2	62.1	97.9	216.3	46
	Lowest (%)	-57.5	-31.7	-100	-100	-35.5
Patuakhali	Highest (%)	216.5	71.9	119.5	251.8	39.4
	Lowest (%)	-68.3	-32.3	-92.8	-100	-29.8
Khepupara	Highest (%)	100.4	28.7	160.4	388.9	27.9
	Lowest (%)	-57.7	-28.8	-93.3	-100	-32.8

Table 9: Seasonal and annual extreme rainfall deviation (%) in recent time.

Trend of 99th Percentile of Seasonal and Annual Rainfall

Trend of 99th percentile of seasonal and annual rainfall in southwestern part of the country is under study. The annual 99th percentile is on a negative trend over Jessore, Barisal, Madaripur and Patuakhali, but positive over Satkhira, Khulna, Mongla and Khepupara.

Highest significant positive trend is observed at Mongla (0.79) at 90% confidence level, while highest negative magnitude is observed at Madaripur (-0.32) (Figures 8i and 8ii).

In pre-monsoon season the trend is negative at maximum stations except Barisal. In case of Khulna and Satkhira it is positive over long period but becomes negative over recent period. In this season highest significant negative magnitude (-0.61) is observed at Madaripur at 95% confidence level. In monsoon trend direction is almost synclined with the annual except Satkhira, where annual trend is positive but becomes negative in monsoon.

In this season highest negative magnitude is observed at Patuakhali (-0.68) and positive at Mongla (0.40) (Table 10).

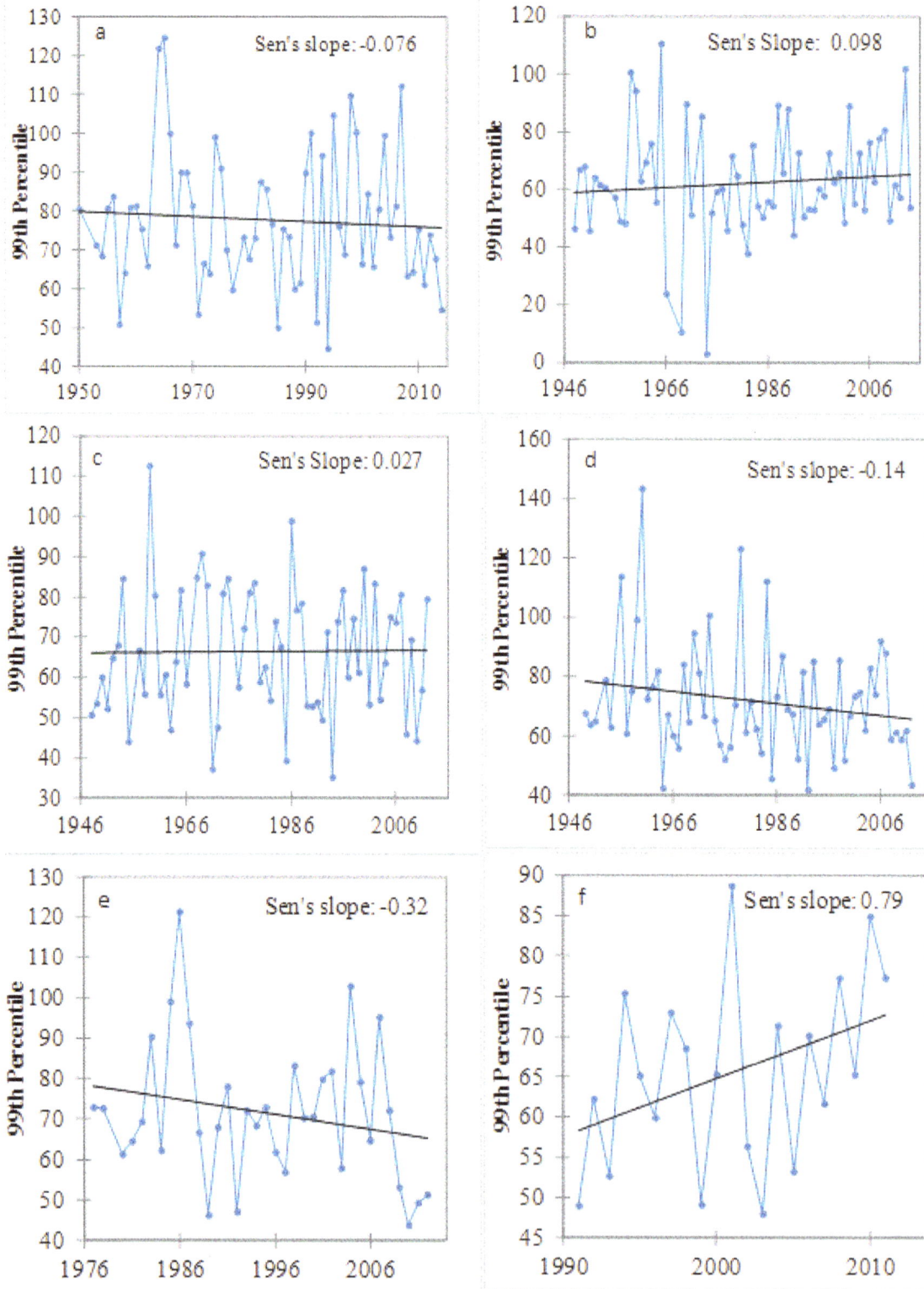

Figure 8i: Trend of 99th Percentile of annual rainfall at (a) Jessore, (b) Satkhira, (c) Khulna, (d) Barisal, (e) Madaripur and (f) Mongla weather stations of BMD.

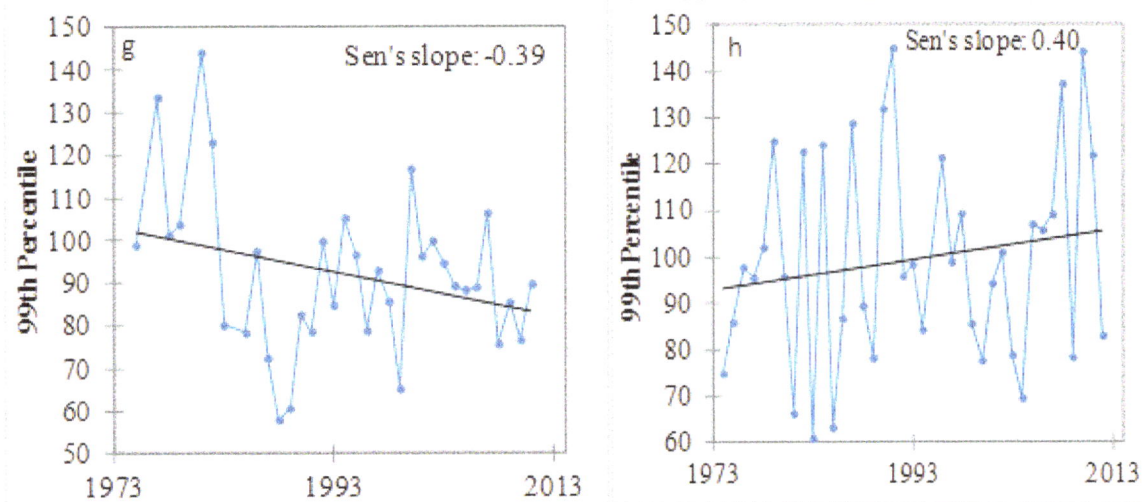

Figure 8ii: Trend of 99th Percentile of annual rainfall at (g) Patuakhali and (h) Khepupara weather stations of BMD.

Station	Time Range	Pre-monsoon	Monsoon	Annual
Jessore	1948-2014	-0.18	-0.61	-0.08
	1981-2014	-0.71	-0.07	-0.22
Satkhira	1948-2014	0.11	-0.67	0.10
	1981-2014	-0.10	-0.19	0.41*
Khulna	1948-2014	0.04	0.01	0.03
	1981-2014	-0.55	0.37	0.41
Barisal	1948-2014	0.05	-0.19	-0.14
	1981-2014	0.07	-0.49	-0.14
Madaripur	1977-2014	-0.61**	-0.32	-0.32
Mongla	1991-2014	-0.23	0.40	0.79*
Patuakhali	1973-2014	-0.39	-0.68	-0.39
Khepupara	1974-2014	-0.20	0.08	0.40

**Significant at 95% confidence level; *Significant at 90% confidence level.

Table 10: Mann-Kendal and Sen's slope test result of 99th percentile of seasonal and annual rainfall at different BMD stations.

Trend of Seasonal and Annual Heavy Rainfall Days

Trend is also studied on seasonal and annual heavy rain days over the study area. Only significant negative trends are observed in Jessore (at 99% confidence level) and Madaripur (90% confidence level) over pre-monsoon, monsoon and annual time period (Tables 10 and 11). In other stations in pre-monsoon and monsoon season no trends are observed. On annual scale slight positive trend is observed in Khulna and slight negative in Barisal and Patuakhali (Figures 9i and 9ii).

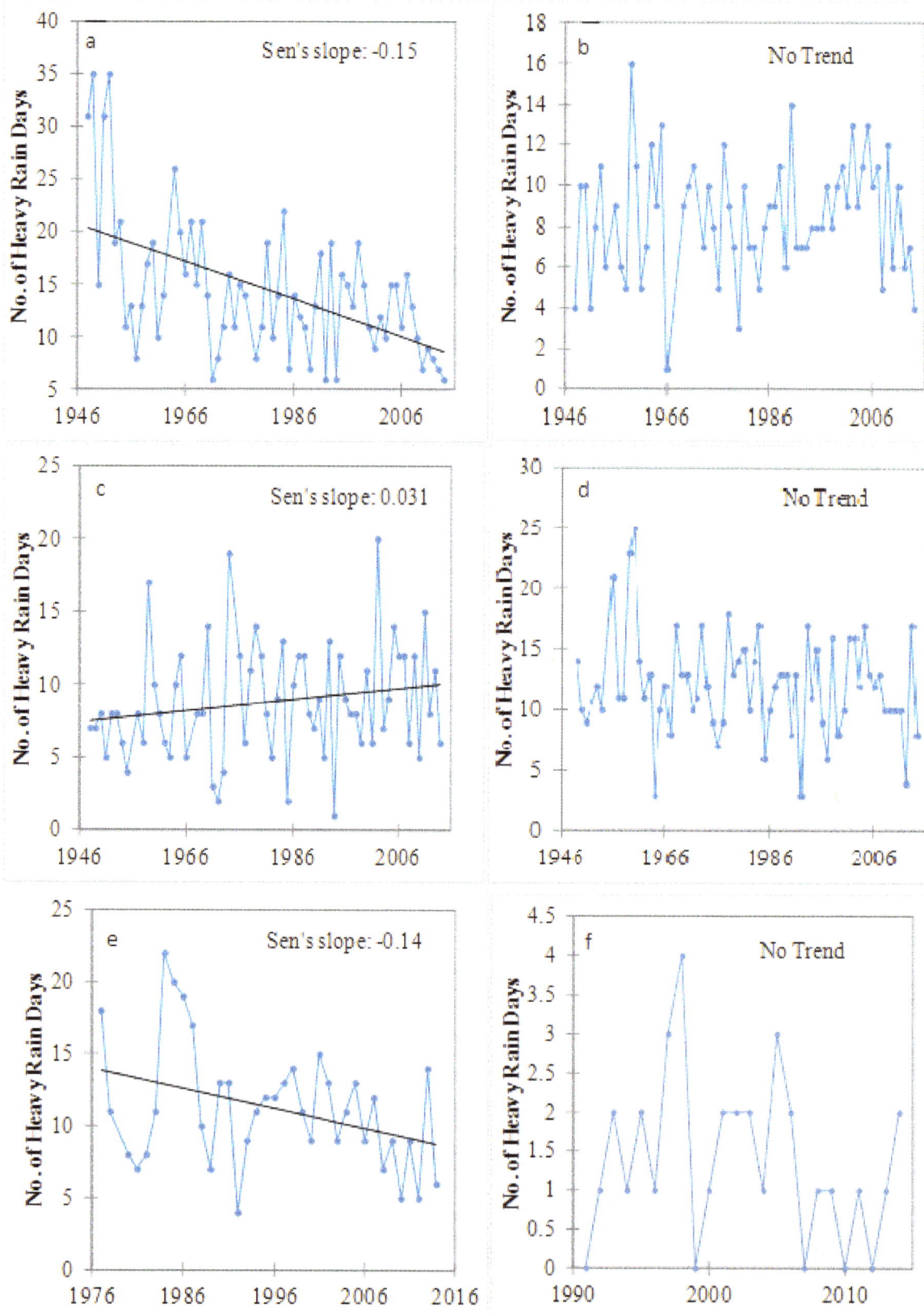

Figure 9i: Trend of annual heavy rain days at (a) Jessore, (b) Satkhira, (c) Khulna, (d) Barisal, (e) Madaripur and (f) Mongla weather stations of BMD.

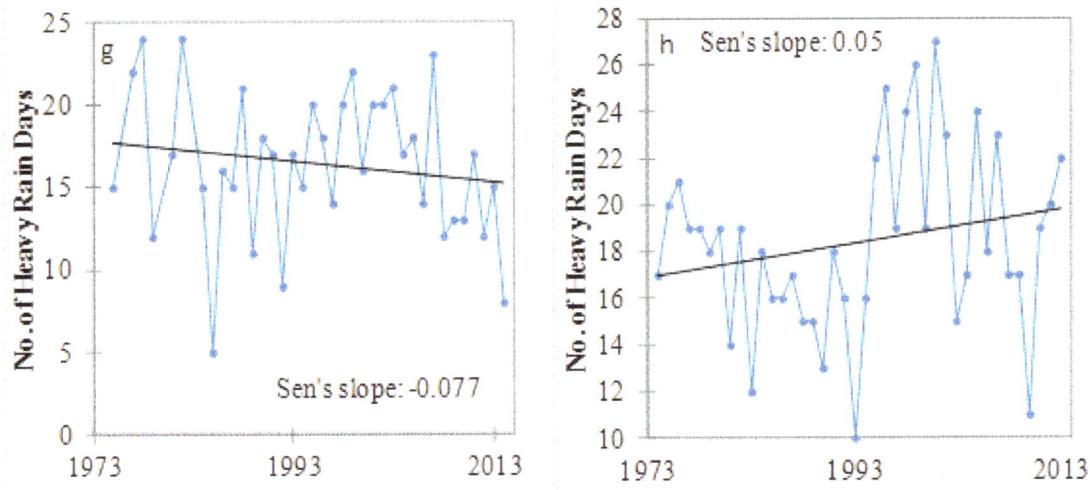

Figure 9ii: Trend of annual heavy rain days at (g) Patuakhali and (h) Khepupara weather stations of BMD.

Station	Time Range	Annual	Pre-monsoon	Monsoon
Jessore	1948-2014	-0.15***	-0.12***	-0.12***
Satkhira	1948-2014	0	0	0
Khulna	1948-2014	0.031	0	0.018
Barisal	1948-2014	-0.023	0	0
Madaripue	1977-2014	-0.14*	-0.059*	-0.066*
Mongla	1991-2014	0	0	-0.063
Patuakhali	1973-2014	-0.077	0	0
Khepupara	1974-2014	0.05	0	0
***Significant at 99% confidence level; *Significant at 90% confidence level.				

Table 11: Mann-Kendal and Sen's slope test result of seasonal and annual heavy rain days at different BMD stations.

Overall Rainfall Scenario over the Southwest Region

Statistical analysis of rainfall variability over the southwestern part of Bangladesh exhibits an overall decreasing scenario, except the Khepupara station which lies in the exposed coastal zone while other stations lies in the interior part. At Khulna and Satkhira, though the rainfall trend is found positive over the long term period but turned into negative (-4.08 mm/year at Khulna and -1.43 mm/year at Satkhira) in recent climatic period. It clearly represents a major turning point in rainfall variability in recent times after 1980s. Magnitude of decreasing trend is highest in the most upper part of southwest region like places in Jessore (-9.4 mm/year at 99% confidence lever) and Madaripur (-15.71 mm/year at 99% confidence level) (Table 3). Besides, some other specific seasonal specialties are also observed. In pre-monsoon period over the places of Khulna, Satkhira and Barisal rainfall trends are positive, but shifted to a negative slopping after 1980s. While in other places of Mongla, Madaripur, and Patuakhali, recent (1981-2014) pre-monsoon trends are significantly negative except the exterior coastal station of

Khepupara. Earlier longer term scenarios are not possible to assess here due to non-availability of long term records. But in monsoon season, rainfall trend is almost similar to annual trend except Khulna, Satkhira and Mongla stations, where though the annual trend is negative but monsoon trend is positive. It indicates monsoon rain intensification in fore southwestern part of the region. In winter season, a major turning point in trend is also prominent. Stations having long term record, e.g. Jessore, Khulna, Satkhira exhibit an increasing trend over long term period but turned to a decreasing one after 1980s (Table 3). Other young stations also exhibit negative trends except Khepupara. It represents a major change in recent climatic condition.

TRMM satellites records also represent significant negative trends (Table 2) with higher magnitudes over recent (1998-2014) times. Annual 99th percentile trend in most of the cases is synclined with the trend of annual rainfall, except Mongla, where rainfall is going on a decreasing trend but 99th percentile is on an increasing trend, i.e., extreme rainfall events are intensifying. In case of Khulna and Satkhira annual rainfall trend is positive over long period (1948-2014) but turned into negative over recent (1981-2014) period. But 99th percentile of these two stations is positive over long period and recent period. Moreover the magnitude is further intensified from 0.10 to 0.41 in Satkhira and 0.03 to 0.42 in Khulna. It reveals that extreme rainfall events are intensifying over Khulna and Satkhira in recent time.

Impact of present rainfall trend on crop agriculture in the study area

The present research identified that the crop agriculture in the study area is greatly influenced by the rainfall pattern as well as the local hydrologic and geomorphic characteristics. The local environmental stress together with rainfall anomaly and changing pattern continues to build up a severe threat to crop agricultural practices.

Impact of monsoon rainfall variability on crop agriculture: As the study area is a part of floodplain delta, there is always a risk of flooding. Most of the land of the study area is between medium high to medium low and thus are flooded to different degrees, from less than 90 cm to more than 180 cm. But it was not a major concern at earlier times. But the current problem is becoming worse is the poor drainage condition (opinion of local people), due to siltation on river bed. It has made rivers lower capacitate to drain water, and thus a little heavy rainfall causes serious flooding condition. And it is becoming intensive day by day. In the current year 2015 rainfall was comparatively higher that has caused severe flooding and crop damage. If we focus on the monsoon rainfall trend (Figure 10) over Khulna district, it has started to increase recently. It is intensifying the risk of similar crop damage. And day by day if the drainage condition becomes poorer and monsoon rain intensified will impose tremendous risk to monsoon season crop cultivation.

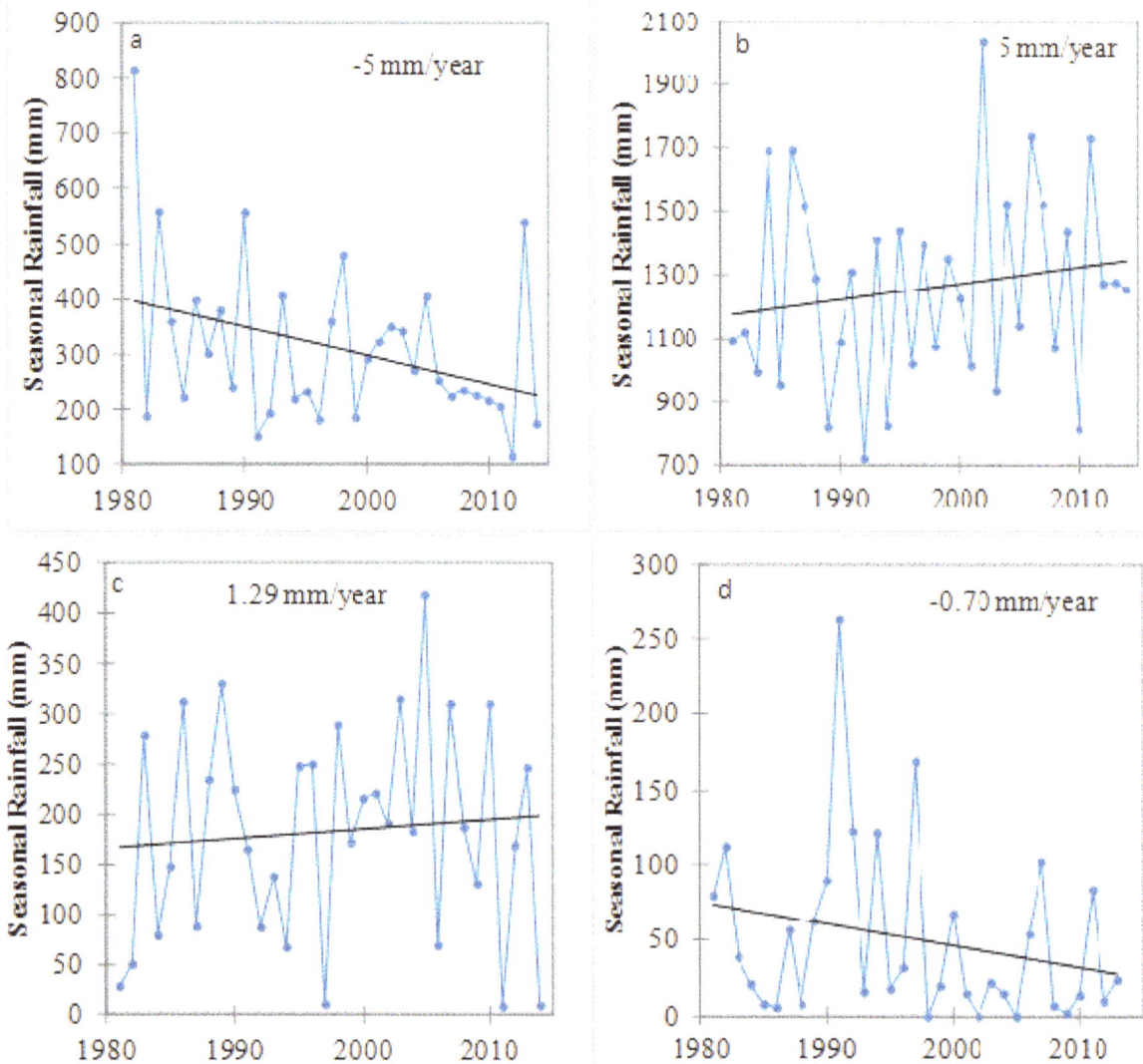

Figure 10: Seasonal Rainfall trend (a) pre-monsoon, (b) monsoon, (c) post-monsoon and (d) winter over Khulna district in recent climatic period (1981-2014).

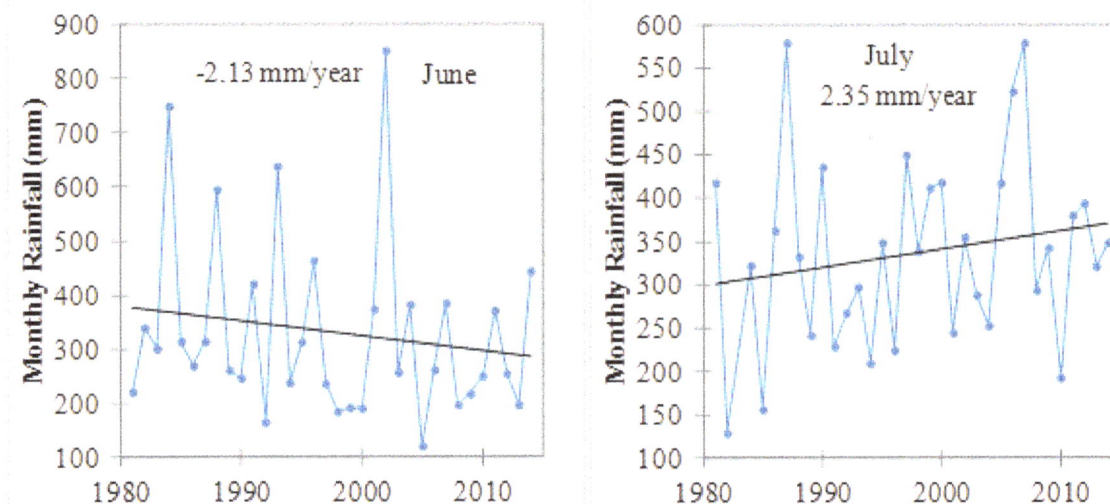

Figure 11: Monthly (June and July) rainfall trend over Khulna district in recent climatic period.

Local people of the study area also opined the observation of anomalous timing of monsoonal onset. Monthly rainfall trend (both Mann-Kendal and linear regression) over the month of June reveals a significant decreasing trend while overall monsoonal rainfall is increasing. Besides, rainfall trend over July is increasing, which represents a delayed onset of monsoon. In case of delayed monsoonal onset, transplantation of aman rice becomes delayed, and farmers have to incur yield loss. Delayed monsoon together with poor drainage and subsequent delayed water recession made loss of land for Boro cultivation. Post-monsoon rainfall also following a increasing trend (Figure 11). In addition to increased post-monsoonal rainfall, the area is under high risk of intense water logging condition.

Impact of winter and Pre-monsoon Rainfall Anomaly

Salinity intensification is a major problem in the study area. Salinity is increasing northward at a high rate. About 0.17 million hactares of new land has been affected by various degrees of soil salinity within 1973-2000 [17]. It is may be due to various reasons like fresh river water withdrawal from upstream, introduction of brackish water shrimp cultivation, faulty management of sluice gates, regular tidal water flooding in unpoldered area, capillary upward movement of soluble salts due to presence of high saline ground water table at shallower depth etc. According to the classification of SRDI the study area lies within slightly saline to highly saline group.

Salinity Status of the Southwest Region

Tidal flooding through a network of tidal creeks and drainage channels connected to the main river system inundates the soil and impregnates them with soluble salts thereby rendering both the top and subsoil saline. Saline water intrusion is highly seasonal. It is at its minimum during the monsoon (June-October) when the main rivers discharge about 80% of the annual fresh water flow. In dry season months, the saline front begins to penetrate inland, and affects the area with sharp rise from 10% in the monsoon to over 40%. The salts enter the soil by flooding with saline river water or by seepage from the rivers, and the salts become concentrated in the surface layers through

evaporation. The saline river water may also cause an increase in salinity of the ground water and make it unsuitable for irrigation. 70% of the 2.35 million hectares within the Khulna and Barisal Divisions is affected by different degree of soil salinity [1]. The decrease in water flow to the Gorai distributaries during the dry season has accentuated the surface water salinity in the southwest region. 1,98,890 ha of cultivable land in the coastal area is under moderately saline with strongly saline class developed by SRDI, where the soil of maximum land (37%) contains high Electrical Conductivity (EC) value of 8.1-12.0 dS/m (Table 3). However 16% land under this class contains extreme EC value of >16.0 dS/m, those are highly detrimental for crop growth.

Conclusion

According to the analysis the following conclusions can be drawn

Amounts of annual rainfall in the southwestern part of Bangladesh are almost equal having very little spatial variation. Significant decreasing trend (Mann-Kendal) of annual rainfall over the region is found (-4.5 mm/year at Mongla, -9.14 mm/year at Jessore, -15.71 mm/year at Madaripur) except Khulna, Satkhira and Khepupara, where the trend is positive over the long period (1948-2014) but it exhibits a decreasing trend during the recent period i.e., after 1980s. It clearly indicates a gradual decrease of the amount of rainfall over southwestern part, which has become most prominent in the recent climatic period of 1981-2014. The trend is more intense at the upper southwestern part, i.e., places like Jessore and Madaripur. Annual rainfall deviates within the range -42.6% to +48.6% over the region. Most wide annual deviation is observed in Khulna from 48.6% to -35.5%.

Like annual rainfall, seasonal rainfall also has anomalous behaviour over the region. Monsoon rainfall at maximum locations are syncline with annual trend, but at Khulna, Satkhira and Mongla though the annual trend is negative but the monsoon trend is positive, it clearly indicates, intensification of rainfall in monsoon period over the fore southwestern part. In pre-monsoon season, overall rainfall trend is significantly negative at maximum places (-8.49 mm/year at Jessore, -2.02 mm/year at Barisal, -7.80 mm/year at Madaripur) over the region except Khulna, satkhira and Khepupara. But in Khulna and Satkhira it is also significantly negative over recent climatic period. Rainfall deviation is higher in seasonal scale than the annual scale. Among the seasons wider deviation is observed in winter (from -100.0% to +586%) and narrower in monsoon (from -43.0% to +62.1%).

The coastal region of southwest Bangladesh has been facing various natural extremes like salinity intensification, drainage congestion, inundation, water logging etc. Anomalous behavior of rainfall in combination with these problems, affecting agricultural crop production in the upazilas under study. The problem is serious in Rabi season, where groundwater irrigation facility is limited for salinity problem. In these areas agriculture is totally dependent on surface water (irrigation canal), which becomes scarce during the month of January, February, March and rainfall is also becoming low in this time (supports by public opinion and rainfall trend analysis) imposing water shortage in crop field and as a result farmers have to incur yield reduction. In monsoon season problem is different. Drainage congestion is a major problem in this time. A little heavy rainfall causes inundation and water logged condition. Besides this problem, increasing trend of rainfall in monsoon is very likely to intensify the risk of inundation. And the farmers of the study area have already faced total damage of Aus crop due to flooding in some years with comparatively high rainfall or fallow due to water logged condition.

Moreover, overall decreasing trend of rainfall is more likely to intensify the risks of salinization due to decreasing upstream flow as well as sedimentation on river bed and consequent poor drainage and water logging. This intensified environmental problem is further likely to intensify the detrimental effects on crop production.

So, changing pattern of rainfall in combination with the local environmental stress is being imposed on risk of agricultural crop production over the study area. It is also occasionally responsible for crop failure over the study area. It is also likely to further intensify the risk for future time.

Salinity status of the southwest region

Tidal flooding through a network of tidal creeks and drainage channels connected to the main river system inundates the soil and impregnates them with soluble salts thereby rendering both the top and subsoil saline. Saline water intrusion is highly seasonal. It is at its minimum during the monsoon (June-October) when the main rivers discharge about 80% of the annual fresh water flow. In dry season months, the saline front begins to penetrate inland, and affects the area with sharp rise from 10% in the monsoon to over 40%. The salts enter the soil by flooding with saline river water or by seepage from the rivers, and the salts become concentrated in the surface layers through evaporation. The saline river water may also cause an increase in salinity of the ground water and make it unsuitable for irrigation. 70% of the 2.35 million hectares within the Khulna and Barisal Divisions is affected by different degree of soil salinity [1]. The decrease in water

flow to the Gorai distributaries during the dry season has accentuated the surface water salinity in the southwest region. 1,98,890 ha of cultivable land in the coastal area is under moderately saline with strongly saline class developed by SRDI, where the soil of maximum land (37%) contains high Electrical Conductivity (EC) value of 8.1-12.0 dS/m (Table 3). However 16% land under this class contains extreme EC value of >16.0 dS/m, those are highly detrimental for crop growth.

References

1. Haque SA (2009) Salinity Problems and Crop Production in Coastal Regions of Bangladesh. Pak J Bot 38: 1359-1365.

2. Intergovernmental Panel on Climate Change (2007) Climate Change 2007-The Physical science Basis; Fourth Assessment Report of the Intergovernmental Panel on Climate Change; Cambridge University Press: Cambridge, UK.

3. Government of Peoples Republic of Bangladesh (2005) Coastal Zone Policy, Ministry of Water Resources, Dhaka.

4. Rafiuddin M, Uyeda H, Islam MN (2010) Characteristics of monsoon precipitation systems in and around Bangladesh. Int J Climatol 30: 1042-1055.

5. Karim Z, Saheed SM, Salauddin ABM, Alam MK, Huq A (1982) Coastal saline soils and their management in Bangladesh. Soils Publication 8: 33.

6. Shahid S, Behrawan H (2008) Drought risk assessment in the western part of Bangladesh. H Nat Hazards 12: 36-42.

7. Ahmed AU (2004) Adaptation to climate change in Bangladesh: learning by doing. UNFCCC Workshop on Adaptation.

8. Ahmed A, Ryosuke S (2000) Climate change and agricultural food production of Bangladesh: an impact assessment using GIS-based biophysical crop simulation model. Center for Spatial Information Science.

9. Climate Change Cell, Who is doing What in Bangladesh? Report on the First Meeting (2006) Comprehensive Disaster Management Programme, Government of Bangladesh: Dhaka.

10. NIDOS (2009) Bangladesh Climate Change Factsheet, Network of International Development Organizations of Scotland.

11. The Center for Environmental and Geographic Information Services (CEGIS) (2011) Final Report on Activity 4: Programmes Containing Measures to Facilitate Adaptation to Climate Change of the Second National Communication Project of Bangladesh.

12. Rahman MR, Salehin M, Matsumoto J (1997) Trends of monsoon rainfall pattern in Bangladesh. Bangladesh J Water Resour 14: 121-138.

13. Rafiuddin M, Uyeda H, Islam MN (2007) Characteristics of Monsoon Precipitation Systems in Bangladesh during 2000–2005.

14. Roy K, Rahaman M (2009) Future Climate Change and Moisture Stress: Impact on Crop Agriculture in South-Western Bangladesh, Climate Change and Development Perspective 1: 1.

15. Center of Excellence for Geospatial Information Science (2013) Vulnerability to Climate Induced Drought Scenario and Impacts, Comprehensive Disaster Management Programme (CDMP II).

16. Karim Z, Hussain SG, Ahmed M (1990) Salinity problems and crop intensification in the coastal regions of Bangladesh. Soils publication 33: 17.

17. Petersen L, Shireen S (2010) Soil and water salinity in the coastal area of Bangladesh. SRDI.

23

A Laboratory Study the Role of Turbulence in Growth of Cloud Droplets in Presence of Environment Aerosols

Pejman M[*], Aliakbari Bidokhti A and Gharaylo M

University of Tehran, Tehran, Iran

[*]**Corresponding author:** Pejman M, Master Level in Field of Meteorology, University of Tehran, Tehran, Iran, E-mail: pejman.moradi@ut.ac.ir

Abstract

In this study, effect of turbulence on cloud droplets growth in the presence of environment aerosols has been studied experimentally. To do this, 3 L of distilled water is poured into the 200 L cloud chamber to provide moisture necessary to create conditions within which air is saturated. After closing the door, the air pressure inside the chamber is increased to 80 mb and after about 25 min air seems to be saturated. Then suddenly the air pressure inside the container is reduced and as a result, warm cloud is formed. Also to create turbulence inside the chamber a propeller attached to an electric motor is used. At the time of the formation of clouds, visible laser beam is scattered by collision with droplets and laser beam signal is reduced. Then the laser signal is gradually increased and returns to its original state after the cloud disappears. The cloud opacity and lifetime are then calculated. The results showed that in the presence of aerosols an increase in turbulence, increased cloud opacity, and cloud lifetime is reduced which indicative of the droplets is getting bigger, hence precipitating faster.

Keywords: Experiments; Turbulence; Warm cloud; Collision coalescence; Cloud opacity; Cloud lifetime

Introduction

For over 80 years, the permeation of turbulence on the process of collision-coalescence of cloud drops has been discussed in the cloud physics meeting. Historically, the first intention to account for the influence of small-scale turbulence on the collisional increase of cloud drops were concentrated on the effect of drop inertia, by Arenberg [1], Gabilly [2], and the most universally by East and Marshall [3]. Later, however, Saffman and Turner [4] consider that this affect is only one possible effect of turbulence, the other being due to variety drop motions with the air. De Almeida extended a method of calculating collision rates on the basis of modeling individual trajectories [5,6]. He found that the effect of weak turbulence fields on collision rates was very strong and impact deeply the expanded of cloud spectra. However, subsequent work [7,8] questioned some of the Almeida's hypothesis, and the results were never fully accepted. More recently, the diffusion equation for a stochastic process was pragmatic to the problem under consideration by Reuter. The role of turbulence in the expansion of clouds remains controversial. For instance, while some have debated that turbulence can lead to droplet clustering, vapour super saturation fluctuations and increased coalescence [9-12]. Turbulence impact not only cloud microphysical processes, such as the collision process and the publication processes rather mixing and entrainment [13]. However, perception techniques have not been sufficiently expanded to recognize the detailed spatio-temporal variability of in-cloud turbulence. It is also challenging to simulate interplay between clouds and turbulent flows in numerical models because simulations that simultaneously discuss both turbulent eddy and cloud systems need large computing resources. Therefore, state of the-art numerical cloud models is used to parameterize the effects of turbulence. Some recent revision has provided the running situation of this topic [13-15]. Direct numerical simulation [12,16,17] and simulations using turbulent statistical models [18] have shown that turbulence growth the collision rate between drops by a few times contrast to the collision rate when only considering gravitational complex, which can result in accelerated and growth area rainfall. By solving the stochastic complex equation, Franklin [19] display that turbulence substantially affects the evolution of the drop size distribution and can reduce the time needed for raindrop formation. Using a Large-Eddy Simulation (LES) model with huge cloud microphysics, Seifert et al. [20] showed that turbulence leads to a vitality enlargement in surface rainfall in warm clouds. Benmoshe et al. [21] study the effects of turbulence on deep convective clouds using a bin microphysics cloud model. They showed that the effects of turbulence are inverse to those of Cloud Condensation Nuclei (CCN): Turbulence-induced collision enlargement accelerates the formation of the first raindrops while leading to reduce in the net accumulated surface rainfall in mixed-phase clouds. Riechelmann et al. [22] expanded a new Lagrangian warm cloud model coupled with an LES model and showed that droplets grow more quickly when the effects of turbulence are contain. Newly, Wyszogrodzki et al. [23] inquire the effects of turbulence-induced collision enlargement under a peculiar range of aerosol concentrations in warm clouds using an LES model with bin microphysics and display an increment in surface rainfall due to turbulence affects. These numerical studies propose that in cloud turbulence plays a significant role in clouds and rainfall. In this research, we test whether the effects of turbulence on clouds and rainfall variety as the aerosol concentration varies, focusing on a single warm cloud using a two-dimensional (2-D) dynamical model with bin microphysics. This examination is anticipated to provide a better understanding of cloud-aerosol interplay and the effects of turbulence on clouds and rainfall.

Method Description and Experimental Setup

In order to analyze the effect of turbulence on the growth of warm cloud droplets, a cylindrical cloud chamber made of Plexiglas with a volume of 200 L is used. In the rubber stopper of this chamber, a few holes used for the passage of temperature sensors, moisture and speed are created. Over the port of the chamber a barometer, an propeller is connected to the electric motor in order to form a turbulence with a variety of intensity with two valves, one connected to air pump and the other to air vents, are embedded. The laser system for determining cloud opacity is placed in a way that the laser beam passes through the chamber and reaches the detector. Laser beam intensity is measured by a power meter connected to the detector and invigorate by an amplifier. The speed of the air is also transferred by speedometer on the spot at a distance of 10 cm. below the propeller and 15 cm wall of the cloud chamber measured and both of them has been transferred to a computer by a board analog digital converter and is recorded in each 0.125 sec. Temperature and moisture inside the cloud chamber is measured by thermometer and barometer with an accuracy of one-tenth is placed and the data related to them is recorded in each second by another computer. The cloud chamber is designed in a way that we can form warm cloud by changing in pressure and adiabatic expansion. The overall layout of the experiment is shown in Figure 1.

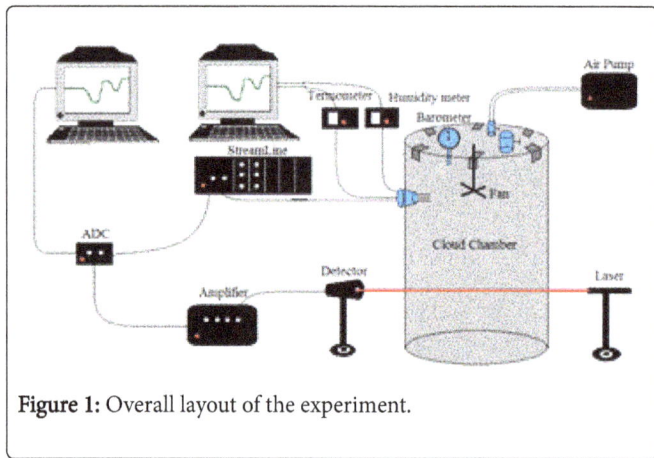

Figure 1: Overall layout of the experiment.

In the present study, the impact of turbulence on the cloud droplets growth is studied in four different turbulence intensities and in the presence of environment aerosols. Three liters of distilled water is poured into the chamber in order to provide the necessary moisture for constructing the supersaturate situation.

After closing the port of the chamber, the air pressure of the inside the chamber is increased by a pump up to 80 mbar. The inside temperature of the chamber is increased because of increase in pressure and the relative moisture of the chamber inside is increased because evaporation of surface water and after about 25 min reaches to the supersaturate phase. Then the air pressure of the inside of the chamber is reduced adiabatically by opening the valve which is embedded on the port which is based on equation of first law of thermodynamic.

$$\frac{dT}{T} = \frac{R}{C_P}\frac{dP}{P} \qquad (1)$$

By reducing the inside pressure of the chamber adiabatically, the temperature environment is reduced adiabatically by condensation over the aerosols of the environment existing in the chamber (condensation nuclears) artificially forms warm cloud. After forming the warm cloud by making turbulence, its effect on the growth of droplets will be discussed. In addition, before forming the warm cloud for recording the data, measuring devices are turning on. When the warm cloud is forming, the laser beam after crossing the chamber by collision toward the droplets will be scattered and appears as a thin line and the intensity of the laser beam will decrease and when drawing its chart causes trough.

Then the laser signals gradually increases and returns to the initial state and the trough will destroy which according to it we can measure the cloud lifetime and cloud opacity (signal depth) and by comparing the experiments in a variety of turbulence intensities, its effect on the growth of the droplets is going to be illustrated. It should be noted that, for decreasing the measurement error of the results, each experiment will be done 4 times, their results are going to be averaged and they move toward a definite quantity and the effect of turbulence on the growth of the droplets display properly.

Results

The speedometer which was placed in order to measure the speed caused by the turbulence of the inside the cloud chamber, recorded the amount of the speed each 0.125 second. The speed variance chart of dimensionless in 3 different turbulences is presented in Figure 2.

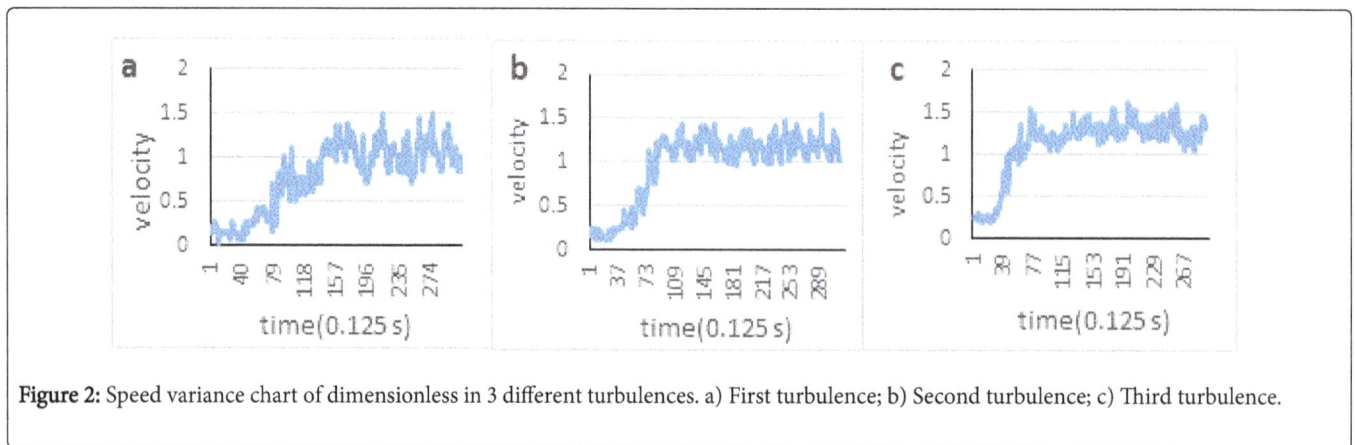

Figure 2: Speed variance chart of dimensionless in 3 different turbulences. a) First turbulence; b) Second turbulence; c) Third turbulence.

The turbulence intensity can be calculated by utilising speed variance which is on a dimensionless quantity.

$$I_u = \frac{\sigma_u}{U} \qquad (2)$$

Which:

$$\sigma_u = \left[\frac{\sum_{i=1}^{i=n}\left(u_i(t) - \bar{U}\right)^2}{n}\right]^{1/2} \qquad (3)$$

$$\mathsf{U} = \overline{\mathsf{U}} - \mathsf{u}_0 \qquad (4)$$

In this equation is turbulence intensity, $u_i(t)$ is speed at any moment of time and $\overline{U}$ is the average speed. Based on the results of the above relations, the turbulence intensity in the first turbulence is 0.16, in the second turbulence is 0.24 and in the third turbulence is 0.3. The definite speed produced by turbulence of 0.16 is also, for turbulence 0.24 is and for the turbulence of 0.3 is. According to the equations 3 and 5:

$$\epsilon = \frac{(\sigma_u)^3}{l} \qquad (5)$$

The amounts of (kinetic energy loss) and also the scales length, speed and time of Kolmogorov in 3 different turbulence intensity is calculated and presented in Table 1.

lu	ε cm²/s³	Tk(S)	η(cm)	ϑk(cm/s)
1.169	0.1453	0.1243	11	0.16
2.821	0.0602	0.0213	373	0.24
4.139	0.041	0.0099	1728	0.3

Table 1: The features of created turbulence inside the chamber in three different intensities.

The speed collision of the droplets is calculated by:

$$V = \frac{2}{9} \frac{g\rho r^2}{\eta} \qquad (6)$$

Equation (24]. In this relation g is the acceleration of gravity, is droplet density, r is droplet radius and is the air viscosity. Now, as the droplets speed collision has a direct relationship with radius, so to the amount the droplet radius is big, the droplets speed collision is big and the cloud's lifetime will be decreased.

Figure 3 shows the variance of cloud opacity in the presence of environmental aerosols for one of the experiments in four turbulence intensities of zero, 0.16, 0.24, and 0.3.

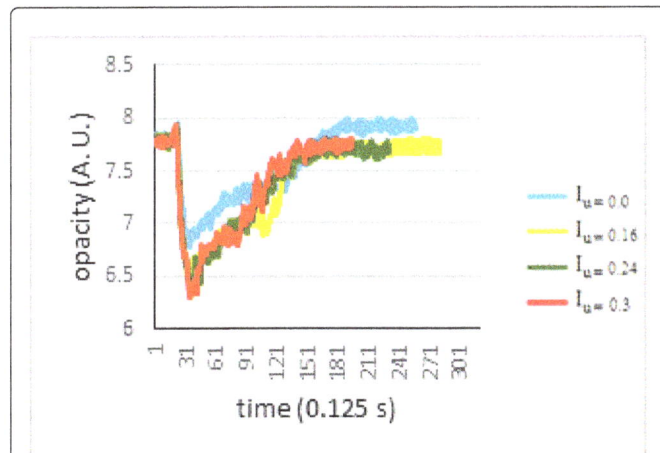

Figure 3: Variance of cloud opacity in the presence of environmental aerosols for one of the experiments in four turbulence intensities of zero, 0.16, 0.24, and 0.3.

Average amount of cloud opacity and cloud lifetime created in different turbulence intensities in the presence of environmental aerosols that shown in Table 2.

lu	t (s)	C (A. U.)
0	18.14	1.09
0.16	17.63	1.16
0.24	17.23	1.44
0.3	17.04	1.64

Table 2: The average amount of cloud opacity and cloud lifetime in different turbulence intensities in the presence of environmental aerosols.

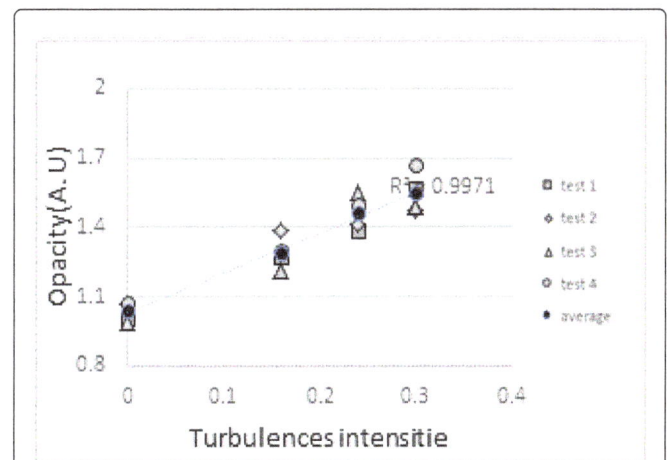

Figure 4: The cloud opacity's scatter chart in relation to turbulence intensity, correlation coefficient, and drawing its trendline.

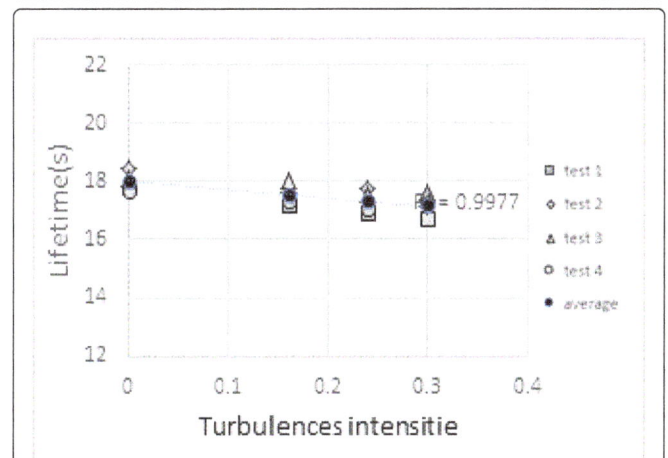

Figure 5: The cloud lifetime's scatter chart in relation to turbulence intensity, correlation coefficient, and drawing its trendline.

As it is clear, cloud opacity is less in no turbulence phase than with turbulence phase and by increasing turbulence, the amount of cloud opacity will be increased. The cloud lifetime, is more in no turbulence phase than with turbulence phase and by increasing the turbulence, the cloud lifetime will be decreased which one can say that turbulence with collision coalescence between cloud droplets, result in the droplets more growth and by increasing radius droplet, the speed of fall increased and concluding its fall time has been decreased.

The cloud opacity and cloud lifetime's scatter chart is shown in relation to turbulence intensity with trend line and the correlation coefficient, in Figures 4 and 5, respectively.

Based on the result, it can be seen that the correlation amount between cloud opacity and cloud lifetime with turbulence intensity are 0.99 that shows the sign of a very much correlation of each quantity with turbulence.

References

1. Arenberg D (1939) Turbulence as the major factor in the growth of cloud droplets. Bull Am Meteor Soc 20: 444-445.

2. Gabilly A (1949) On the role that turbulence can play in the coalescence of cloud droplets. Ann Geophys 5: 232-234.

3. East TWR, Marshall JS (1954) Turbulence in clouds as a factor in precipitation. Quart J Roy Meteor Sos 80: 26-47.

4. Saffman P, Turner J (1956) On the collision of drops in turbulent clouds. J Fluid Mech 1: 16-30.

5. De Almeida FC (1976) The collisional problem of cloud droplets moving in a turbulent environinent-part I: A method of solution. J Atmosph Sci 33: 1571-1578.

6. De Almeida FC (1979) The collisional problem of cloud droplets moving in a turbulent environment-part II: turbulent collision efficiencies. J Atmosph Sci 36: 1564-1576.

7. Pruppacher HR, Klett JD (1980) Microphysics of cloud and precipitation. Reidel p: 714.

8. Lomaya VA, Mazin IP, Neizvestnyy AI (1990) Effect of turbulence on the coagulation efficiency of cloud droplets. Izv Atmos Oceanic Phys 26: 595-600.

9. Shaw RA, Reade WC, Collins LR, Verlinde J (1998) Preferential concentration of cloud droplets by turbulence: effects on the early evolution of cumulus cloud droplet spectra. J Atmos Sci 55: 1965-1976.

10. Pinsky M, Khain A (2003) Fine structure of cloud droplet concentration as seen from the Fast-FSSP measurements. Part II: results of in situ observations. J Appl Meteor 42: 65-73.

11. Shaw RA (2003) Particle-turbulence interactions in atmospheric clouds. Annu Rev Fluid Mech 35: 183-227.

12. Ayala O, Rosa B, Wang LP (2008) Effects of turbulence on the geometric collision rate of sedimenting droplets. Part II: theory and parameterization. New J Phys 10: 075016.

13. Grabowski WW, Wang LP (2013) Growth of cloud droplets in a turbulent environment. Annu Rev Fluid Mech 45: 293-324.

14. Khain AP, Pinsky M, Elperin T, Kleeorin N, Rogachevskii I (2007) Critical comments to results of investigations of drop collisions in turbulent clouds. Atmos Res 86: 1-20.

15. Devenish BJ, Bartello P, Brenguier JL, Collins LR, Grabowski WW, et al. (2012) Droplet growth in warm turbulent clouds. QJR Meteorol Soc 138: 1401-1429.

16. Zhou Y, Wexler AS, Wang LP (2001) Modelling turbulent collision of bidisperse inertial particles. J Fluid Mech 433: 77-104.

17. Franklin CN, Vaillancourt PA, Yau MK, Bartello P (2005) Collision rates of cloud droplets in turbulent flows. J Atmos Sci 62: 2451-2466.

18. Pinsky M, Khain AP, Krugliak H (2008) Collisions of cloud droplets in a turbulent flow. Part V: Application of detailed tables of turbulent collision rate enhancement to simulation of droplet spectra evolution. J Atmos Sci 65: 357-374.

19. Franklin CN (2008) A warm rain microphysics parameterization that includes the effect of turbulence. J Atmos Sci 65: 1795-1816.

20. Seifert A, Nuijens L, Stevens B (2010) Turbulence effects on warm-rain autoconversion in precipitating shallow convection. QJR Meteorol Soc 136: 1753-1762.

21. Benmoshe N, Pinsky M, Pokrovsky A, Khain A (2012) Turbulent effects on the microphysics and initiation of warm rain in deep convective clouds: 2-D simulations by a spectral mixed-phase microphysics cloud model. J Geophys Res 117: D06220.

22. Riechelmann T, Noh Y, Raasch S (2012) A new method for large-eddy simulations of clouds with Lagrangian droplets including the effects of turbulent collision. New J Phys 14: 065008.

23. Wyszogrodzki AA, Grabowski WW, Wang LP, Ayala O (2013) Turbulent collision-coalescence in maritime shallow convection. Atmos Chem Phys 13: 8471-8487.

24. Wallace JM, Hobbs PV (2006) Atmospheric science: an introductory survey.

Weekly Seasonal Agro Climatic Potentialities of India and Climate Change in Extremities

Sivaram B[1][*] and Sarma AALN[2]

[1]*Department of Mathematics, Sri Venkateswara College of Engineering and Technology, Chittoor, India*

[2]*Andhra University, Visakhapatnam, India*

[*]**Corresponding author:** Sivaram B, Department of Mathematics, Sri Venkateswara College of Engineering and Technology, Chittoor, India
E-mail: sivaramboppe@yahoo.co.in

Abstract

The Ocean atmospheric interaction such as El Nino and Southern Oscillation (ENSO) and LaNina and Southern Oscillation (LNSO) plays an important role in monsoon performance and in turn it affects the agro climatic potentialities. The present investigation addresses the impact of the ENSO and LNSO in the periods 1951-1980, 1981-1990 and 1992-1998 on the agro climatic potentialities such as precipitation, potential and actual evapotranspiration, surplus, indices such as moisture, humid and aridity, moisture adequacy, soil wetness of India and selected stations from the climate spectrum of India in different seasons in dry and wet years along with ENSO and LNSO situations on a mean weekly basis. The climate status of All India enriched to fourth humid (B4) and third humid (B3) status during the monsoon seasons of the study periods 1981-1990 and 1992-1998 respectively. During the retrieval of the monsoon All India climate status depleted to moist sub humid (C2) in the years 1992-1998. In the cold weather season of 1992-1998 the climate status depleted to dry sub humid (C1) and experienced aridity in the years 1981-1991. All India climate status enriched to semi aridity (D) during the hot weather season of 1992-1998 compared to the normal status of the same season.

Keywords: Humidity; Aridity; Moisture indices; Soil moisture; Adequacy

Introduction

The past and current anthropogenic land use changes have lead the way for the significant variations of land surface dynamic parameters such as surface albedo, surface resistance, vegetation index and vegetation fractional coverage. These variations resulted in large changes in surface energy and water balance that modulate atmospheric circulation over the region. In understanding the atmospheric and climatic changes, it is very important to understand these variations. India being an agrarian country, the climate change plays an important role in monitoring the crop performance and Indian economy.

Gadgil [1] reported the climatic changes and its impact on agricultural production in India.

The activity of summer monsoon rainfall has a strong bearing on many climate teleconnections. In 1982/83 El Niño resulted in severe drought and flood damage in several countries. This year, first season crops have been affected by drought in most Central American and some Caribbean countries [2].

The ENSO and LNSO effects the economics of the agriculture due to shortening and enlargement of monsoon duration and its performance. The abnormal loss to the agriculture by these two events is reported by Richard et al. [3].

Das and Goankar [4] investigated the effects of the climatic environment on three different varieties of paddy using the crop coefficient in different stages of growth, the consumptive uses and radiation use efficiency.

Kashyaji [5] reported the influence of meteorological parameters such as temperature, rainfall, relative humidity, and bright sunshine hours on the possible growing season and crop performance in rainfed cropping system in a given area. Deng [6] discussed about the impacts of ENSO on rice production.

Sarma and Vizayabhaskar [7] studied and reported that ENSO reduces the agriculture production and yields of major crops, while LNSO increases the agriculture production and yields during the Kharif season.

The influence of seasonal teleconnections such as El Nino and Southern Oscillations patterns on cryosphere of Antarctica and Himalayan regions have been discussed by Krishna et al. [8].

Various aspects of Indian summer monsoon activity, El Nino and Southern Oscillation (ENSO) and LaNina & SO (LNSO) relationship have been investigated by several research workers to name a few [9-13].

Bhalme et al. [14], Sarker [15], Chaudhury and Mhasawade [16] have investigated association between El Nino, monsoon rainfall and droughts in India.

The present investigation addresses the climate change, variations in precipitation, potential and actual evapotranspiration, surplus, indices such as moisture, humid and aridity, moisture adequacy, soil wetness, of India and selected stations from the climate spectrum of India in different seasons during the dry and wet situations accompanied by ENSO and LNSO in the periods 1951-1980, 1981-1990 and 1992-1998, which gives a better resolution of effects of these events on the agriculture sector of India on a weekly basis.

Material and Methodology

The revised water balance concept of Thornthwaite and Mather [17] is followed in computing the water budget elements for 90 stations that are drawn from varied geographical settings of India. The Normals of Agro climatic Observatories India (India Meteorological Department) and the data supplied by the India Meteorological Department, PUNE on a weekly basis for the standard monsoon period (22nd week to 39th week) [18], retrieval of monsoon, cold and hot weather seasons for the periods 1951-1980, 1981-1990 and 1992-1998 have been used to arrive at the agro climatic indices.

The mean weekly climatic indices such as humidity (Iwh %), aridity (Iwa %), moisture indices (Iwm %) are derived from the basic water budget elements. Mean weekly humidity and aridity indices in four seasons-monsoon, retrieval of monsoon, cold and hot weather seasons of All India as well as selected stations will reflect the seasonal mean weekly affectivity of moisture. These indices have been obtained for the study of moist and dry climates respectively in extreme dry and wet years accompanied by ENSO and LNSO. The lowest and highest moisture indices of All India are the two ends of the moisture status. A year in which moisture index is low will be considered as a dry year which may be a consequence of poor performance of the south west monsoon. Hence these dry and wet years are the two extremities of the moisture status, which may be accompanied by El Nino and Southern Oscillation (ENSO) or La Nina and Southern Oscillation (LNSO). The absence of moisture leads to drought. The year in which the moisture index is high will be a wet year that may be due to the good amount of rainfall. The presence of moisture is necessary for the growth of plants and cultivation.

In determining the moisture status of All India, the revised expression of moisture index [19] is followed and is obtained for four seasons of the year on mean weekly climate concept.

The maximum water holding capacity of a location will depend upon the structure and texture of the soil of the location. The maximum water holding capacity of a soil against the pull of gravity is called field capacity (Fc mm). Field capacity values are assigned depending upon the root zone of the crops and soil characteristics of the selected stations.

The soil wetness of All India as well as the selected stations from the climate spectrum of India is estimated as the percentage ratio of water storage (St mm) to field capacity (Fc mm).

Soil wetness (Swt) (%)=(St/Fc) × 100

The index of moisture adequacy is the ratio of actual evapotranspiration to potential evapotranspiration expressed as a percentage.

In climate classification moisture index plays an important role. So much of information is available for the climate classification, but the scheme which was given by Thornth waite was extended for climate classification on a weekly basis during the Hot weather season (March to May), South west monsoon season (June to September), retreating south west monsoon season (October to November) and Cold Weather Season (December to February) of the study, which is presented here under:

$$I_m = I_h - I_a$$

Where I_h=Humidity index, I_a=Aridity index, all being percentages.

Results and Discussion

Mean weekly seasonal climate change of India

It is evident from the present investigation that the normal (1951-1980) hydrologic regime of All India is first humid climate (B1), during the monsoon period with a mean weekly precipitation of 57 mm and potential evapotranspiration of 36 mm that could result in a moderate water surplus of 14 mm with a water deficit of 1 mm. The soils are at 98% moisture adequacy with a 74% of wetness in general (Table 1). The climate status of All India during the retrieval of monsoon is shifted to dry sub humid (C1) with a mean weekly precipitation of 17 mm and potential evapotranspiration of 26 mm which resulted a water deficit of 2 mm. The soils are at 92% adequate with 80% wetness. Climate status has become semi-arid (D) during cold weather season (December to February) with a dip in the rainfall to 4 mm, water need to 12 mm, water surplus to 0 mm accompanied with an increase in water deficit to 7 mm per week. The soils are 48% adequate with 20% wetness during this period.

During the hot weather season (March-May), with a mean weekly precipitation of 11 mm and potential evapotranspiration of 38 mm that resulted a deficit of 28 mm with no water surplus, the All India climate depleted to aridity (E) with 27% soil moisture adequacy, accompanied by 1% wetness.

The hydrological regime of All India during the monsoon period of the decade 1981-1991 is fourth humid (B4) with mean weekly rainfall of 73 mm and potential evapotranspiration of 35 mm that could result in a surplus of 30 mm with no water deficit. The soils are adequate at 99% and wet by 82%. During the retrieval of monsoon, India experienced dry sub humid climatic (C1) status with a mean weekly precipitation of 18 mm and a potential evapotranspiration of 26 mm that resulted a water deficit of 1 mm, 95% adequacy and 86% wetness.

In the cold weather conditions with a rainfall of 4 mm and water need of 13 mm per week resulted to a soil moisture adequacy of 64%, with 44% soil wetness. As a result All India climate shifted to semi-arid (D) in cold weather season. A mean weekly precipitation of 9 mm with a potential water need of 39 mm which resulted to deficit of 27 mm accompanied by an adequacy of 31% and a wetness of 10% forced the All India climatic status to shift to aridity (E).

During the study period of 1992-1998, in the monsoon season All India witnessed a mean weekly precipitation of 66 mm, potential water need of 33 mm that result a deficit of 1 mm and water surplus of 25 mm with a soil moisture adequacy of 98% and 85% wetness of soils. As a result, All India climate witnessed third humid (B3). All India climate shifted to moist sub humid (C2) in the retrieval of monsoon period with 26 mm rainfall and potential evapotranspiration, 5 mm water surplus resulting 98% adequacy and 91% soil wetness. During the cold weather season All India experienced dry sub humid (C1) climatic status due to a rainfall of 6 mm and 13 mm water need.

Though the rainfall is depreciated by 60 mm from the monsoon period the soils are 70% adequate with 52% wetness. In the hot weather season a precipitation of 12 mm with water need 36 mm resulted a deficit of 21 mm and forced the climate of All India to semiarid (D). During this season the soils are wet by 12% with an adequacy of 42%.

Period	Season	P (mm)	PE (mm)	AE (mm)	D (mm)	S (mm)	Iwh (%)	Iwa (%)	Iwm (%)	Wetness (%)	Adequacy (%)
	June-Sept	57	36	35	1	14	40	2	38 (B1)	74	98
	Oct-Nov	17	26	24	2	0	0	8	-8(C1)	80	92
	Dec-Feb	4	12	5	7	0	0	52	-52(D)	20	48
1951-1980	Mar-May	11	38	11	28	0	0	73	-73 (E)	1	27
	June-Sept	73	35	34	0	30	90	1	89 (B4)	82	99
	Oct-Nov	18	26	25	1	0	0	5	-5 (C1)	86	95
	Dec-Feb	4	13	8	5	0	0	36	-36(D)	44	64
1981-1991	Mar-May	9	39	12	27	0	0	69	-69 (E)	10	31
	June-Sept	66	33	33	1	25	76	2	74(B3)	85	98
	Oct-Nov	26	26	25	0	5	16	2	14(C2)	91	98
1992-1998	Dec-Feb	6	13	9	4	0	0	30	-30(C1)	52	70

Table 1: Seasonal water budget elements on mean weekly basis-all India extremities.

From the above analysis it can be concluded that the All India climate is normally first humid (B1), which is enriched to fourth (B4) and third humid (B3) during the monsoon season of 1981-1991 and 1992-1998 respectively. In the retrieval of monsoon season, All India experienced dry sub humid (C1) climate in the years 1951-1980 and 1981-1991, while the climate status improved to moist sub humid (C2) in 1992-1998. The All India climate depleted to semiarid (D) during the cold weather seasons of 1951-1980 and 1981-1991, enriched to dry sub humid (C1) in 1992-1998. The climate status enriched to semiarid (D) in the hot weather seasons of 1981-1991 and 1992-1998 compared to the normal status of aridity (E).

Mean weekly seasonal climate change of per humid zone (A) in extremities

Tocklai: It is evident that the water surplus of Tocklai that hails from per humid zone, during the monsoon season of the extreme wet year (1987) is only 1 mm more than the normal, with 47% depreciation during the dry year (1988). As a consequence, the climate in the extreme wet (1987) and dry (1988) years has taken a shift down ward into humid (B) climates with a moisture status of fourth humid (B4) and third humid (B2) respectively. The soil moisture adequacy and wetness depreciated by 1%, and 6% respectively in the dry year, but adequacy maintained its normality with a 2% decrease in wetness in the extreme wet year (1987). It is also observed that the climate status taken a shift to dry sub humid (C1) from the normal status of moist sub humid, during the monsoon retrieval period in dry and wet years due to an increase in potential water need by 4% with a depreciation in rainfall by 68% from the normal. In this season soil moisture adequacy and wetness reduced by 15% and 19% respectively during the dry year, 25% and 33% depreciation of adequacy and wetness in the wet year.

During the cold weather season, Tocklai maintained its normal status of dry sub humid (C1) during the dry and wet years. But the soil moisture adequacy and wetness showed a depreciation of 20% in each during dry year. The wet year experienced a depreciation of 17% adequacy and 8% wetness. During the hot weather season Tocklai

maintained its normal moist sub humid (C2) status in both the extremities.

Karjat: It is observed that during the monsoon season of a dry year (1987), which is also an ENSO year, there is 43% depreciation in the water surplus, which is due the 35% shortage of rainfall from the normal. As a result 8% depreciation in soil moisture adequacy and 2% decrease in wetness are observed. During the retrieval of monsoon season due to a depreciation of 29% in rainfall accompanied with an increase in potential water need of 9%, resulted a drop in the soil moisture adequacy by 25% and soil wetness by 29% from the normal. As a result climatic status of Karjat forced to shift to dry sub humid (C1). In the cold weather and hot weather seasons the rain fall showed further depreciation from the normal and hence climate experienced aridity (E), though this station hails from the per humid zone. As a consequence, during these seasons adequacy showed a fall of 21%, 80% and wetness dropped by 37%, 50% in cold and hot weather seasons respectively.

During the summer monsoon season of the wet year (1983), Karjat has experienced a water surplus of 173 mm, which is 20% increase, with a rise in the rainfall by 14% from the normal. Witnessed an 8% increase in the water surplus with a rise in rainfall by 6% from the normal (Table 2). In this season, soil wetness depleted by 5% in association with 10% drop in soil moisture adequacy. Karjat experienced moist sub humid status (C2) in the retrieval of monsoon season due to less amount of rainfall and more water need.

Karjat maintained its Arid (D) status during the cold and hot weather seasons due to a drop in the rainfall that results no surplus. The soil moisture adequacy and wetness dropped by 4% and 11% in the cold weather seasons. The variations in the water surplus might be due to the variations in the monsoon activity. It is interesting to observe that during the monsoon period of dry year, coupled with ENSO there is a considerable compression in the wetness as well as the amount of rainfall. In the rest of the extremities the wetness prevailed throughout the monsoon period, which might be due to the higher amounts of rainfall caused by the active monsoon circulation pattern. Throughout the period of study the moisture status of Karjat is consistent and maintained its per humid (A) status in monsoon season but shifted to moist sub humid (C2) and dry sub humid (C1)in the retrieval and then dropped to Arid (E) in the cold and hot weather seasons during the Dry and Wet years.

Station	Period	Season	P (mm)	PE (mm)	AE (mm)	D (mm)	S (mm)	Iwh (%)	Iwa (%)	Iwm (%)	Wetness (%)	Adequacy (%)
Tocklai	1951-1980	June-Sept	75	37	37	0	38	107	0	107(A)	100	100
		Oct-Nov	19	24	23	1	3	8	6	3(C2)	88	94
		Dec-Feb	5	8	7	1	0	0	11	-11(C1)	66	89
		Mar-May	41	28	27	1	7	20	3	17(C2)	77	97
	Dry year (1988)	June-Sept	55	38	37	0	20	54	1	53(B2)	94	99
		Oct-Nov	19	25	22	3	4	13	15	-1(C1)	81	85
		Dec-Feb	7	9	6	2	0	0	29	-29(C1)	53	71
		Mar-May	42	29	27	2	6	16	6	10(C2)	69	94
	Wet year (1987)	June-Sept	76	37	37	0	39	99	0	99(B4)	98	100
		Oct-Nov	6	25	20	5	0	0	25	-25(C1)	67	75
		Dec-Feb	4	9	6	2	0	0	27	-27(C1)	61	73
		Mar-May	35	26	26	1	3	8	4	4(C2)	84	96
Karjat	1951-1980	June-Sept	185	34	33	1	144	439	2	437(A)	87	98
		Oct-Nov	17	32	26	6	2	6	20	-14(C1)	70	80
		Dec-Feb	1	18	4	14	0	0	77	-77(E)	19	23
		Mar-May	1	44	2	42	0	0	95	-95(E)	2	5
		June-Sept	121	36	32	4	81	238	10	229(A)	85	90
	Dry & ENSO year (1987)	Oct-Nov	12	35	21	13	3	7	40	-33(C1)	50	60
		Dec-Feb	3	20	3	16	0	0	82	-82(E)	12	18
		Mar-May	0	45	0	44	0	0	99	-99(E)	1	1
	Wet year (1983)	June-Sept	211	34	30	4	173	542	12	530(A)	83	88
		Oct-Nov	18	27	20	6	10	30	28	3(C2)	66	72
		Dec-Feb	0	17	4	13	0	0	78	-78(E)	21	22
		Mar-May	0	41	1	40	0	0	96	-96(E)	3	4

Table 2: Seasonal water budget elements on weekly basis-per humid zone (A)–extremities.

Mean weekly seasonal climate change of humid zone (B) in extremities

Adhartal: In the dry year (1996), the water surplus is decreased by 20 mm (69%) with an increase in water deficit by 10 mm (250%) from the normal due to sub normal rainfall, which was responsible for the shift of the station's climatic regime to the dry sub humid climate (C1) type. Soils are adequate by 90% with wetness 70% during the monsoon season. The dry sub humid climate (C1) prevailed during the retrieval and cold weather seasons and further depreciated to arid type (E) in the hot weather season.

In the wet year (1994), there was such a heavy rainfall which is nearly double the normal that caused to record an 80 mm water surplus, which is 51 mm (176%) more than the normal. As a result, the station's moisture status enriched to per humid (A) type. The soils are 75% wet with an adequacy of 85%. During the retrieval of the monsoon, cold and hot weather seasons there is a depreciation of 100% surplus, due to the drop in rainfall by 100%, 75% and 0% in the respective seasons, which forced the climate status to shift Arid (E).

The soil wetness reduced by 51%, 62% and 40% in these seasons from their normal. As a result, the soil moisture adequacy also showed depletion by 55%, 73% and 50% from the corresponding normal in the retrieval, cold and hot weather seasons.

Station	Period	Season	P (mm)	PE (mm)	AE (mm)	D (mm)	S (mm)	Iwh (%)	Iwa (%)	Iwm (%)	Wetness (%)	Adequacy (%)
Adhartal	1951-80	June-Sept	68	37	33	4	29	91	10	81(B4)	70	90
		Oct-Nov	8	21	15	6	0	0	31	-31(C1)	55	69
		Dec-Feb	4	7	5	2	0	0	30	-30(C1)	37	70
		Mar-May	2	41	6	35	0	0	84	-84(E)	10	16
	Dry year (1996)	June-Sept	37	38	24	14	9	29	34	-5(C1)	50	66
		Oct-Nov	15	20	15	5	0	0	29	-29(C1)	62	71
		Dec-Feb	9	7	6	1	1	6	19	-13(C1)	69	81
		Mar-May	1	42	11	31	0	0	70	-70(E)	25	30
	Wet year (1994)	June-Sept	114	35	30	6	80	247	15	232(A)	75	85
		Oct-Nov	0	21	7	14	0	0	69	-69(E)	27	31
		Dec-Feb	1	8	2	6	0	0	81	-81(E)	14	19
		Mar-May	2	42	3	39	0	0	92	-92(E)	3	8
Rudrur	1951-80	June-Sept	48	34	31	3	9	31	8	23(B1)	64	92
		Oct-Nov	11	24	19	4	0	0	21	-21(C1)	64	79
		Dec-Feb	2	15	5	10	0	0	61	-61(D)	28	39
		Mar-May	4	45	5	39	0	0	88	-88(E)	4	12
	Dry year (1984)	June-Sept	37	34	25	10	9	25	27	-3(C1)	51	73
		Oct-Nov	10	23	13	10	0	0	47	-47(D)	43	53
		Dec-Feb	0	15	3	13	0	0	80	-80(E)	18	20
		Mar-May	0	45	1	44	0	0	98	-98(E)	2	2
	Wet year (1989)	June-Sept	81	31	28	3	45	150	9	141(A)	75	91
		Oct-Nov	0	22	13	9	0	0	45	-45(D)	50	55
		Dec-Feb	3	14	4	10	0	0	71	-71(E)	23	29
		Mar-May	4	42	5	37	0	0	84	-84(E)	4	16

Table 3: Seasonal water budget elements on weekly basis-humid zone (B)-extremities.

It is very interesting to note that in the extreme dry year, Adhartal that hails from humid zone shifted to dry sub humid (C1) in monsoon season, and remained in the same status during retrieval of monsoon period and cold weather season. Climate status depleted to Aridity (E) in the hot weather season. During the wet year, Adhartal improved its moisture status to per humid (A) but witnessed aridity (E) during the retrieval, cold and hot weather seasons.

Rudrur: The mean weekly water budget elements in extremities are presented in the Table 2. It is evident that in the dry year (1984), there is an increase in the water deficit by 233% with depreciation in the rainfall by 23% from the normal that has forced the moisture status of Rudur to dry sub humid (C1) type from first humid (B1) status. There

is a depreciation of 21% in soil moisture adequacy and 20% depreciation in soil wetness in monsoon season. In the retrieval of monsoon period, due to 100% depreciation in water surplus the moisture regime of Rudrur witnessed semi aridity (D), soil wetness and soil moisture adequacy decreased by 33% each from the respective normal. In the cold and hot weather seasons the climate status dropped to aridity (E) due to a drop in water surplus by 100%, as a result the soil wetness and adequacy reduced by 36%, 49% and 50%, 83% respectively in the respective seasons of dry year (1984).

There is an increase in the water surplus by 36 mm (400%) with a rise in the rainfall by 33 mm (69%) and depreciation in the water need by 3 mm (9%) from the respective normalcy, which have forced the moisture status to per humid (A), improved the soil wetness by 15% and adequacy by 1% from their respective normal during the monsoon season of the wet year (1989). A 100% drop in the water surplus due to 100% fall in the rain fall and 8% increase in the water need from the normal forced the climate status to shift to semi aridity (D), 22% decline in soil wetness and 30% decrease in soil moisture adequacy during the retrieval of monsoon season. Rudrur experienced Aridity (E) during the cold and hot weather seasons of wet year due to a drop in the rainfall with the absence of water surplus.

As a result, soil wetness and adequacy were dropped by 18%, 26% respectively in the cold weather season. There is a 33% increase in soil moisture adequacy during the hot weather season of the wet year.

It is observed that during the monsoon season of normal, dry and wet years Rudrur witnessed first humid (B1), dry sub humid (C1) and per humid(A) status respectively. During retrieval of monsoon season the climate status dropped to semi aridity (D), and during cold and hot weather seasons it has experienced aridity (E), both in dry and wet years respectively, which are due to the fluctuations in rainfall and surplus in the respective seasons.

Mean weekly seasonal climate change of moist sub humid zone (C2) in extremities

Agro climatic potentialities of Junagarh and Sabour representing the Moist sub humid zone (C2) in extreme situations such as dry and wet years in the monsoon, retrieval of monsoon, cold weather and hot weather seasons on a weekly basis are presented in the Table 3.

Junagarh: It is observed that in the extreme dry year (1991), there is no water surplus due to sub normal rainfall and as a result there is a climatic shift to semiarid (D) type. The soil wetness and moisture adequacy depreciated by 41% and 36% from the respective normal during the monsoon season. In the retrieval of monsoon season also there is no water surplus due to subnormal rainfall, as a result Junagarh shifted its climate status to Aridity (E), with a decrease in soil wetness by 67% and soil moisture adequacy by 74% from the corresponding normal. During the cold weather season, the aridity (E) prevailed with soil wetness and adequacy of 6% and 7%.

In the hot weather season also the climate status depleted to aridity (E) with 1% soil wetness and moisture adequacy. On the other hand, in the extreme wet year (1988), which is also happened to be the LNSO year, the water surplus is increased by 33 mm with an increase in the rainfall by 32 mm (71%) from the normal that resulted an improvement in the moisture status to third humid (B3) type accompanied by an increase in soil wetness by 6% and a decrease in adequacy by 13% from the corresponding normality during the monsoon season. In the retrieval of monsoon season though there is no water surplus due to absence of rainfall, the soils improved their wetness by 24% with no change in the moisture adequacy. The moisture status of this station dropped to semi aridity (D). In the rest of the seasons the climate status further dropped to aridity (E), but soil wetness showed an improvement of 11% with depreciation in the adequacy by 21% during the cold weather season and a decrease of 50% adequacy during the hot weather season of wet year and LNSO year (1988).

It is also evident from the Table 4 that in the dry year, the water deficit is 18 mm, which is 12 mm (200%) more than the normal (16

mm) and in the wet year, the water deficit is 10 mm, which is 4 mm (67%) more than the normal. The variations in the water surplus and water deficits are due to the strong or weak monsoon circulation pattern over that region. It is worth mentioning that during the wet year there is a high amount of monsoon rainfall, which might be due to the impact of LNSO on the southwest monsoon circulation pattern. As a result, the climate status of Junagarh enriched to third humid (B3).

Sabour: In the dry year (1990), Sabour registered a water surplus of 6 mm, though there is depletion in rainfall and an increase in water need from the respective normalcy during the monsoon season. The water deficits are increased by 5 mm during the monsoon season of the dry year. The soil wetness showed an improvement by 29% with a rise in adequacy by 17% during the monsoon season of the dry year. The climate status depleted to dry sub humid (C1) with 66% soil wetness and 82% moisture adequacy. During the retrieval of monsoon season, cold and hot weather seasons of dry year, the moisture status further depleted to semiarid (D).

The monsoon season of the extreme wet year (1987), experienced a water surplus of 58 mm due to abnormal rainfall of 103 mm, which is twice that of normal rainfall, has enriched the station's moisture regime to per humid (A) (Table 4). The water deficit decreased by 1 mm from the normality. An increase in soil wetness by 41% in association with a rise in adequacy by 1% during the wet year from their corresponding normality is observed. In all other seasons such as retrieval of monsoon, cold and hot weather seasons of the wet year the moisture regime of this station experienced semi aridity (D).

The soils are wet by 46%, 41% and 19% during the retrieval of monsoon season, cold and hot weather seasons of wet years (1987) respectively. From the studies, it is evident that the climate status of Sabour enriched to Per humid (A) during the monsoon period of wet year (1987) and depleted to dry sub humid (C1) during the monsoon season of dry year (1990).

Mean weekly seasonal climate change of Drysub humid zone (C1) in extremities

From the Dry sub humid zone (C1) the agro climatic potentialities of Solapur and Rajendranagar have been studied during the monsoon, retrieval of monsoon, cold and hot weather season in the extreme dry and wet years along with normal.

Solapur: The climatic agro climatic potentialities of Solapur representing the Dry sub humid zone (C1) showed a distribution of rainfall that produced an 8 mm water deficiency, with no water surplus in the monsoon season during 1951-80. As a result the normal moisture status maintained its dry sub humid status with 10% soil wetness and 79% adequacy. In the retrieval of monsoon season, Solapur recorded 14 mm rainfall, which produced no water surplus, which forced the moisture status to shift semi aridity (D), with a soil wetness of 36% and adequacy of 66%.

During the cold weather season 1 mm rainfall with a 13 mm water deficit, pushed the moisture status to Aridity (E), with soil wetness of 14% and adequacy 20%. The Aridity (E) prevailed during the hot weather season with 40 mm water deficit, 2% wetness and 11% adequacy (Table 5).

Station	Period	Season	P (mm)	PE (mm)	AE (mm)	D (mm)	S (mm)	Iwh (%)	Iwa (%)	Iwm (%)	Wetness (%)	Adequacy (%)
Junagarh	1951-1980	June-Sept	45	37	31	6	8	22	15	6(C2)	68	85
		Oct-Nov	7	35	17	18	0	0	53	-53(D)	33	47
		Dec-Feb	1	13	2	11	0	0	86	-86(E)	9	14
		Mar-May	0	43	1	43	0	0	98	-98(E)	1	2
	Dry year (1991)	June-Sept	22	37	19	18	0	1	46	-45(D)	40	54
		Oct-Nov	0	31	4	27	0	0	88	-88(E)	11	12
		Dec-Feb	0	11	1	10	0	0	93	-93(E)	6	7
		Mar-May	0	43	1	43	0	0	99	-99(E)	1	1
	Wet & LNSO year (1988)	June-Sept	77	38	28	10	41	100	26	74(B3)	64	74
		Oct-Nov	0	29	15	14	0	0	53	-53(D)	41	47
		Dec-Feb	0	15	1	13	0	0	89	-89(E)	10	11
		Mar-May	0	45	0	45	0	0	99	-99(E)	1	1
Sabour	1951-1980	June-Sept	50	39	37	2	3	7	5	2(C2)	51	95
		Oct-Nov	14	25	21	3	3	10	19	-9(C1)	76	81
		Dec-Feb	2	6	4	2	0	0	30	-30(C2)	49	70
		Mar-May	7	43	14	29	0	0	68	-68(E)	16	32
	Dry year (1990)	June-Sept	42	40	33	7	6	15	18	-3(C1)	66	82
		Oct-Nov	9	25	16	9	0	0	41	-41(D)	43	59
		Dec-Feb	2	7	4	3	0	0	52	-52(D)	36	48
		Mar-May	12	36	16	20	0	0	56	-56(D)	21	44
	Wet year (1987)	June-Sept	103	39	37	1	58	165	4	162(A)	87	96
		Oct-Nov	3	26	15	11	0	0	46	-46(D)	46	54
		Dec-Feb	1	7	3	4	0	0	53	-53(D)	41	47
		Mar-May	8	38	14	24	0	0	63	-63(D)	19	37

Table 4: Seasonal water budget elements on weekly basis-moist sub humid zone (C2)-extremities.

In the monsoon season of the dry year (1994), there is no water surplus, but an increase in the water deficiency by 11 mm, a moderate depreciation in the rainfall by 15 mm (52%) from the normal have forced the climate to take a shift down ward to semiarid (D) type of climate. It is observed that there is no change in the soil wetness, but 43% depreciation in moisture adequacy. During the retrieval of monsoon season, though there is an increase in water deficit by 63%, with no water surplus could not change its moisture status of its monsoon season, i.e., semi aridity (D). But soil wetness and adequacy showed a depletion of 42% and 30% from their corresponding normality. During the cold weather season, a decrease in water deficit by 15%, with the absence of rainfall changed its moisture status to Aridity (E), with an increase in soil wetness by 92% and adequacy by 50%. In the hot weather season also Solapur experienced the aridity

(E) due to very less amount of rainfall and high water need. It is observed that soils are 5% wet with an adequacy of 12%.

On the other hand, in the monsoon season of the wet year (1987), the water surplus showed an increase of 15 mm due to a fall in the water need by 22 mm (65%) from the corresponding normal and as a result the station's climatic status was improved to per humid (A) type. The soil wetness increased by 650% accompanied by a rise in the soil moisture adequacy by 11% (Table 5). In the retrieval of monsoon season an increased precipitation of 521% and a decrease in water need by 65% resulted a high amount of water surplus of 77 mm which have forced the station's moisture status to per humid (A) type and soils have become wet by 99% with an adequacy of 100%. In the cold weather season of the wet year (1987), Solapur experienced semiarid (D) climate status due to an increase in water deficit by 10 mm. The soils dried to 40% with adequacy of 43%. Solapur experienced aridity (E), due to a depreciation in water deficit by 75% with a reduction in water need by 71% from the corresponding normalcy. The soil wetness and adequacy improved by 250%, 118% respectively from the respective normality. From the aforesaid results it is evident that the moisture status of this station enriched to per humid (A) during the monsoon and retrieval of monsoon seasons of the wet year (1987) and depleted to aridity (E) during the hot weather seasons of both dry and wet years.

Station	Period	Season	P (mm)	PE (mm)	AE (mm)	D (mm)	S (mm)	Iwh (%)	Iwa (%)	Iwm (%)	Wetness (%)	Adequacy (%)
Solapur	1951-1980	June-Sept	31	34	27	8	0	0	21	-21(C1)	10	79
		Oct-Nov	14	26	18	8	0	0	34	-34(D)	36	66
		Dec-Feb	1	17	3	13	0	0	80	-80(E)	14	20
		Mar-May	4	45	5	40	0	0	89	-89(E)	2	11
	Dry year (1994)	June-Sept	16	34	16	19	0	0	55	-55(D)	10	45
		Oct-Nov	15	27	14	13	0	0	54	-54(D)	21	46
		Dec-Feb	0	15	5	11	0	0	70	-70(E)	27	30
		Mar-May	3	44	5	39	0	0	88	-88(E)	5	12
	Wet year (1987)	June-Sept	31	12	10	1	15	145	12	133(A)	75	88
		Oct-Nov	87	9	9	0	77	894	0	894(A)	99	100
		Dec-Feb	2	14	3	10	0	7	57	-50(D)	40	43
		Mar-May	5	13	3	10	0	0	76	-76(E)	7	24
Rajendranagar	1951-1980	June-Sept	32	38	30	8	0	0	20	-20(C1)	17	80
		Oct-Nov	10	31	15	17	0	0	56	-56(D)	20	44
		Dec-Feb	1	17	2	15	0	0	90	-90(E)	2	10
		Mar-May	5	43	5	38	0	0	88	-88(E)	1	12
	Dry year (1984)	June-Sept	20	34	17	17	2	8	48	-40(D)	26	52
		Oct-Nov	10	22	10	12	0	0	58	-58(D)	26	42
		Dec-Feb	0	14	0	14	0	0	96	-96(E)	4	4
		Mar-May	4	43	4	39	0	0	90	-90(E)	1	10
	Wet year (1985)	June-Sept	55	34	23	11	31	104	32	71(B3)	41	68
		Oct-Nov	8	19	8	12	0	0	65	-65(D)	20	35
		Dec-Feb	0	16	1	15	0	0	92	-92(E)	4	8
		Mar-May	3	44	3	41	0	0	92	-92(E)	1	8

Table 5: Seasonal water budget elements on weekly basis-dry sub humid zone (C1) extremities.

Rajendranagar: The mean weekly agro climatic potentialities of Rajendranagar in the monsoon season, retrieval of monsoon season, cold and hot weather seasons during the long term, dry and wet situations have been presented in Table 5. It is evident that there is an increase in water deficit by 9 mm and water surplus by 2 mm due to depreciation in rainfall by 12 mm and water need by 4 mm from the normality during the monsoon season of Dry year (1984). As a result the moisture status of this status depleted to semiarid (D), accompanied by an improvement in soil wetness by 53% and decrease in moisture adequacy by 35%. In the monsoon withdrawal season, depreciation in water deficit by 29% accompanied by depreciation in water need by 33% from their normalcy forced the moisture status to shift to semi aridity (D), but soils improved their wetness by 30% and moisture adequacy dropped by 5%. In the cold weather season of dry year (1984) Rajendranagar experienced Aridity (E), with 100% increase in soil wetness and 60% decrease in soil moisture adequacy from the corresponding normality. In the hot weather season of the dry year moisture status depleted to aridity. Due to high water deficits and water needs the moisture adequacy depleted to 10% in the hot weather season of dry year.

In the monsoon season of the wet year (1985), experienced a rainfall which is almost twice that of the normal. As a result, the water surplus rose by 31 mm and the water need depreciated by 4 mm (11%) from the normal and these were responsible for the climatic shift of the station to the third humid (B3) type, and enrichment in soil wetness by 141% with depletion in adequacy by 15%. In the monsoon withdrawal season, moisture status shifted to semi aridity (D) due to subnormal rainfall that resulted to absence of water surplus. As a result soil wetness maintained its normality, but 20% depreciation in soil moisture adequacy is observed. During the cold and hot weather seasons, Rajendranagar experienced Aridity (E) due to the depreciation in rainfall that could not meet the water demand, which resulted an increase in water deficit.

From the aforesaid discussion it can be concluded that the moisture status of this station enriched to third humid (B3) during the monsoon season of wet year (1985). The climate status depleted to aridity (E) during the cold and hot weather seasons of long term, dry and wet years.

Mean weekly seasonal climate change of semiarid zone (D) in extremities

The mean weekly water budget elements for the four seasons in the long term, extreme dry and wet years are presented here for the two stations Jodhpur and Aduthurai representing the semiarid zone, from the climatic spectrum of India.

Jodhpur: It is evident from the water balance studies (Table 6) that during the monsoon seasons of all the extremities the rainfall distribution is subnormal and could not meet the potential water need, except in the wet year (1990). In the extreme dry situation (1987), which is also the ENSO year, there is no water surplus. The depreciation of rainfall by 16 mm, accompanied by an increase in potential water need by 3 mm increased the water deficit by 20 mm from the respective normality. As a consequence the moisture status depleted to aridity (E). The soils are wet by 1% and adequate by 8%. In the cold and hot weather seasons of the Dry & ENSO year (1987), the moisture status depleted to aridity (E), due to subnormal rainfall.

The monsoon season of the wet year (1990), witnessed a water surplus of 23 mm due to the heavy rainfall of 44 mm, which is 24 mm (120%) higher than the normal. As a result, the moisture status improved to dry sub humid (C1) type of climate. The soil wetness and moisture adequacy showed an increase by 725% and 4% respectively in the wet year from their normality. During the post monsoon season, cold and hot weather seasons of the wet year (1990), Jodhpur witnessed subnormal rainfall, which could not meet the potential water need. As a result, the climate status depleted to aridity (E), with soil wetness by 1%. From this analysis it can be concluded that moisture status improved to dry sub humid (C1) and depreciated to aridity (E)

during the monsoon season of Wet year and dry years respectively. In all other seasons of long term, dry & ENSO and Wet years the climate status dropped to aridity (E).

Aduthurai: From the water balances of Aduthurai (Table 6), it is clear that the distribution of rainfall is subnormal, during the monsoon period of dry year (1994) that could not meet the demand. As a result, the moisture status depleted to aridity (E), with soil wetness of 1% and adequacy 9%. During the retrieval of monsoon season of dry year (1994)), the climate status enriched to first humid (B1), due to huge amount of rainfall of 54 mm, 33 mm water need and an increase in water surplus by 20 mm, along with a depreciation in soil wetness by 18% and soil moisture adequacy by 38% from the corresponding normalcy.

During the cold weather season of Dry year, the climate status of Aduthurai depleted to dry sub humid (C1). Soil wetness decreased to 73%, with an adequacy of 79% which may be due to the depreciation in the rainfall by 5 mm from the corresponding normality, which increased the water deficit.

The climate status further depleted to aridity (E) in the hot weather season of dry year (1994), which may be due to less amount of rainfall and high water need.

The monsoon season of wet year (1996) recorded a water surplus of 5 mm, due to an amount of rainfall 31 mm. As a consequence the moisture status was improved and the climate was shifted into dry sub humid (C1) type. The soil wetness and soil moisture adequacy were improved by 37% and 29% respectively from the corresponding normality.

During the retrieval of monsoon season of this year, this station registered a rainfall of 49 mm with a water need 36 mm, as a result the water surplus increased by 2 mm from its normality. As a result the climate is force to shift to dry sub-humid (C2) with a soil wetness of 41% and soil moisture adequacy 77%.

In the cold weather season of the wet year (1996), Aduthurai experienced Per humid (A) status, with 46% soil wetness and 48% adequacy due to an increase in the rainfall by 28 mm from its normality, with stationary potential water need, which resulted 43 mm water surplus.

During the hot weather season of the wet year, Aduthurai witnessed aridity (E), due to the depreciation in precipitation and increase in water need that resulted to absence of water surplus and increase in water deficit. The soil wetness and moisture adequacy reduced to 3% and 16% respectively.

From the above discussion it can be concluded that the climate status of Jodhpur enriched to dry sub humid (C1) during the monsoon season of the wet year (1990) and depleted to aridity (E) in all the seasons of dry and wet years. On the other hand, Aduthurai shifted its climate status to first humid (B1) and moist sub humid (C2) in the retrieval of monsoon season of dry and wet years. During cold weather season of dry year (1994), Aduthurai experienced dry sub humid (C1) and improved its climate status to per humid (A) in the wet year (1996). Both the stations that hail from the semiarid zone shifted their climate status to aridity (E), in the hot weather season of the long term, dry and wet years.

Station	Period	Season	P (mm)	PE (mm)	AE (mm)	D (mm)	S (mm)	Iwh (%)	Iwa (%)	Iwm (%)	Wetness (%)	Adequacy (%)
Jodhpur	1951-1980	June-Sept	20	41	20	21	0	0	50	-50(D)	4	50
		Oct-Nov	1	29	1	29	0	0	98	-98(E)	1	2
		Dec-Feb	1	7	1	7	0	0	93	-93(E)	1	7
		Mar-May	1	42	1	41	0	0	96	-96(E)	1	4
	Dry & ENSO year (1987)	June-Sept	4	44	4	41	0	0	92	-92(E)	1	8
		Oct-Nov	0	32	0	32	0	0	100	-100(E)	1	0
		Dec-Feb	1	7	1	6	0	0	84	-84(E)	2	16
		Mar-May	3	44	3	41	0	0	94	-94(E)	1	6
	Wet year (1990)	June-Sept	44	41	21	20	23	47	48	0(C1)	33	52
		Oct-Nov	0	14	1	12	0	0	88	-88(E)	10	12
		Dec-Feb	2	9	2	7	0	0	83	-83(E)	3	17
		Mar-May	1	42	2	40	0	0	96	-96(E)	1	4
Aduthurai	1951-1980	June-Sept	18	38	18	20	0	0	51	-51(D)	1	49
		Oct-Nov	58	33	32	1	7	25	2	23(B1)	49	98
		Dec-Feb	19	20	17	3	9	47	14	32(B1)	80	86
		Mar-May	8	42	12	30	0	0	71	-71(E)	12	29
	Dry year (1994)	June-Sept	4	38	4	34	0	0	91	-91(E)	1	9
		Oct-Nov	54	33	18	15	20	75	40	35(B1)	40	60
		Dec-Feb	14	20	16	4	3	15	21	-6(C1)	73	79
		Mar-May	4	42	10	33	0	0	77	-77(E)	13	23
	Wet year (1996)	June-Sept	31	37	23	14	5	15	37	-22(C1)	37	63
		Oct-Nov	49	36	27	9	9	27	23	4(C2)	41	77
		Dec-Feb	47	20	9	11	43	245	52	193(A)	46	48
		Mar-May	6	42	7	35	0	0	84	-84(E)	3	16

Table 6: Seasonal water budget elements on weekly basis-semi arid zone (D)–extremities.

Summary and Conclusion

In the monsoon period, during the three study periods 1951-1980, 1981-1991, and 1992-1998 the All India climate witnessed first, fourth and third humid (B1, B4 and B3) status respectively. In the retrieval of monsoon season, All India moisture status depleted to dry sub humid (C1)in the years 1951-1980 and 1981-1991, whereas in 1992-1998 All India witnessed moist sub humid (C2) climate. Among the three study periods the All India climate changed to semiarid (D) during the cold weather seasons of 1951-1980 and 1981-1991, with a shift to dry sub humid (C1) in 1992-98. The moisture status depreciated to arid (E) during the hot weather seasons of 1951-1980, 1981-1991 and semiarid (D) in 1992-1998.

Tocklai that hails from per humid climate zone depleted to second and fourth humid climate during the monsoon season of dry and wet years respectively. In the retrieval of monsoon season of long term, dry and wet years the climate status shifted to moist sub humid (C2) and dry sub humid (C1). In the cold and hot weather seasons of the study period, this station's moisture status shifted to dry sub humid (C1) and moist sub humid (C2) respectively.

Karjat is consistent and maintained its per humid (A) status in monsoon season but shifted to moist sub humid (C2) and dry sub humid (C1)in the retrieval and then dropped to Arid (E) in the cold and hot weather seasons during the Dry and Wet years.

In the extreme dry year, Adhartal that hails from humid zone shifted to dry sub humid (C1) in monsoon season, and remained in the same status during retrieval of monsoon period and cold weather season, shifted to Arid (E) status in the hot weather season. During the wet year Adhartal improved its moisture status to per humid (A) but witnessed aridity during the retrieval, cold and hot weather seasons.

Rudrur a representative of humid zone depleted to dry sub humid (C1) and enriched to per humid (A) during the monsoon season of dry and wet years. During retrieval of monsoon season the climate status dropped to semi aridity (D), and during cold and hot weather seasons of dry and wet years, its climate status depleted to aridity (E).

Junagarh's moisture status depleted to semiarid (D) and enriched to third humid (B3), during the monsoon season of dry and wet years respectively. In the monsoon retrieval season the moisture regime shifted to semiarid (D) during the long term and wet years of study and experienced aridity (E) during the same season of dry year. In the rest of the seasons of the study period Junagarh's moisture status depleted to aridity (E). Sabour improved its moisture regime to per humid (A) and dropped to dry sub humid (C1) in the monsoon period of wet and dry years respectively. In the rest of the seasons of the dry and wet years Sabour experienced semi aridity (D).

The moisture status of Jodhpur improved to dry subs humid (C1) and depreciated to aridity (E) during the monsoon season of wet year and dry years respectively. In all other seasons of long term, dry & ENSO and wet years the climate status dropped to aridity (E).

Aduthurai shifted its climate status to first humid (B1) and moist sub humid (C2) in the retrieval of monsoon season of dry and wet years. During cold weather season of dry year (1994), Aduthurai experienced dry sub humid (C1) and improved its climate status to per humid (A) in the wet year (1996).

Acknowledgements

Authors acknowledge Additional Director General of Meteorology (Research), PUNE for supplying the meteorological data for the study period.

References

1. Gadgil S (1996) Climate change and agriculture-An Indian perspective. In Climate Variability and Agriculture pp: 1-18.
2. Food and Agriculture Organization (FAO) (1997) Impact of El Niño on Agriculture, Fisheries and Forestry a Report.
3. Richard MA, Chi-Chung C, Bruce A McCarl, Rodney F (1999) The economic consequences of ENSO events for agriculture. Clim Res 13: 165-172.
4. Das HP, Goankar SB (2000) Impact of weather on low land rice cultivation and planning decisions over humid tropical India. Mausam 51: 261-268.
5. Kashyaji A (2002) Influence of Meteorological parameters on performance of rain fed Cropping Systems. Mausam 53: 1.
6. Deng X (2010) Impacts of El Nino-Southern Oscillation events on China's rice production. J Geographys Sci 20: 3-16.
7. Sarma AALN, Vizayabhaskar (2013) Monsoon season soil wetness over India-ENSO and LNSO Events. J Ind Geophys Union 17: 383-402.
8. Krishna K, Singh GP, Sekhar MS (2015) The Influence of seasonal teleconnection patterns on the cryosphere of Antarctic and Northern Himalayas. Int J Earth Atmos Sci 2: 80-89.
9. Sikka DR (1980) Some aspects of large scale fluctuations of summer monsoon rainfall over India in relation to fluctuations in the planetary and regional scale circulation parameters. Proc Indian Acad Sci 89: 179-195.
10. Krishnamurthy TN (1985) Summer Monsoon Experiment: A review. Mon Weather Rev 113: 1590-1626.
11. Mooley DA, Parthasarathy B, Pant GB (1986) Relationships between India summer rainfall and location of the ridge at 500 hPa level along 75° E. J Cli App Met 25: 633-640.
12. Shukla J (1987) Interannual variability of monsoon "In Monsoons". J Weley & Sons pp: 339-463.
13. Sarma AALN, Srinivas S, Karthikeya A (2005) Studies on Aberrations in Climate Impacts-Water Balance Model. Ind Geophys Union 9: 205-214.
14. Bhalme HN, Mooley DA, Jadhav SK (1984) Fluctuations in drought/flood area over India and relationships with the Southern Oscillations. Monthly Weather Rev 111: 86-94.
15. Sarker RP (1987) Proc Int Conf On tropical micro met and air pollution, Indian Institute of Technology, New Delhi, India.
16. Chaudhury A, Mhasawade SV (1991) Variations in meteorological floods during summer monsoon over India. Mausam 42: 167-170.
17. Thornthwaite CW, Mather JR (1955) The Water Balance. Publ in Clim Drexel Instt Tech 8: 1.
18. Gore PG, Thapliyar V (1999) Occurrence of Dry and Wet weeks over Maharashtra. Mausam 51: 1.
19. Carter DB, Mather JR (1966) Climatic Classification for Environmental Biology. Publi Incline Inst 19: 341- 352.

PERMISSIONS

LIST OF CONTRIBUTORS

P. Alfredini
Polytechnic School of São Paulo University, Department of Hydraulic and Environmental Engineering, Harbour and Coastal Area of the Hydraulic Laboratory, Avenida Professor Luciano Gualberto, Travessa 3, n. 380, Cidade Universitaria, CEP 05508-010, São Paulo, Brazil

E. Arasaki
Polytechnic School of São Paulo University, Department of Hydraulic and Environmental Engineering, Harbour and Coastal Area of the Hydraulic Laboratory, Avenida Professor Luciano Gualberto, Travessa 3, n. 380, Cidade Universitaria, CEP 05508-010, São Paulo, Brazil National Institute of Space Research, Av. Dos Astronautas n. 1758, CEP 12227-010, São José dos Campos, São Paulo, Brazil

M. Rosso and A. Pezzoli
Polytechnic of Torino, Engineering Faculty, Department of Environment, Land and Infrastructure Engineering, CorsoDucadegli Abruzzi n. 24, 10129, Torino, Italy

W. C. de Sousa Jr.
Aeronautic Technology Institute, Civil Engineering Division, Department of Water Resources and Environmental Sanitation; PraçaMarechal Eduardo Gomes n. 50, Vila das Acacias, CEP 12228-900, São Jose dos Campos, São Paulo, Brazil

Mequaninta F, Mitikub R and Shimelesc A
Climate and Geospatial Research, Ethiopian Institute of Agricultural Research, Addis Ababa, Ethiopia

Uthaya Siva M, Selvakumar P and Sakthivel A
CAS in Marine Biology, Faculty of Marine Science, Annamalai University, Parangipettai-608502, India

Eludoyin OM
Department of Geography and Planning Sciences, AdekunleAjasin University, Akungba-Akoko, Ondo State, Nigeria

Sintayehu Legesse Gebre
Department of Natural Resources Management, Jimma University, P.o.box 307, Ethiopia

Fulco Ludwig
Department of Earth System Science, Wageningen University, P.o.box 47 6700AA, The Netherlands

Mohamed Aoubouazza
Centre de la Recherche Forestière, BP. 763 Agdal-Rabat, Morocco

Rachid Rajel and Rachid Essafi
Direction de la Recherche et de la Planification de l'Eau, Agdal-Rabat, Morocco

Kachaje O and Chavula G
University of Malawi-Polytechnic, Chichiri, Blantyre, Malawi

Kasulo V
Mzuzu University, Mzuzu, Malawi

Sagar A. Marathe and Shankar Murthy
National Institute of Industrial Engineering, Mumbai, India 400 087

Paulus AW and Shanas SP
State College Meteorology and Geophysics, Indonesia Meteorological and Geophysical Agency, Jakarta, Indonesia

Sukumar Lala, Nabojit Chakraborty and Milan Kanti Das
Regional Meterological Centre, Kolkata, India

Adigun AO and Ayolabi E A
Department of Geosciences, Faculty of Science, University of Lagos, Lagos State, Nigeria

Jae-Won Choi, Yumi Cha and Jeoung-Yun Kim
National Institute of Meteorological Sciences, Seohobuk-ro, Jeju, 63568, Korea

Bilel Fathalli
Université de Tunis El Manar, Ecole Nationale d'Ingénieurs de Tunis, Tunis, Tunisie

Benjamin Pohl and Thierry Castel
Centre de Recherches de Climatologie, Biogéosciences, CNRS / université de Bourgogne Franche-Comté, Dijon, France

Mohamed Jomâa Safi
Université de Tunis El Manar, Ecole Nationale d'Ingénieurs de Tunis, Tunis, Tunisie

Raneesh KY
Ex-research Scholar, Department of Civil Engineering, National Institute of Technology, Calicut - 673601, Kerala, India

Thampi SG
Associate Professor, Department of Civil Engineering, National Institute of Technology, Calicut - 673601, Kerala, India

Belay TT
Department of Urban Environment and Climate Change Management, Ethiopian Civil Service University, Ethiopia

Kailash Chand Pandey
Climatologist in G.E.A.G. India, Gorakhpur Environmental Action Group, Gorakhpur, India

Abuloye AP, Nevo AO and Popoola KS and Awotoye OO
Institute of Ecology and Environmental Studies, Obafemi Awolowo University, Ile-Ife, Nigeria

Eludoyin OM
Adekunle Ajasin University, Akungba-Akoko, Ondo State, Nigeria

Abdulkadir G
Food and Agriculture Organization of the United Nations, Koddbur, Hargeisa, Somaliland

Ruchita S and Rohit S
Department of Sciences, Pandit Deendayal Petroleum University, Gandhinagar, Gujarat, India

Mugume I, Basalirwa C, Bamutaze Y, Twinomuhangi R and Waiswa D
Department of Geography, Geo-Informatics and Climatic Sciences, Makerere University, Uganda

Mesquita MDS
Uni Research Climate, Bergen, Norway

Bjerknes Centre for Climate Research, Bergen, Norway

Reuder J
Geophysical Institute, University of Bergen, Norway

Tumwine F and Sansa Otim J
Department of Networks, Makerere University, Kampala, Uganda

Jacob Ngailo T
Department of General Studies, Dar er Salaam Institute of Technology, Tanzania

Ayesiga G
Uganda National Meteorological Authority, Uganda

Frazier AE and Hemingway B
Department of Geography, Oklahoma State University, Stillwater, OK, 74074, USA

Brasher J
Department of Geography, University of Tennessee, Knoxville, TN, 37996, USA

Kabir H
International Islamic University Malaysia, Kulliyyah of Architecture and Environment Design (KAED), Malaysia

Golder J
Department of Geography and Environment, Faculty of Earth and Environmental Sciences, University of Dhaka, Dhaka

Pejman M, Aliakbari Bidokhti A and Gharaylo M
University of Tehran, Tehran, Iran

Sivaram B
Department of Mathematics, Sri Venkateswara College of Engineering and Technology, Chittoor, India

Sarma AALN
Andhra University, Visakhapatnam, India

Index

www.ingramcontent.com/pod-product-compliance
Lightning Source LLC
Chambersburg PA
CBHW080652200326
41458CB00013B/4830